METHODS IN MOLECULAR BIOLOGY™

AF813243

Series Editor
John M. Walker
School of Life Sciences
University of Hertfordshire
Hatfield, Hertfordshire, AL10 9AB, UK

For further volumes:
http://www.springer.com/series/7651

Methods in Molecular Biology

Biolistic DNA Delivery

Methods and Protocols

Edited by

Stephan Sudowe

Ganzimmun Diagnostics AG, Mainz, Germany

Angelika B. Reske-Kunz

Department of Dermatology, University Medical Center of the Johannes Gutenberg-University, Mainz, Germany

Humana Press

Editors
Stephan Sudowe
Ganzimmun Diagnostics AG
Mainz, Germany

Angelika B. Reske-Kunz
Department of Dermatology
University Medical Center of the Johannes Gutenberg-University
Mainz, Germany

ISSN 1064-3745 ISSN 1940-6029 (electronic)
ISBN 978-1-4939-6068-2 ISBN 978-1-62703-110-3 (eBook)
DOI 10.1007/978-1-62703-110-3
Springer New York Heidelberg Dordrecht London

© Springer Science+Business Media, LLC 2013
Softcover re-print of the Hardcover 1st edition 2013
This work is subject to copyright. All rights are reserved by the Publisher, whether the whole or part of the material is concerned, specifically the rights of translation, reprinting, reuse of illustrations, recitation, broadcasting, reproduction on microfilms or in any other physical way, and transmission or information storage and retrieval, electronic adaptation, computer software, or by similar or dissimilar methodology now known or hereafter developed. Exempted from this legal reservation are brief excerpts in connection with reviews or scholarly analysis or material supplied specifically for the purpose of being entered and executed on a computer system, for exclusive use by the purchaser of the work. Duplication of this publication or parts thereof is permitted only under the provisions of the Copyright Law of the Publisher's location, in its current version, and permission for use must always be obtained from Springer. Permissions for use may be obtained through RightsLink at the Copyright Clearance Center. Violations are liable to prosecution under the respective Copyright Law.
The use of general descriptive names, registered names, trademarks, service marks, etc. in this publication does not imply, even in the absence of a specific statement, that such names are exempt from the relevant protective laws and regulations and therefore free for general use.
While the advice and information in this book are believed to be true and accurate at the date of publication, neither the authors nor the editors nor the publisher can accept any legal responsibility for any errors or omissions that may be made. The publisher makes no warranty, express or implied, with respect to the material contained herein.

Printed on acid-free paper

Humana Press is a brand of Springer
Springer is part of Springer Science+Business Media (www.springer.com)

Preface

Gene transfer into eukaryotic cells has been established as an important tool in basic research for the study of gene regulation, the analysis of protein expression and function within somatic cells/tissues, as well as for the production of transgenic organisms. Furthermore, it allows a variety of valuable applications in the clinics such as the implementation of gene therapy strategies for therapeutic purposes. Since the term transfection ("infection by transformation") has been introduced to describe the procedure of artificial introduction of foreign nucleic acids (DNA or RNA) into the nucleus of eukaryotic cells using nonviral methods, considerable progress has been made in the development of effective gene transfer technologies, which are broadly classified into chemical-based and physical-based transfection methods.

Biolistic transfection represents a direct physical gene transfer approach in which nucleic acids are precipitated on biologically inert high-density microparticles (usually gold or tungsten) and delivered directly through cell walls and/or membranes into the nucleus of target cells by high-velocity acceleration using a ballistic device such as the gene gun. Originally invented by Sanford and Klein in the late 1980s and designed to introduce foreign DNA into plant cells (as it accomplishes gene transfer across cell walls), the gene gun technique has continuously advanced to allow for the transfection of animal cells and organs. It is now appreciated as a technology that is potentially applicable in vitro to a wide range of cell and tissue types as well as in vivo to various organisms, including plants, nematodes, mammals, and ultimately even to humans.

We made allowance for the multifaceted applications of biolistic transfection in the conceptual design of this volume of the *Methods in Molecular Biology* series. This comprehensive collection of detailed protocols is intended to provide the definitive practical guide for the novice as well as the advanced user in the field of gene transfer on how to introduce nucleic acids into eukaryotic cells using the biolistic technique. Therefore, the volume is divided into a total of six parts, each highlighting a separate field of research activity. Parts I to IV cover biolistic gene transfer into plants (Chaps. 1, 2, 3, 4, 5, and 6), nematodes (Chaps. 7, 8, and 9) or mammalian cells in vitro (Chaps. 10, 11, 12, and 13) and in vivo (Chaps. 14, 15, 16, 17, and 18). Part V, in which the use of gene gun-mediated DNA vaccination in various experimental animal models of human diseases is introduced (Chaps. 19, 20, 21, 22, 23, 24, 25, and 26), is supplemented by a non-protocol chapter that provides an overview of the basic safety issues which have been raised with respect to biolistic DNA vaccines (Chap. 27). Part VI, which focuses on the description of biolistic delivery of molecules other than nucleic acids (Chaps. 28 and 29), completes the volume.

In summary, the compendium *Biolistic DNA Delivery: Methods and Protocols* brings together the knowledge and the experience of leading experts in the field of gene transfer, who work with the methods on a regular basis. The chapters provide step-by-step instructions to reproducibly implement biolistic transfection immediately into scientific practice. Each chapter is supported by a helpful "Notes" section, which offers valuable advices and tricks for the respective transfection procedure described. Thus, this methods book represents a useful resource for students, postdoctoral fellows, as well as principal investigators who work on gene transfer as basic researchers in plant science or molecular/cell biology or as preclinical/clinical investigators in biomedical disciplines such as immunology, virology, or neurology.

Finally, we sincerely thank the authors of each chapter for their excellent contributions to this volume and their straightforward cooperation. We are also very grateful to the series editor, Dr. John Walker, for his guidance in the development of this book and to Humana Press for giving us the opportunity to make this project possible.

Mainz, Germany

Stephan Sudowe, Ph.D.
Angelika B. Reske-Kunz, Ph.D.

Contents

Contributors

GUSTAF AHLÉN • *Department of Laboratory Medicine, Karolinska Institutet, Karolinska University Hospital Huddinge, Stockholm, Sweden*
H. OYA ALPAR • *Division of Clinical Microbiology, Department of Pharmaceutics, School of Pharmacy, University of London, London, UK*
VERONICA A. ALVAREZ • *Section on Neuronal Structure, National Institute on Alcohol Abuse and Alcoholism, National Institutes of Health, Bethesda, MD, USA*
KANDAN ARAVINDARAM • *Agricultural Biotechnology Research Center, Academia Sinica, Nankang, Taipei, Taiwan, ROC*
DIRK BAUER • *Department of Ophthalmology, Ophtha-Lab, St. Franziskus Hospital, Münster, Germany*
EUGENE BEREZIKOV • *Hubrecht Institute and University Medical Center Utrecht, Utrecht, The Netherlands*
ELKE S. BERGMANN-LEITNER • *Division of Malaria Vaccine Development, United States Military Malaria Vaccine Program, Walter Reed Army Institute of Research, Silver Spring, MD, USA*
MURRAY R. BOASE • *The New Zealand Institute for Plant and Food Research Limited, Palmerston North, New Zealand*
MARIANNE BRYAN • *Department of Immunology, University of Pittsburgh, School of Medicine, Pittsburgh, PA, USA; Department of Immunology, University of Washington, Seattle, WA, USA*
LIQUAN CAI • *Department of Urology, University of Pittsburgh, Pittsburgh, PA, USA*
KLAUS CICHUTEK • *Division of Medical Biotechnology, Paul-Ehrlich-Institut, Langen, Germany*
VITALY CITOVSKY • *Department of Biochemistry and Cell Biology, State University of New York, Stony Brook, NY, USA*
KEVIN M. DAVIES • *The New Zealand Institute for Plant and Food Research Limited, Palmerston North, New Zealand*
SIMON C. DEROLES • *The New Zealand Institute for Plant and Food Research Limited, Palmerston North, New Zealand*
YING DOU • *Department of Anesthesiology, Pharmacology and Therapeutics, University of British Columbia, Vancouver, BC, Canada*
DAVID FEDIDA • *Department of Anesthesiology, Pharmacology and Therapeutics, University of British Columbia, Vancouver, BC, Canada*
ANNABEL A. FERGUSON • *Division of Geriatric Medicine, Department of Medicine, University of Pittsburgh, Pittsburgh, PA, USA*
ALFRED L. FISHER • *Division of Geriatric Medicine, Department of Medicine, University of Pittsburgh, Pittsburgh, PA, USA*
LARS FRELIN • *Division of Clinical Microbiology, Department of Laboratory Medicine, Karolinska Institutet, Karolinska University Hospital Huddinge, Stockholm, Sweden*
DEBORAH H. FULLER • *Department of Microbiology, University of Washington, Seattle, WA, USA*

James T. Fuller • *Department of Microbiology, University of Washington, Seattle, WA, USA*
Victor Gaba • *Department of Plant Pathology and Weed Science, The Volcani Center, Agricultural Research Organization, Bet Dagan, Israel*
Amit Gal-On • *Department of Plant Pathology and Weed Science, The Volcani Center, Agricultural Research Organization, Bet Dagan, Israel*
Caixia Gao • *The State Laboratory of Plant Cell and Chromosome Engineering, Institute of Genetics and Developmental Biology, Chinese Academy of Sciences, Beijing, China*
Ragisha Gopalakrishnan • *Albany Medical College, Albany, NY, USA*
Siobhan Guyach • *Department of Immunology, University of Pittsburgh, School of Medicine, Pittsburgh, PA, USA*
Arnd Heiligenhaus • *Department of Ophthalmology, Ophtha-Lab, St. Franziskus Hospital, Münster, Germany; Department of Ophthalmology, University of Duisburg-Essen, Essen, Germany*
Tarig B. Higazi • *Department of Biological Science, Ohio University, Zanesville, OH, USA*
Yvonne Höhn • *Institute for Laboratory Animal Science and Experimental Surgery, RWTH-Aachen, Aachen, Germany*
Don A. Hunter • *The New Zealand Institute for Plant and Food Research Limited, Palmerston North, New Zealand*
Meltem Isik • *Hubrecht Institute and University Medical Center Utrecht, Utrecht, The Netherlands*
Ahmad Jalili • *Division of Immunology, Allergy and Infectious Diseases, Department of Dermatology, Medical University of Vienna, Vienna, Austria*
Luv Kashyap • *Division of Geriatric Medicine, Department of Medicine, University of Pittsburgh, Pittsburgh, PA, USA*
Yukio Koide • *Department of Infectious Diseases, Hamamatsu University School of Medicine, Hamamatsu, Shizuoka, Japan*
Benoît Lacroix • *Department of Biochemistry and Cell Biology, State University of New York, Stony Brook, NY, USA*
Ming-Derg Lai • *Department of Biochemistry and Molecular Biology, College of Medicine, Center of Infectious Disease and Signaling Research, National Cheng Kung University, Tainan, Taiwan, ROC*
Barbara Langer • *Division of Medical Biotechnology, Paul-Ehrlich-Institut, Langen, Germany*
Moshe Lapidot • *Department of Vegetable Research, The Volcani Center, Agricultural Research Organization, Bet Dagan, Israel*
Wolfgang W. Leitner • *Basic Immunology Branch, Division of Allergy, Immunology and Transplantation, National Institute of Allergy and Infectious Diseases, National Institute of Health, Bethesda, MD, USA*
Xiongwei Li • *Department of Pharmaceutics, School of Pharmacy, University of London, London, UK*
Mengji Lu • *Department of Virology, University of Duisburg-Essen, Essen, Germany*
Sarah C.R. Lummis • *Department of Biochemistry, University of Cambridge, Cambridge, UK; Division of Neurobiology, MRC Laboratory of Molecular Biology, Cambridge, UK*
Shimpei Magori • *Department of Biochemistry and Cell Biology, State University of New York, Stony Brook, NY, USA*

JOACHIM MAXEINER • *Asthma Core Facility, Research Center Immunology, University Medical Center of the Johannes Gutenberg University Mainz, Mainz, Germany*
PETER MERTENS • *Institute for Animal Health, Pirbright Laboratory, Pirbright, UK*
TOSHI NAGATA • *Department of Health Science, Hamamatsu University School of Medicine, Hamamatsu, Shizuoka, Japan*
KLAUS K. NIELSEN • *Research Division, DLF-TRIFOLIUM A/S, Store Heddinge, Denmark*
ARTO NIEMINEN • *The Instrument Center, University of Helsinki, Helsinki, Finland*
KAREN A. NORRIS • *Department of Immunology, University of Pittsburgh, School of Medicine, Pittsburgh, PA, USA*
JOHN A. O'BRIEN • *Division of Neurobiology, MRC Laboratory of Molecular Biology, Cambridge, UK*
STEVE PASCOLO • *Department of Oncology, University Hospital of Zürich, Zürich, Switzerland*
VERENA RAKER • *Clinical Research Unit Allergology, Department of Dermatology, University Medical Center of the Johannes Gutenberg University Mainz, Mainz, Germany*
MATTHIAS RENNER • *Division of Medical Biotechnology, Paul-Ehrlich-Institut, Langen, Germany*
ANGELIKA B. RESKE-KUNZ • *Department of Dermatology, University Medical Center of the Johannes Gutenberg-University, Mainz, Germany*
UWE RITTER • *Department of Immunology, University of Regensburg, Regensburg, Germany*
MATTI SÄLLBERG • *Division of Clinical Microbiology, Department of Laboratory Medicine, Karolinska Institutet, Karolinska University Hospital Huddinge, Stockholm, Sweden*
SANDRA SCHEIBLHOFER • *Division of Allergy and Immunology, Department of Molecular Biology, University of Salzburg, Salzburg, Austria*
JÜRGEN SCHERER • *Division of Medical Biotechnology, Paul-Ehrlich-Institut, Langen, Germany*
SILKE SCHÜLE • *Division of Medical Biotechnology, Paul-Ehrlich-Institut, Langen, Germany*
KATHY E. SCHWINN • *The New Zealand Institute for Plant and Food Research Limited, Palmerston North, New Zealand*
NYSSA SHERAZEE • *Section on Neuronal Structure, National Institute on Alcohol Abuse and Alcoholism, National Institutes of Health, Bethesda, MD, USA*
SIDONA SIKORSKAITE • *Department of Agricultural Sciences, University of Helsinki, Helsinki, Finland*
MICHAL ŠMAHEL • *Department of Experimental Virology, Laboratory of Molecular Oncology, Institute of Hematology and Blood Transfusion, Prague, Czech Republic*
CRISTY STAGNAR • *Division of Infectious Diseases, Wadsworth Center, New York State Department of Health, Albany, NY, USA*
DAVID F. STEELE • *Department of Anesthesiology, Pharmacology and Therapeutics, University of British Columbia, Vancouver, BC, Canada*
JULIA STEITZ • *Institute for Laboratory Animal Science, University Hospital of RWTH Aachen, Aachen, Germany*
STEPHAN SUDOWE • *Ganzimmun Diagnostics AG, Mainz, Germany*
MULPURI SUJATHA • *Directorate of Oilseeds Research, Rajendranagar, Hyderabad, India*

JOSEF THALHAMER • *Division of Allergy and Immunology, Department of Molecular Biology, University of Salzburg, Salzburg, Austria*
THOMAS TÜTING • *Department of Dermatology, Laboratory of Experimental Dermatology, University of Bonn, Bonn, Germany*
MASAKI UCHIDA • *Department of Pharmaceutics, School of Pharmacy, University of London, London, UK; Josai University, Sakado, Saitama, Japan*
SHOKO UEKI • *Institute of Plant Science and Resources, Okayama University, Kurashiki, Okayama, Japan*
THOMAS R. UNNASCH • *GHIDR Program, College of Public Health, University of South Florida, Tampa, FL, USA*
JARI P.T. VALKONEN • *Department of Agricultural Sciences, University of Helsinki, Helsinki, Finland*
K.B.R.S. VISARADA • *Directorate of Sorghum Research, Rajendranagar, Hyderabad, India*
ANSSI L. VUORINEN • *Department of Agricultural Sciences, University of Helsinki, Helsinki, Finland*
SUSANNE WASMUTH • *Department of Ophthalmology, Ophtha-Lab, St. Franziskus Hospital, Münster, Germany*
RICHARD WEISS • *Division of Allergy and Immunology, Department of Molecular Biology, University of Salzburg, Salzburg, Austria*
ERIC J. YAGER • *Department of Arts and Sciences, Albany College of Pharmacy and Health Sciences, Albany, NY, USA*
NING-SUN YANG • *Agricultural Biotechnology Research Center, Academia Sinica, Nankang, Taipei, Taiwan, ROC*
MENG-CHI YEN • *Department of Biochemistry and Molecular Biology, College of Medicine, Center of Infectious Disease and Signaling Research, National Cheng Kung University, Tainan, Taiwan, ROC*
SHU-YI YIN • *Agricultural Biotechnology Research Center, Academia Sinica, Nankang, Taipei, Taiwan, ROC*

Part I

Biolistic DNA Delivery in Plant Science and Agricultural Biotechnology

Chapter 1

Comparison Between *Agrobacterium*-Mediated and Direct Gene Transfer Using the Gene Gun

Caixia Gao and Klaus K. Nielsen

Abstract

Agrobacterium-mediated transformation and direct gene transfer using the gene gun (microparticle bombardment) are the two most widely used methods for plant genetic modification. The *Agrobacterium* method has been successfully practiced in dicots for many years, but only recently have efficient protocols been developed for grasses. Microparticle bombardment has evolved as a method delivering exogenous nucleic acids into plant genome and is a commonly employed technique in plant science. Here these two systems are compared for transformation efficiency, transgene integration, and transgene expression when used to transform tall fescue (*Festuca arundinacea* Schreb.). The tall fescue transformation protocols lead to the production of large numbers of fertile, independent transgenic lines.

Key words: Tall fescue (*Festuca arundinacea* Schreb.), Grass transformation, Embryogenic callus, Particle bombardment, *Agrobacterium* transformation, Bialaphos selection

1. Introduction

Transformation of plant tissues and plants is a powerful research tool for studying gene function and is a key component of agricultural biotechnology. The most commonly used methods for integrating foreign genes into plant cells are *Agrobacterium*-mediated transformation or direct gene transfer using the gene gun (1). In the former, soil bacteria of the genus *Agrobacterium* have been harnessed for their natural ability to act as genetic engineers. Since the mid-1990s the use of super virulent strains of the bacteria has resulted in successful generation and recovery of transgenic plants in all the major cereal crops and in a range of other crops which were previously transformable by particle bombardment only (2–4). The direct gene transfer method involves direct mechanical

Stephan Sudowe and Angelika B. Reske-Kunz (eds.), *Biolistic DNA Delivery: Methods and Protocols*,
Methods in Molecular Biology, vol. 940, DOI 10.1007/978-1-62703-110-3_1, © Springer Science+Business Media, LLC 2013

integration of the DNA into the host genome using metal beads as carriers. Particle bombardment has proved to be a highly successful technique for gene transfer to plant cells and has been widely adapted by plant biotechnologists since the early 1990s (5).

Much debate has revolved around the various merits of particle bombardment and *Agrobacterium*, and which is the preferred method for the production of genetically transformed plants. The ability of *Agrobacterium* to generate single-copy insertions compared with the prevalence of multicopy insertions and broken transgene integration when using particle bombardment is an attractive feature of the former method. However, several advantages make microparticle bombardment the method of choice for engineering crop species: (a) transformation of recalcitrant species; engineering of important agronomic crops such as wheat, maize, soybean, and cotton etc., is restricted to a few non-commercial genotypes when *Agrobacterium*-mediated transformation method is used. Limited host range still presents a barrier in some species, and for these, particle bombardment remains the only reliable method for the production of transgenic plants; (b) co-transformation with multiple transgenes; co-transformation is an attractive technology because it facilitates the introduction of multiple genes of interest in one simultaneous transformation step, requiring only one selectable marker (6). Multiple gene transfer is necessary for sophisticated genetic manipulation strategies, such as stacking of transgenes regulating different agronomic traits, introduction of several enzymes acting sequentially in a metabolic pathway, or expression of a target protein and one or more enzymes required for specific types of post-translational modification. Although this can be achieved by single gene transformation followed by crossing of the plant lines carrying different transgenes, it is much quicker and more straightforward to introduce all the necessary genes simultaneously, and, in many cases, they will integrate into the same locus; (c) "clean gene" technology; in *Agrobacterium*-mediated transformation, the transgene must be placed between T-DNA repeats, and the T-DNA is naturally excised from the vector during the transformation process. This frequently leads to the integration of vector backbone sequences into the plant genome. In contrast, vectors are not required for particle bombardment. Fu et al. (7) devised a clean DNA strategy in which all vector sequences were removed prior to particle loading. This minimal, linear cassette was then coated onto the metal particles before transformation. Many results proved that the cassette transformation method was at least as efficient as with whole plasmids (8–11).

Tall fescue is an out-crossing, perennial, cool-season polyploid grass widely used for forage and turf. The majority of tall fescue transformation studies utilized either particle bombardment (12–17) or *Agrobacterium*-mediated transformation system (18–20).

This paper provides complete protocols that we have developed and used to produce transgenic tall fescue plants via both *Agrobacterium*-mediated transformation and particle bombardment. Average transformation efficiency of 10.5% for *Agrobacterium*-mediated transformation and 11.5% for particle bombardment are routinely obtained. Similar transgene integration patterns and co-integration frequencies of *bar* and *uidA* are observed in both gene transfer systems. However, while GUS activity is detected in leaves of 53% of the *Agrobacterium* transformed lines, only 20% of the bombarded lines show GUS activity (the *uidA* gene of *Escherichia coli* that encodes the enzyme beta-glucuronidase (GUS)). This indicates that for the efficient production of transgenic tall fescue plants with stable and predictable transgene transcription and translation, the *Agrobacterium*-mediated method offers advantages over particle bombardment.

2. Materials

2.1. *Agrobacterium*-Mediated Transformation

1. *Agrobacterium* strain and vector: The super-virulent *Agrobacterium tumefaciens* strain AGL1 with the standard binary vector pDM805 (21), which contains the *Act1:uidA:rbcS* reporter gene at the left border and the *Ubi1:bar:nos* selectable marker gene at the right border.
2. Bacterium culture medium: The basic bacterium culture medium is MG/L (21), which contains 5.0 g/L tryptone, 5.0 g/L mannitol, 2.5 g/L yeast extract, 1.0 g/L L-glutamic acid, 250 mg/L KH_2PO_4, 100 mg/L NaCl, 100 mg/L $MgSO_4 \cdot 7H_2O$. The medium is adjusted to pH 7.2 with KOH. Add biotin at 1 µg/L after autoclaving. For the preparation of plates, 15 g/L agar is added.

2.2. Particle Bombardment

1. PDS-1000/He particle gun [BIO-RAD].
2. Particle bombardment plasmid: Plasmid pDM803, which expression cassette comprises the *Act1:uidA:rbcS* reporter gene and the *Ubi1:bar:nos* selectable marker gene. It is constructed in a similar way as the binary vector pDM805. Thus, plasmids containing the same expression cassettes are used for both *Agrobacterium*-mediated and particle bombardment transformations.
3. Osmotic treatment medium: The osmotic treatment medium contains 4.3 g/L Murashige and Skoog medium including vitamins (Duchefa, M0222), 3 mg/L 2,4-D, 45.55 g/L sorbitol, 45.55 g/L mannitol, 3.75 g/L Gelrite.

2.3. Plant Tissue Media

Three different basic plant tissue culture media are used during the transformation and regeneration process: the callus induction, regeneration, and rooting media. During all selection stages, bialaphos is added to the media at 2 mg/L (Wako, Japan) (see Note 1). The antibiotic Augmentin (amoxicillin Na/K 5:1 clavulanate) is also applied at 200 mg/L to all the *Agrobacterium* experiments. Additional copper (0.6 mg/L $CuSO_4$) is added during callus induction and regeneration stages. Unless otherwise stated, all media components are supplied by Duchefa and all media and stocks made up using milli Q water.

1. Callus induction: 4.3 g/L Murashige and Skoog medium including vitamins, 30 g/L sucrose, 500 mg/L casein hydrolysate, 5.0 mg/L 2,4-D (Sigma), 3.75 g/L Gelrite. The medium is adjusted to pH 5.8 with KOH.
2. Regeneration: 4.3 g/L Murashige and Skoog medium including vitamins, 30 g/L sucrose, 0.2 mg/L kinetin (Sigma), 3.75 g/L Gelrite. The medium is adjusted to pH 5.8 with KOH.
3. Rooting: 2.15 g/L Murashige and Skoog medium including vitamins, 30 g/L sucrose, 3.75 g/L Gelrite. The medium is adjusted to pH 5.8 with KOH.

2.4. Supplements for Tissue Media

1. Biotin (1 mg/L): Dissolve in water. Filter sterilize and store at –20°C in 0.5 mL aliquots.
2. 2,4-Dichlorophenoxyacetic acid (2,4-D, 1 mg/mL): Dissolve powder in a small volume of 1 M KOH and make up to volume with water. Filter sterilize and store at –20°C in 1 mL aliquots.
3. $CuSO_4$ stock (1 mg/mL): Dissolve in water. Filter sterilize and store at 4°C.
4. Bialaphos (8 mg/mL): Dissolve in water. Filter sterilize and store at –20°C in 1 mL aliquots.
5. Augmentin (amoxicillin Na/K 5:1 clavulanate, 200 mg/mL): Dissolve augmentin in water. Filter sterilize and store at 20°C in 0.5 mL aliquots.
6. Kinetin (1 mg/mL): Dissolve powder in a small volume of 1 M KOH and make up to volume with water, and mix well. Filter sterilize and store at –20°C in 1 mL aliquots.
7. Rifampicin (50 mg/mL): Dissolve 50 mg of rifampicin to 1 mL of DMSO. Store at –20°C in 0.5 mL aliquots.
8. Tetracycline (5 mg/mL): Dissove 5 mg of tetracycline to 1 mL of methanol. Store at –20°C in 0.5 mL aliquots.

2.5. Reagents

1. Spermidine (0.1 M): Dissolve in water. Filter sterilize and store at –20°C in 1 mL aliquots.

2. TE buffer: 1 mL of 1 M tris HCL (pH 8.0) and 0.2 mL EDTA (0.5 M) and make up with distilled water up to 100 mL.
3. Silwet L-77 (1% v/v): Dissolve in water. Filter sterilize and store at 4°C in 0.5 mL aliquots.
4. X-Gluc (5-bromo-4-chloro-3-indolyl-β-D-glucuronic acid, 1 mg/mL): dissolve 10 mg X-Gluc in 100 μL dimethylformamide under a fume hood. Add to 10 mL of 100 mM sodium-phosphate buffer, pH7.0. Wrap the tube in aluminum foil to protect from light and store at −20°C.
5. Bleach: 15% sodium hypochlorite.
6. Tween 20.
7. 30% Aqueous glycerol.
8. Ethanol solution (70%, 96%, 99%v/v): make up with de-ionized water.

3. Methods

3.1. Isolation of Tall Fescue Mature Embryos

1. Sterilization of mature seeds: Sterilize the mature seeds in a 50-mL tube by washing in 70% ethanol for 30 s followed by three washes in sterile distilled water. This is followed by sterilization in 30 mL of bleach (15% sodium hypochlorite)+ 1 drop of Tween 20 during incubation for 45–60 min and invert on ice. Wash six times in sterile distilled water. Drain off most of the water, and incubate the seeds in the water for 1–3 days at 4°C.
2. Isolation of mature embryos: Perform all operations in a laminar flow hood under sterile conditions. Hold the seed firmly with a pair of fine forceps, and use a second pair of fine forceps to expose the mature embryo under a dissecting microscope. Then place the embryo scutellum side up on callus induction medium. Place twenty-five embryos on each 9-cm plate and store at 23–24°C in the dark for 8–20 weeks.

3.2. Induction of Embryogenic Callus

After 4 weeks on callus induction medium, friable, white to yellowish embryogenic calluses (see Fig. 1a) develop. At this stage, some of the calluses break up naturally when handing.

1. Place the callus pieces near each other in a marked area so that all the materials derived from a single embryo can be tracked.
2. Grow embryogenic calluses derived from single embryos, representing individual genotypes, separately and subculture every 4 weeks.

Fig. 1. Transgenic tall fescue (*Festuca arundinacea* Schreb.) plants obtained from *Agrobacterium*-mediated transformation and particle bombardment. (**a**) Embryogenic calluses. (**b**) Bialaphos resistant callus clones 4 weeks after transformation. (**c**) Shoot differentiation under bialaphos selection. (**d**) Transgenic plants in rooting medium. (**e**) Transgenic plants in greenhouse. (**f**) GUS activity in leaves of different transgenic lines. Figure reprinted with kind permission of Springer Science + Business Media from: Gao CX et al (2008) Comparative analysis of transgenic tall fescue (*Festuca arundinacea* Schreb.) plants obtained by *Agrobacterium*-mediated transformation and particle bombardment. Plant Cell Rep 27:1601–1609. Copyright by Springer Science + Business Media.

3. To evaluate if the age of callus affects regeneration efficiency, transfer part of the calluses from each callus clone to regeneration medium periodically and measure the percentage of calluses producing green shoots after 3–6 weeks incubation in the light (24°C, under a 16/8-h photoperiod light/dark, white fluorescent light, 100 μmol/m²/s).

3.3. *Agrobacterium* Preparation, Inoculation and Co-cultivation

1. Use a single colony of *Agrobacterium* AGL1, containing the appropriate pDM805 vector, to inoculate 10 mL of MG/L medium with 20 μg/mL Rifampicin and 10 μg/mL tetracycline (see Note 2). Incubate at 28°C and shake at 180 rpm for 40 h.
2. Add 10 mL of sterile 30% aqueous glycerol to the bacterial culture and mix by inverting several times.
3. Place aliquots of 400 μL of the standard inoculum into 0.5-mL Eppendorf tubes and maintain at room temperature for 2 h mixing by inversion every 30 min.
4. Store standard inoculums at −80°C ready for use.
5. Take a 400 μL glycerol stock aliquot, one per 10 mL MG/L liquid medium, and grow overnight (17–20 h) in a 50-mL sterile disposal tube with the appropriate antibiotics at 28°C

on an orbital shaker (250 rpm) (17–20 h), until an OD 1.0–1.5 approximately (A600) is reached.

6. Pellet *Agrobacterium* cells by centrifugation at 4,500 × *g* for 10 min, dispose the supernatant, and resuspend the cells in 4 mL of liquid callus induction medium.
7. Put the 50-mL tube with 4 mL of callus induction medium plus *Agrobacterium* cells back in the shaker for 1–3 h at 28°C in the dark (see Note 3). When the culture is ready, take the tube from the shaker and add 60 μL of 1% (v/v) Silwet to make a final concentration of 0.015%.
8. Place small pieces of embryogenic calluses (0.5–1.0 mm in diameter, aging from 4 to 8 months old) in *A. tumefaciens* AGL1/pDM805 suspension for 35 min (see Notes 4 and 5).
9. Remove excess bacteria after the incubation and drain the infected calluses on filter paper and co-cultivate on callus induction medium (see Note 6).
10. Carry out co-cultivation at 22–23°C in a culture room in the dark (see Note 7).

3.4. Osmotic Treatment

Distribute embryogenic calluses (approx. 60–90 small pieces of callus) as a 1.5-cm diameter monolayer in a 5-cm Petri dish containing osmotic treatment medium for a 4-h osmotic treatment prior to bombardment.

3.5. Preparation of Gold Particle Solution

1. Place 20 mg BIO-RAD sub-micron gold particles (0.6 μm) and 1 mL 99% ethanol in a 1.5 mL Eppendorf tube.
2. Pulse spin in a microfuge for 3 s and remove the supernatant. Repeat this ethanol wash twice more.
3. Add 1 mL sterile water, pulse spin in a microfuge for 3 s and remove the supernatant. Repeat this step.
4. Resuspend gold in 1 mL sterile water, divide into 50 μL aliquots and store at −20°C.
5. When ready to precipitate DNA onto gold, allow a 50 μL aliquot of prepared gold to thaw at room temperature.
6. Add 5 μL DNA (1 mg/mL in TE or water) or water for bombarded-only controls and vortex briefly.
7. Place 50 μL 2.5 M $CaCl_2$ and 20 μL 0.1 M spermidine into the Eppendorf tube and vortex into the gold + DNA solution.
8. Pellet the DNA-coated particles by centrifugation and discard the supernatant.
9. Add 150 μL 99% ethanol to wash the particles and resuspend. Again, pellet the particles and discard the supernatant. Resuspend fully in 85 μL 99% ethanol and maintain on ice.

3.6. Delivery of DNA-Coated Gold Particles

1. The following settings are recommended for this procedure: gap 3.0 cm, stopping plate aperture 0.8 cm, target distance 6.0 cm, vacuum 88–94 kPa, vacuum flow rate 5.0, vent flow rate 7–8.
2. Sterilize the gun's chamber and component parts with 70% (v/v) ethanol. Sterilize macro-carrier holders, macro-carriers, stopping screens, and rupture discs by dipping in 99% ethanol and allow to evaporate completely.
3. Briefly vortex the coated gold particles, take 5 μL, place centrally onto the macro-carrier membrane and allow to dry.
4. Load a rupture disc (1,100 psi) into the rupture disc retaining cap and tighten the screw into place firmly. Place a stopping screen in the fixed nest.
5. Invert the macro-carrier holder containing macro-carrier + gold particles/DNA and place it over the stopping screen in the nest and maintain its position using the retaining ring.
6. Mount the fixed nest assembly onto the second shelf from the top to give a gap of 3.0 cm. Place a sample on the target stage, draw a vacuum of 88–94 kPa and fire the gun (see Note 8).

3.7. Selection and Regeneration of Transformed Calluses and Identification of Transgenic Plants

1. After co-cultivation with *Agrobacterium*, rinse the infected calluses 2–3 times in a solution containing 200 mg/L Augmentin, drain on filter paper and transfer to callus selection medium supplemented with 2 mg/L bialaphos and 200 mg/L Augmentin. Incubate the infected calluses at 24°C in the dark for the first 2–3 weeks, and then cultivate under a 16/8-h (light/dark) photoperiod.
2. Incubate bombarded calluses on the same osmotic treatment medium overnight and then transfer to callus induction medium supplemented with 2 mg/L bialaphos. Incubate the bombarded calluses at 24°C under a 16/8-h (light/dark) photoperiod.
3. Putative bialaphos resistant callus clones after 3–6 weeks selection are healthy, yellowish, compact, and friable, whereas susceptible calluses show marked necrosis (Fig. 1b, c). Transfer the selected clones to regeneration medium supplemented with 2 mg/L bialaphos, with the regeneration medium used for the *Agrobacterium* experiments also containing 200 mg/L Augmentin (see Note 9).
4. When healthy resistant clones are transferred to regeneration medium, approximately 6 weeks are needed to regenerate bialaphos resistant green plantlets.
5. Transfer fully recovered plantlets to containers for further root development (Fig. 1d) and finally transfer green plants to soil in a greenhouse (Fig. 1e) (see Note 10).

6. Once plants are established in soil, leaf samples can be collected for further analysis to confirm the presence of the introduced genes (see Note 11).
7. GUS activity can be examined histochemically as described by Jefferson et al. (22) in calli one day after bombardment, in 2-month-old resistant calli and in leaves of putative transgenic plants. Prior to examination, incubate tissues overnight at 37°C in a phosphate-buffered solution containing X-Gluc (1 mg/mL). After incubation, bleach leaves in 96% ethanol overnight. Figure 1f and Table 1 show the expression of the *uidA* gene in leaf samples from plants transformed with pDM805. Figure 2 and Table 1 show the expression of *bar* and *uidA* in transgenic lines as detected by Southern blot analysis (23).
8. Transformation efficiency is defined as the number of independent transformed plants as a percentage of the original number of calluses initially bombarded or infected. When more than one confirmed transgenic plant is produced from one piece of callus, they are treated as only one independent event (see Note 12).

4. Notes

1. Bialaphos used as the selective agent in this protocol has the *bar* gene as the selectable marker.
2. The antibiotics used depend on the selectable markers in the *Agrobacterium* strain and binary vectors used. For the AGL1 strain used in this protocol, rifampicin (20 mg/L) is used and pDM805 is selected with tetracycline (10 mg/L).
3. The tube should be in the shaker for at least 30 min after resuspension but before use.
4. Most of the calluses will stay attached to the medium while submerged in the *Agrobacterium* suspension. If some calluses float, use a little force to submerge them, as they need to be in contact with *Agrobacterium* cells.
5. The 30-min incubation time is not a critical maximum; it could be extended up to 3 h, but be aware that calluses will be damaged by long-term wetness, and too much *Agrobacterium* growth is also damaging.
6. Do not leave calluses too wet; remove as much *Agrobacterium* suspension as possible before transfer. Use filter paper to dry the calluses. This is a crucial step, as too much *Agrobacterium* growth during the co-cultivation stage will damage the calluses.

Table 1
Expression and estimated minimum copy numbers of *bar* and *uidA* genes in transgenic tall fescue lines derived from *Agrobacterium* and particle bombardment transformation

Agrobacterium-mediated transformation					Particle bombardment				
	Marker gene activity		Minimum copy number			Marker gene activity		Minimum copy number	
Lines	Herbicide resistance	GUS activity in leaves	*bar*	*uidA*	Lines	Herbicide resistance	GUS activity in leaves	*Bar*	*uidA*
A-087	+	++	>3	3	P-154	+	–	>2	3
A-013	+	–	1	0	P-136	+	–	2	2
A-084	+	++	>3	2	P-138	+	–	1	1
A-027	+	–	2	1	P-143	+	–	1	1
A-073	+	+++	>4	4	P-124	+	–	2	0
A-099	+	+	1	1	P-157	+	–	1	0
A-044	+	++	>4	5	P-130	+	+	3	3
A-091	+	–	1	0	P-147	+	–	2	3
A-035	+	–	1	0	P-141	+	–	1	0
A-105	+	++	>1	2	P-156	+	++	1	1
A-095	+	++	1	1					
A-079	+	++	>1	3					
A-075	+	–	1	0					
A-018	+	–	4	1					

A-052	+	–	1	1	
A-059	+	–	1	0	
A-021	+	–	1	0	
A-037	+	++	3	1	
A-061	+	++	2	1	
Co-integration of *bar* and *uidA*		68%			70%
GUS activity		53%			20%
Lines with 1 or 2 *uidA* transgene copies *with a size greater than 5.0 kb*		47%			30%

For herbicide resistance, + denotes that plants grew healthy after spraying the herbicide Basta. For GUS activity, +++ denotes very strong GUS activity (dark blue staining), ++ denotes strong activity (blue staining), + denotes weak activity (light blue staining) and – denotes no detectable activity (no blue staining). Table reprinted with kind permission of Springer Science + Business Media from: Gao CX et al (2008) Comparative analysis of transgenic tall fescue (***Festuca arundinacea*** Schreb.) plants obtained by ***Agrobacterium***-mediated transformation and particle bombardment. Plant Cell Rep **27**:1601–1609. Copyright by Springer Science + Business Media

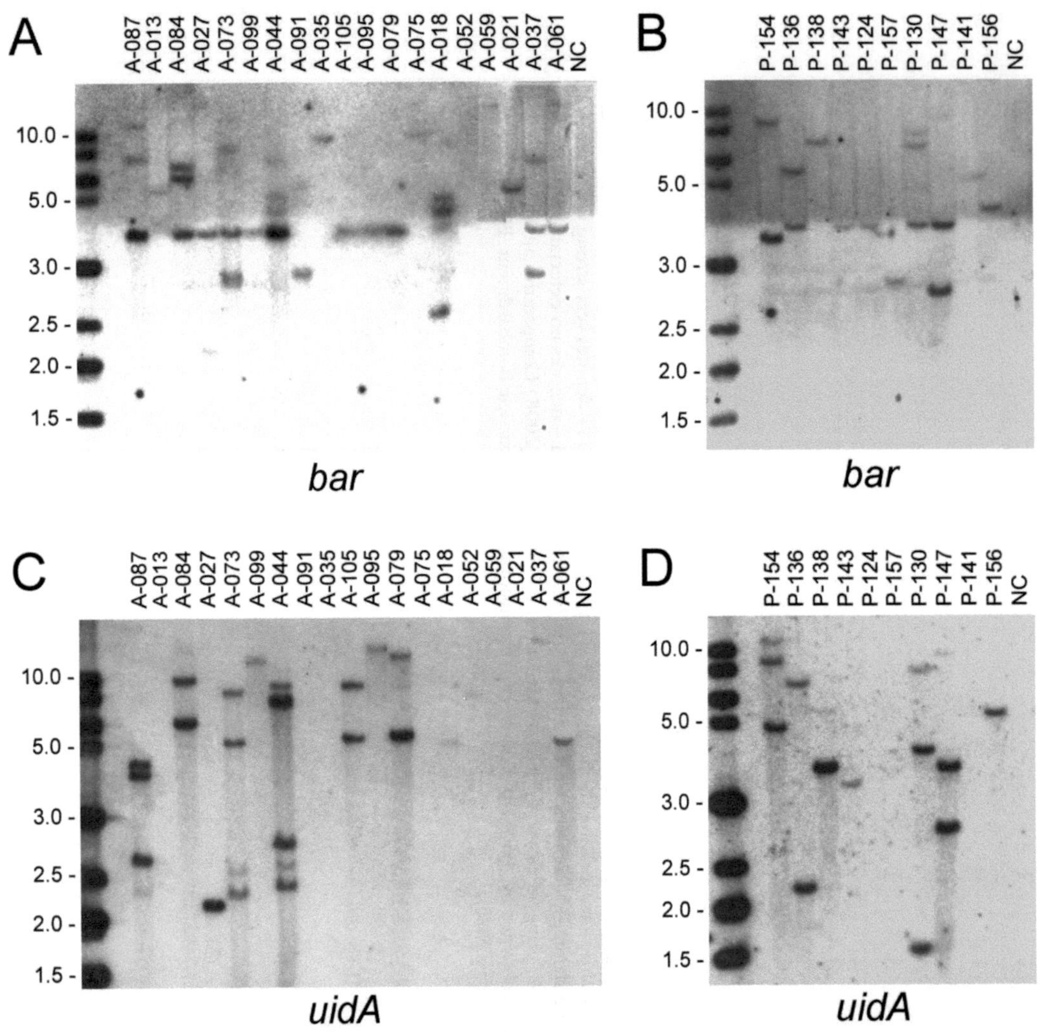

Fig. 2. Southern blot analysis of genomic DNA from tall fescue plants transformed with *Agrobacterium* (**a** and **c**) and particle bombardment (**b** and **d**). For each sample, 20 μg of genomic DNA is digested with *Bam*HI and probed with a 0.36 kb PCR product from the coding region of *bar* and a 0.32 kb PCR product from the coding region of *uidA*. (**a**) *Agrobacterium*-mediated transformation, the *bar* probe. (**b**) Particle bombardment, the *bar* probe. (**c**) *Agrobacterium*-mediated transformation, the *uidA* probe. (**d**) Particle bombardment, the *uidA* probe. *NC* non-transformed control plant. Fragment sizes in kb are indicated to the left. Figure reprinted with kind permission of Springer Science + Business Media from: Gao CX et al (2008) Comparative analysis of transgenic tall fescue (*Festuca arundinacea* Schreb.) plants obtained by *Agrobacterium*-mediated transformation and particle bombardment. Plant Cell Rep 27:1601–1609. Copyright by Springer Science + Business Media.

7. Always prepare control plates. Controls are treated in exactly the same way, i.e., for bombardment, the gold particle solution prepared without plasmid DNA is used to bombard the control plate; for the *Agrobacterium* transformation, the callus induction liquid medium plus silwet without *Agrobacterium* cells is used as the inoculum. These controls are used to monitor the quality of the calluses, and the liquid and solid media.

8. The PDS-1000/He particle gun [BIO-RAD] delivery system involves the use of high pressure gas to accelerate particles to high velocity. Appropriate safety precautions described by the manufacturer should be taken when operating the gun.
9. If callus growth looks good after 3 weeks of selection, then it is possible to omit the final 3 weeks selection on callus induction medium and to transfer to regeneration medium at this stage. If in any doubt, continue with selection on callus induction medium, as this is likely to yield better results. Severely damaged calluses without any sign of growth during the induction period can be removed at this stage.
10. If the first leaf is not healthy, it is unlikely that the plantlet is tolerant to selection pressure. Usually, if a plantlet has developed strong roots deep in the selection medium, it is a good indication that it may be transgenic.
11. If the plantlets are big and strong enough at this stage, leaf samples can be taken for DNA extraction and PCR reaction, enabling an early confirmation of transgenic plants. If the binary vector has *uidA* gene, a segment of leaf can be taken for a GUS activity assay.
12. The transformation efficiencies for *Agrobacterium* (10.5%) and bombardment transformation (11.5%) have been obtained by bombarding or infecting calluses.

References

1. Danford JC (1988) The biolistic process. Trends Biotechnol 6:299–302
2. Ishaida Y et al (1996) High frequency transformation of maize (*Zea mays* L.) mediated by *Agrobacterium tumefaciens*. Nat Biotechnol 14:745–750
3. Hiei Y, Komari T, Kubo T (1997) Transformation of rice mediated by *Agrobacterium tumefaciens*. Plant Mol Biol 35:205–218
4. Gonzalez A et al (1998) Regeneration of transgenic plants of cassava (*Manihot esculenta* Crantz.) through *Agrobacterium*-mediated transformation of embryogenic suspension cultures. Plant Cell Rep 17:827–831
5. Altpeter F et al (2005) Particle bombardment and the genetic enhancement of crops: myths and realities. Mol Breed 15:305–327
6. Francois I, Broekaert W, Cammue B (2002) Different approaches for multi-transgene-stacking in plants. Plant Sci 163:281–295
7. Fu XD et al (2000) Linear transgene constructs lacking vector backbone sequences generate low-copy-number transgenic plants with simple integration patterns. Transgenic Res 9:11–19
8. Breitler JC et al (2002) Efficient microprojectile bombardment mediated transformation of rice using gene cassettes. Theor Appl Genet 104:709–719
9. Loc NT et al (2002) Linear transgene constructs lacking vector backbone sequences generate transgenic rice plants which accumulate higher levels of proteins conferring insect resistance. Mol Breed 9:231–244
10. Romano A et al (2003) Transgene organization in potato after particle bombardment-mediated (co)transformation using plasmids and gene cassettes. Transgenic Res 12: 461–473
11. Romano A et al (2005) Expression of poly-3-(R)-hydroxyalkanoate (PHA) polymerase and acyl-CoA-transacylase in plastids of transgenic potato leads to the synthesis of a hydrophobic polymer, presumably medium-chain-length PHAs. Planta 220:455–464
12. Spangenberg G et al (1995) Transgenic tall fescue (*Festuca arundinacea*) and red fescue (*F. rubra*) plants from microprojectile bombardment of embryogenic suspension cells. J Plant Physiol 145:693–701

13. Cho MJ, Ha CD, Lemaux PG (2000) Production of transgenic tall fescue and red fescue plants by particle bombardment of mature seed-derived highly regenerative tissues. Plant Cell Rep 19:1084–1089
14. Bai Y, Qu RD (2001) Genetic transformation of elite turf-type cultivars of tall fescue. Int Turf Soc Res J 9:129–136
15. Chen L, Auh C, Dowling P (2003) Improved forage digestibility of tall fescue (*Festuca arundinacea*) by transgenic down regulation of cinnamyl alcohol dehydrogenase. Plant Biotechnol J 1:437–449
16. Wang ZY et al (2003) Inheritance of transgenes in tall fescue (*Festuca arundinacea*). In Vitro Cell Dev Biol-Plant 39:277–282
17. Chen L et al (2004) Transgenic down regulation of caffeic acid o-methyltransferase (COMT) led to improved digestibility in tall fescue (*Festuca arundinacea*). Funct Plant Biol 31:235–245
18. Wang ZY, Ge YX (2005) *Agrobacterium*-mediated high efficiency transformation of tall fescue (*Festuca Arundinacea*). J Plant Physiol 162:103–113
19. Dong SJ, Qu RD (2005) High efficiency transformation of tall fescue with *Agrobacterium tumefaciens*. Plant Sci 168:1453–1458
20. Gao CX et al (2008) Comparative analysis of transgenic tall fescue (Festuca arundinacea Schreb.) plants obtained by *Agrobacterium*-mediated transformation and particle bombardment. Plant Cell Rep 27:1601–1609
21. Tingay S et al (1997) *Agrobacterium tumefaciens*-mediated barley transformation. Plant J 11:1369–1376
22. Jefferson RA, Kavanagh TA, Bevan MW (1987) GUS fusions: β-glucuronidase as a sensitive and versatile gene fusion marker in higher plants. EMBO J 6:3901–3907
23. Sambrook J, Fritsch EF, Maniatis T (1989) Molecular cloning: a laboratory manual, 2nd edn. Cold Spring Harbor Laboratory Press, Cold Spring Harbor, New York, pp 9.31–9.62

Chapter 2

Transient Gene Expression in Epidermal Cells of Plant Leaves by Biolistic DNA Delivery

Shoko Ueki, Shimpei Magori, Benoît Lacroix, and Vitaly Citovsky

Abstract

Transient gene expression is a useful approach for studying the functions of gene products. In the case of plants, *Agrobacterium* infiltration is a method of choice for transient introduction of genes for many species. However, this technique does not work efficiently in some species, such as *Arabidopsis thaliana*. Moreover, the infection of *Agrobacterium* is known to induce dynamic changes in gene expression patterns in the host plants, possibly affecting the function and localization of the proteins to be tested. These problems can be circumvented by biolistic delivery of the genes of interest.

Here, we present an optimized protocol for biolistic delivery of plasmid DNA into epidermal cells of plant leaves, which can be easily performed using the Bio-Rad Helios gene gun system. This protocol allows efficient and reproducible transient expression of diverse genes in *Arabidopsis*, *Nicotiana benthamiana* and *N. tabacum*, and is suitable for studies of the biological function and subcellular localization of the gene products directly *in planta*. The protocol also can be easily adapted to other species by optimizing the delivery gas pressure.

Key words: Plant, Biolistic gene delivery, Transient expression, Bio-Rad Helios gene gun system, Leaf epidermis

1. Introduction

Ectopic gene expression in living organisms is an important mean for studying the function of the gene products. To this end, it would be ideal to obtain transgenic plants that stably express a gene of interest. However, production of such transgenic plants requires a significant time investment, and transgene loci are often silenced transcriptionally and/or post-transcriptionally. Moreover, the location of transgene insertions in the genome and the resulting gene expression variability may confound data interpretation (1). Thus, it is crucial to develop an efficient, reproducible, and rela-

Stephan Sudowe and Angelika B. Reske-Kunz (eds.), *Biolistic DNA Delivery: Methods and Protocols*, Methods in Molecular Biology, vol. 940, DOI 10.1007/978-1-62703-110-3_2, © Springer Science+Business Media, LLC 2013

tively simple methodology for transient gene expression in plant tissues.

Currently, at least four types of basic approaches are available for transient gene expression in plants: polyethylene glycol (PEG)-mediated or electroporation-mediated transformation of protoplasts, infiltration of *Agrobacterium tumefaciens* (agroinfiltration), and biolistic bombardment. PEG-mediated and electroporation-mediated transformation of protoplasts work efficiently in some plant species (2, 3), but both are time-consuming and only allow for studies in isolated protoplasts, which notoriously do not reflect the biology of plant tissues. Transient gene expression by agroinfiltration represents a relatively non-invasive and cost-effective method that enables fine tuning of the transgene expression levels by changing the concentration of the *Agrobacterium* cell inoculum (4, 5), and is a favored technique for several plant species, such as tobacco (*Nicotiana tabacum*) or *Nicotiana benthamiana*. However, *Agrobacterium* infection induces changes in gene expression pattern of specific sets of genes, including defense-related genes (6–8), and also interferes with host RNA silencing pathways (9), introducing a potential bias into the experiments' outcome and interpretation. Moreover, this technique does not work well in leaves of many plants, including *Arabidopsis*, which is the most widely used model species for plant biology research.

An alternative approach to DNA delivery for transient gene expression, which circumvents many shortcomings of agroinfiltration, is microbombardment (10–12). Here, we describe a protocol for delivery of plasmid DNA into the epidermis of plant leaves by microparticle bombardment, which can be easily achieved using the Bio-Rad Helios gene gun system. Our technique is characterized by its high efficiency, reproducibility, and suitability for transient expression of functional proteins with diverse biological activities and different patterns of subcellular localization in *Arabidopsis*, *N. benthamiana*, and *N. tabacum*. The technique can easily be adapted to other species by optimizing the delivery gas pressure.

Figure 1 illustrates expression levels and localization patterns of different proteins expressed by this procedure. Using microbombardment with 0.6-μm gold particles prepared with the protocol described here, on average 4 ~ 8 cells expressing the unfused YFP are observed under 10× objective lens in 600 μm × 600 μm area (panel A), exemplifying the transformation efficiency of the technique. We also demonstrated the application of this technique to study protein localization *in planta*. Agrobacterium VirE2, VirE3, and VirF are previously demonstrated to localize to cell nucleus, and the *Arabidopsis* protein VirE2-interacting protein 1 (VIP1) is shown to be required for the targeting of VirE2 (13, 14). The biolistic bombardment technique was successfully utilized to analyze the localization of CFP-VirF fusion protein (panels B, C). Furthermore,

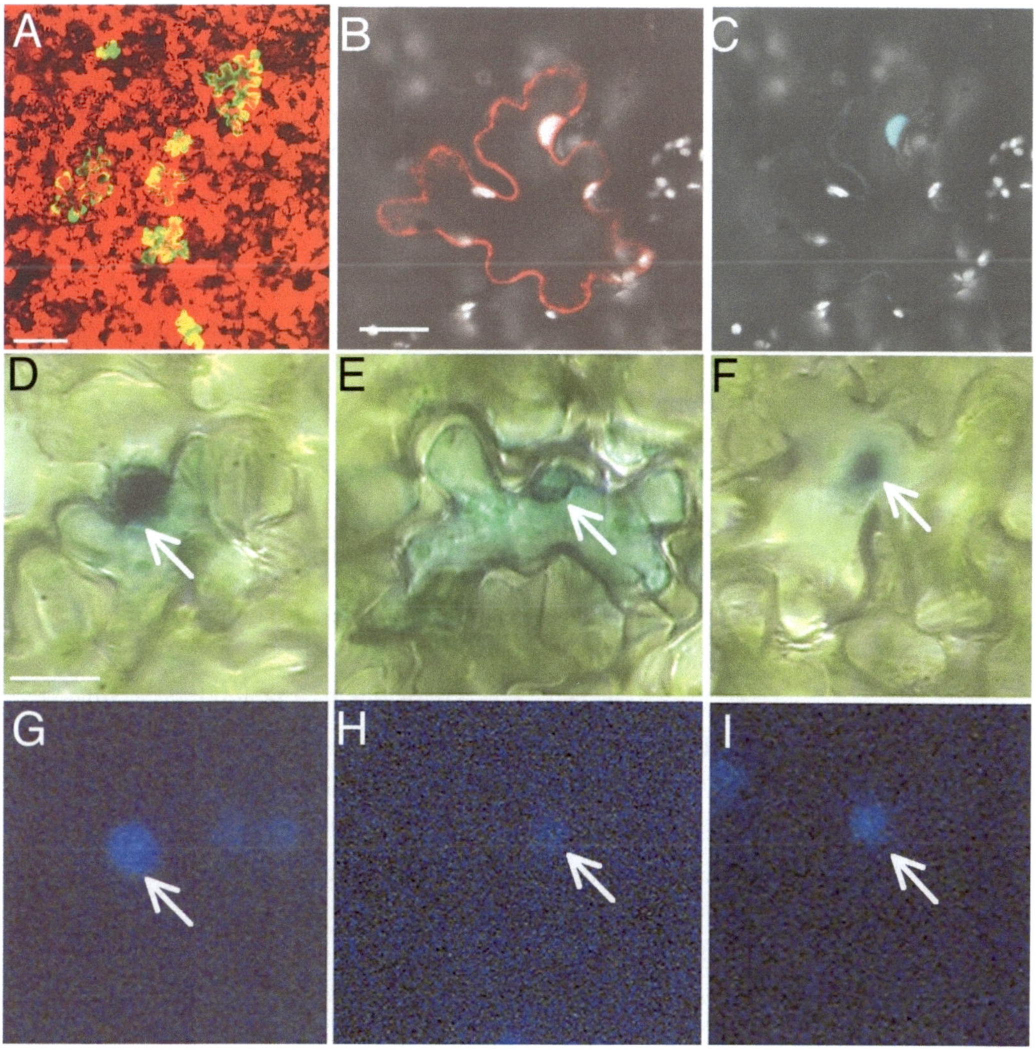

Fig. 1. Examples of transient gene expression in leaf tissues following microbombardment. (**a**) Expression of a tandem repeat of the YFP gene in *N. tabacum* cv. Turk leaf (bar = 100 μm), YFP expressing cells in *yellow* and chlorophyll autofluorescence in *red*. (**b**, **c**) Expression of free DsRed and CFP-VirF fusion in a leaf cell of *N. benthamiana* (bar = 20 μm), showing merged image with DsRed in *red*, CFP-VirF in *blue*, and chlorophyll autofluorescence in *white* (**b**) and CFP-VirF alone (**c**). For (**a–c**), observations were performed under a confocal microscope (Zeiss, LSM5 Pa), 24 h after microbombardment. (**d–i**) Localization of β-glucuronidase (GUS)-VirE2 fusion in *A. thaliana* leaves (bar = 10 μm). GUS-VirE2 is targeted to the nucleus in wild-type *N. tabacum* (**d**, **g**), whereas it is essentially cytoplasmic in *vip1*-antisense *N. tabacum* (**e**, **h**); in double transgenic *vip1*-antisense plants expressing VirE3, GUS-VirE2 nuclear localization is restored (**f**, **i**). Panels **D–F** represent GUS staining, and panels **G–I** represent DAPI staining. *Arrows* indicate cell nuclei. Histochemical GUS assay was done 24 h after microbombardment, and observation were performed after 3 h staining (as described in (13)).

a functional assay allowed us to show that β-glucuronidase (GUS)-tagged VirE2, which accumulates in the cell nucleus in wild-type tobacco (panels D, G), is localized in the cytoplasm in a *vip*1 antisense background (panels E, H), and that co-expression of VirE3 restores the GUS-VirE2 nuclear localization (panels F, I) (14).

In addition, this method was used with *Arabidopsis thaliana* to demonstrate cell-to-cell movement of the *Tobacco mosaic virus* movement protein (MP) tagged with YFP (11, 12). These data indicate that the protocol described in this article can be efficiently utilized to demonstrate intracellular localization and function of different proteins in plant tissues.

2. Materials

1. Eppendorf tubes.
2. Microcentrifuge.
3. Ultrasonic cleaner (see Note 1).
4. Vortex mixer with adjustable speed.
5. Bio-Rad Helios Gene.
6. Bio-Rad Helios cartridge preparatory station (see Note 2).
7. Bio-Rad Tubing cutter.
8. Tank with Helium gas with Bio-Rad Helium gas regulator.
9. Tank with dry N_2 gas.
10. Window screen mesh, cut into 10×10 cm squares.
11. Flat Styrofoam surface (e.g., a lid of a Styrofoam box).
12. Epifluorescence or confocal microscope.
13. Plant growth chamber.
14. Pro-Mix BX.
15. *A. thaliana* plants (4–6-week old) *N. benthamiana*, or *N. tabacum* (7–10 weeks) plants.
16. Plasmid DNA for expression of the gene of interest (*at* >0.5 μg/μL in H_2O, up to 50 μg total), (see Note 3).
17. Gold microparticles, 0.6 or 1.0 μm diameter (Bio-Rad or other brand).
18. Absolute ethanol (see Note 4).
19. Bio-Rad Tefzel tubing.
20. Bio-Rad Polyvinylpyrolidone (PVP), MW 360,000 (included in the Tefzel tubing kit).
21. 5 mL Syringe (without needle).
22. Double-distilled water (ddH_2O), autoclaved.
23. Spermidine stock solution: 3.0 M spermidine in ddH_2O, stored at −20°C.

24. PVP stock solution: PVP in absolute ethanol at 20 mg/mL, stored at −20°C.
25. $CaCl_2$ solution: 1 M $CaCl_2$. Autoclave and store at room temperature.
26. Cotton balls.
27. Scintillation vials.
28. Drierite.
29. Whatmann filter paper.
30. Petri dishes.
31. Parafilm.

3. Methods

3.1. Plant Growth

Grow plants in environmental chamber with appropriate photoperiod cycle and humidity.

1. For *Arabidopsis*, grow one or two plants on Pro-Mix BX in a pot (10×10×10 cm) in an environment-controlled chamber with a short photoperiod (8 h of 130–150 μE/m^2 s light at 23°C/16 h dark at 20°C), and 40–65% relative humidity for 6–8 weeks (15) (see Note 5).
2. For *N. benthamiana* and *N. tabacum*, grow one plant on Pro-Mix BX in a pot (20 cm×20 cm×20 cm) in an environment-controlled chamber with a long photoperiod (16 h of 130–150 μE/m^2 s light at 23°C/8 h dark at 20°C) and 40–65% relative humidity for 7–10 weeks.
3. Supplement plants occasionally with commercially available fertilizers following manufacturers' instructions (see Note 6).

3.2. Preparation of Working Solutions

Prepare spermidine working solution (50 mM in ddH_2O) and PVP working solution (50 μg/mL in ethanol) from stock solutions. These solutions need to be prepared fresh just prior to experiments.

3.3. DNA Precipitation Onto the Surface of Gold Microparticles

1. Weigh 12 mg of gold microparticles and transfer into a 1.5 mL microcentrifuge tube.
2. Add 100 μL of spermidine working solution.
3. Sonicate the mixture for 10 s, then vortex the tube vigorously for 10 s in order to disperse gold particles (see Note 7).
4. Add 25–50 μg of plasmid DNA, in a maximal volume of 100 μL (ideally 50 μL), to the gold microparticle suspension

(when more than one plasmid are used, mix them thoroughly before adding plasmid DNA to the gold particles).

5. Sonicate the tube for 10 s, then vortex vigorously for 5 s at full speed (see Note 8). Lower the speed of vortex.
6. Open the tube, while continuing to mix the gold microparticle suspension (you must vortex with the lid open without spilling the suspension from the tube), add 100 μL of 1.0 M $CaCl_2$ slowly, drop by drop, waiting 5 s between each drop (see Note 9).
7. Allow the suspension to settle at room temperature for about 10 min.
8. Meanwhile, connect the Tefzel tubing to N_2 flow and dry the inner wall of the tubing for at least 5 min.
9. Centrifuge at 10,000 rpm for 30 s in a microcentrifuge to collect the gold microparticles. Remove the supernatant without disturbing the pellet.
10. Add 1 mL of absolute ethanol to the gold microparticles, resuspend, and centrifuge at 10,000 rpm for 30 s to wash the particles.
11. Repeat the wash twice more (a total of three washes) and remove the supernatant completely.
12. Resuspend the gold microparticles in 0.5 mL PVP-ethanol solution and transfer the microparticle suspension to a 15-mL conical tube. Wash the microcentrifuge tube with 0.5 mL PVP-ethanol solution to collect the microparticles as much as possible and add them to the conical tube.
13. Adjust the total volume of microparticle suspension in the conical tube to 3.0 mL with PVP-ethanol.
14. Sonicate the resulting mixture for 10 s to disperse the gold microparticles before proceeding to next step (see Note 10).

3.4. Cartridge Preparation with Tubing Prep Station

1. Close the N_2 flow after drying the Tefzel tubing.
2. Remove the tubing from the Tubing Prep Station, and load it with the DNA-coated gold microparticle suspension using a 5-mL syringe connected to the Tefzel tubing via a short segment of flexible Tygon tubing (see Note 11).
3. Place the Tefzel tubing horizontally in the Tubing Prep Station immediately after loading.
4. Allow the Tefzel tubing with gold microparticle suspension to lie for 5 min for 1-μm gold or for 15 min for 0.6-μm gold, to settle the microparticles on the inner surface of the Tefzel tubing (see Note 12).

5. Remove ethanol from the tubing, using the 5-mL syringe. After ethanol removal, the gold microparticles must remain on the inner surface of the Tefzel tubing.
6. Turn the tubing 180°, wait for 5 s, then rotate the tubing at a speed of 60 rpm for 30 s.
7. Open the N_2 flow for 10 min to dry the tubing.
8. Cut the microparticle-loaded Tefzel tubing into 1-cm-long segments (cartridges) using the Tubing Cutter supplied with the Tubing Prep Station. A 70-cm-long Tefzel tubing loaded with gold microparticle prepared using this protocol should yield approximately 50 cartridges.
9. The cartridges can be kept at −20°C in a scintillation vial containing drying agent, such as silica gel or Drierite. Overlay the drying agents with a cotton ball to secure the drying agent particles to the bottom of the container, place the prepared cartridges on top of the cotton ball, and tightly close the vial with its lid. The cartridges can be stored for several months in dry environment at −20°C.

3.5. Microbombardment

1. Select well-expanded leaves from plants. Leaves with size larger than 50 mm × 70 mm for *N. benthamiana*, 100 mm × 125 mm for *N. tabacum* (these length measurements do not include petiole)or 15 × 35 mm for *A. thaliana* (the length measurement includes petiole) work well.
2. Remove the selected leaves with a sharp razor blade and immediately place them with the abaxial sides facing up onto a flat Styrofoam surface. The abaxial side of the leaf represents a better substrate for bombardment because of its lower trichome density and thinner cuticle (see Note 13).
3. Adjust the pressure of the gene gun at 100–120 psi for 1-µm gold microparticles, and 160–180 psi for 0.6-µm gold for *N. benthamiana* and *N. tabacum*, and at 90–110 psi for 1-µm gold and 140–160 psi for 0.6-µm gold for *Arabidopsis* (see Note 14).
4. Load the cartridge into the gun and shoot. Hold the tip of the barrel liner of the gene gun as close as possible to the leaf tissue, and aim to the center of the leaf mid-rib. Note that this procedure is aimed at transformation of the epidermal cell layers of the leaf (see Note 15).
5. Place the leaves into a Petri dish over three layers of wet Whatman filter paper, seal the Petri dish with Parafilm, and leave it in the dark at room temperature for 16–48 h to allow expression of the delivered transgene (see Note 16).
6. Analyze the transformed tissue under a confocal microscope (Fig. 1).

4. Notes

1. Ultrasonic cleaner for jewelry or glasses, such as Misonix ultrasonic cleaner, frequency 40 Hz.
2. Helios Gene Gun system and Tubing Prep Station are set up and used according to the manufacturer's instructions.
3. Use plasmid purified using common commercial kits, such as Qiagen.
4. The ethanol bottle has to be opened freshly before each experiment, since ethanol that has absorbed moisture from air tends to give poor results.
5. The short photoperiod is required to obtain larger leaves from *Arabidopsis* plants; growing plants under long photoperiod conditions will result in much smaller leaves, which are less convenient for the experiments.
6. Ensure that the plants are healthy and well maintained. Leaves harvested from plants grown under inappropriate conditions yield poor transformation efficiency.
7. Dispersing the gold microparticles by sonication and vortexing is required for uniform DNA coating of the particle surface.
8. Poor quality of DNA-coated microparticles leads to low expression level. To achieve high transformation efficiency, DNA solution with a concentration higher than 0.5 mg/mL should be used.
9. The 1.0 M $CaCl_2$ solution has to be added slowly, while the microparticle suspension is constantly mixed, for an even binding of DNA on the surface of the gold particles.
10. Sonication at this step is required for efficient particle loading into the Tefzel tubing in step 9.
11. Minimize the handling time for Subheading 3.3, 3.4, 3.5, and 4.
12. Insufficient amount of microparticles loaded into the cartridge may lead to low expression levels. The Tefzel tubing must remain stationary after the microparticle loading precisely as described in this Step.
13. For easier handling of small leaves (e.g., leaves from *Arabidopsis*), cover the leaves with a piece of window screen mesh and secure the mesh with pushpins to the Styrofoam surface. Maintaining leaves flat using the window screen mesh increases the efficiency of the particle delivery and minimizes the damage to the tissue during the bombardment.
14. The weight of the 0.6-μm microparticle is as ~22% of 1-μm microparticles, so the same pressure using the microparticle

with different size results in different speed of the particles reaching the epidermis. Therefore, for transformation of other plant species with thin leaves, using 0.6-μm microparticle is preferable.

15. Even well-prepared cartridges could give low expression levels because of inappropriate bombardment conditions. Use the pressure values indicated in the text for *N. benthamiana* and *Arabidopsis*. For other plants, the pressures should be determined empirically. Too low pressure will give poor transformation whereas too high pressure will damage the cells.
16. The time period between the bombardment and the microscopy/activity assay should be determined empirically. For example, for imaging GFP and its different spectral variants, 24–36 h is usually sufficient, whereas for imaging of DsRed2, expression/protein maturation time of 48 h may be required.

Acknowledgements

We apologize to colleagues whose original works have not been cited due to the lack of space. The work in our laboratory is supported by grants from NIH, NSF, USDA NIFA, and BARD to VC.

References

1. Janssen BJ, Gardner RC (1990) Localized transient expression of GUS in leaf discs following cocultivation with *Agrobacterium*. Plant Mol Biol 14:61–72
2. Abel S, Theologis A (1994) Transient transformation of Arabidopsis leaf protoplasts: a versatile experimental system to study gene expression. Plant J 5:421–427
3. Yoo SD, Cho YH, Sheen J (2007) Arabidopsis mesophyll protoplasts: a versatile cell system for transient gene expression analysis. Nat Protoc 2:1565–1572
4. Yang B et al (2000) The virulence factor AvrXa7 of *Xanthomonas oryzae* pv. *oryzae* is a type III secretion pathway-dependent nuclear-localized double-stranded DNA-binding protein. Proc Natl Acad Sci USA 97:9807–9812
5. Goodin MM et al (2002) pGD vectors: versatile tools for the expression of green and red fluorescent protein fusions in agroinfiltrated plant leaves. Plant J 31:375–383
6. Ditt RF, Nester EW, Comai L (2001) Plant gene expression response to Agrobacterium tumefaciens. Proc Natl Acad Sci USA 98: 10954–10959
7. Ditt RF et al (2006) The Arabidopsis thaliana transcriptome in response to Agrobacterium tumefaciens. Mol Plant Microbe Interact 19:665–681
8. Zaltsman A et al (2010) Agrobacterium induces expression of a host F-box protein required for tumorigenicity. Cell Host Microbe 7:197–209
9. Dunoyer P, Himber C, Voinnet O (2006) Induction, suppression and requirement of RNA silencing pathways in virulent Agrobacterium tumefaciens infections. Nat Genet 38:258–263
10. Taylor NJ, Fauquet CM (2002) Microparticle bombardment as a tool in plant science and agricultural biotechnology. DNA Cell Biol 21:963–977
11. Ueki S et al (2009) Functional transient genetic transformation of Arabidopsis leaves by biolistic bombardment. Nat Protoc 4:71–77
12. Ueki S et al (2010) A cell-to-cell macromolecular transport assay in planta utilizing biolistic bombardment. J Vis Exp 42:2208

13. Tzfira T, Vaidya M, Citovsky V (2004) Involvement of targeted proteolysis in plant genetic transformation by Agrobacterium. Nature 431:87–92
14. Lacroix B et al (2005) The VirE3 protein of Agrobacterium mimics a host cell function required for plant genetic transformation. EMBO J 24:428–437
15. Rivero-Lepinckas L, Crist D, Scholl R (2006) Growth of plants and preservation of seeds. Methods Mol Biol 323:3–12

Chapter 3

Transformation of Nuclear DNA in Meristematic and Embryogenic Tissues

Mulpuri Sujatha and K.B.R.S. Visarada

Abstract

Particle bombardment/biolistic delivery is a very popular method of genetic transformation of diverse targets including cells and intact tissues. Delivery of DNA through particle bombardment is genotype and species independent, nevertheless, an efficient protocol for large-scale generation of transgenic plants through embryogenic tissues with a high (≥80%) shoot regeneration efficiency is a prerequisite. Young embryogenic tissues or multiple shoot buds in early stages of induction are the most suited target tissues for recovery of transgenic plants. We describe the protocol for delivery of foreign genes using particle delivery system (Biorad gene gun, PDS-1000/He) in to the meristematic tissues of embryonic axes derived from mature seeds of castor. With the optimized physical and biological parameters, putative transformants were obtained at a frequency of 1.4% through particle gun bombardment of castor embryo axes. Also, transformation of embryogenic calli of sorghum using particle inflow gun (PIG) is described.

Key words: Biolistics, Castor, Embryo axes, Embryogenic calli, Gene gun, Microprojectile bombardment, Plant transformation, Sorghum, Transgenics

1. Introduction

Biolistics or particle bombardment is a physical method for gene delivery which involves helium pressure for acceleration of microscopic metal particles called microcarriers directly into a broad range of cells and intact tissues (1). Biolistics has no host range limitation and is not restricted to certain specific tissues or cell types, and thus, proved to be a valuable technique for delivering DNA into diverse targets, such as, the cells of plants, animals, microbial species as well as into subcellular organelles. Particle bombardment is one of the most important and efficient method for delivery of DNA into plant cells and has been successfully used

Stephan Sudowe and Angelika B. Reske-Kunz (eds.), *Biolistic DNA Delivery: Methods and Protocols*, Methods in Molecular Biology, vol. 940, DOI 10.1007/978-1-62703-110-3_3, © Springer Science+Business Media, LLC 2013

in genetic transformation of economically important crop plants like cotton, maize, soybean, rice, wheat, sorghum, barley, pine, papaya, etc. The procedure has its own advantages of employing clean gene cassettes, more simplified plasmid constructs, is reproducible, eliminates false positives that are common with vector-mediated transformation, and facilitates simultaneous multiple gene transformations. According to Birch and Bower (2), the ideal components of a system for production of transgenic plants by particle bombardment are, (1) to ensure that the target cells possess competence for regeneration and transformation, (2) minimize cell damage while maintaining a high frequency transfer of an appropriate DNA load, and, (3) efficient selection for transformants. Thus, an optimized regeneration protocol and an efficient selection system are prerequisites for successful transformation of embryogenic tissues.

In embryogenic cultures, plants are regenerated from somatic embryos either through direct embryogenesis or indirectly through a callus step. For establishment of embryogenic cultures, the choice of explant is critical and the cells should have competence for regenerability and transformability. The most widely used targets are embryogenic calli and embryogenic cell suspensions derived from immature zygotic embryos, young leaves, immature inflorescences, pollen, scutellar tissue, immature cotyledons, etc. Embryogenic calli are of two types and both types of calli can be used to generate transgenic plants (3). Type I callus is formed by hard, compact and yellowish tissue, which usually is able to regenerate plants. Type II callus is soft, friable, highly embryogenic and able to regenerate a higher number of plants than type I callus. As particle delivery system facilitates DNA transfer to intact tissues, plant transformation through meristem-based shoot proliferation has also been exploited for production of transgenic plants.

In this chapter we present the protocols for genetic transformation of embryo axes of castor (*Ricinus communis* L.) through particle gun bombardment method and transformation of embryogenic callus cultures of sorghum (*Sorghum bicolor* L.) through particle inflow gun (PIG) method. Each instrument has its own advantages and disadvantages but in most cases the results obtained are the same. While particle delivery system (Biorad PDS-1000/He) gives very high pressures, gene delivery through PIG is simple, cost-effective and requires less time for clean up and set up for each bombardment. The method of using embryo axes bypasses the genotype-dependent tissue culture regeneration procedures involving callus cultures thus enabling the rapid recovery of putative transformants in agronomically desirable cultivars. Particle gun bombardment with embryonic axes as target tissues has been used as a method for production of transgenic plants in soybean (4) and peanut (5) while in our lab we had used it for successful production of stable transgenics in castor (6, 7). For obtaining maximum transient and stable expression events, it is essential to optimize several physical and biological parameters. The physical parameters include the size and type of microparticles, number of microprojectiles used for bombardment, the acceleration parameters, such as, the helium pressure, gap distance (the distance between the rupture disk and macrocarrier), the macrocarrier travel distance (the distance between the microprojectile and the stopping screen), the target distance (distance between the stopping screen and the biological target), concentration of $CaCl_2$, concentration of spermidine used to adsorb DNA to the microprojectiles. The biological parameters include the gene construct, genotype, correct type and stage of target cells, size of the target tissue, cell density, selection system, regeneration capability of the target explant, influence of growth regulators in the medium, the osmolarity of the bombardment medium, the treatment of the target cells/tissues post-bombardment, etc. Regardless of the method of bombardment, each experiment should include at least three types of controls: (1) target tissues without bombardment but subjected to selection (selection control—to determine the selection stringency particularly for meristematic tissues), (2) bombardment with uncoated microcarriers, but without selection (bombardment control—to determine the damage effects due to bombardment), and, (3) without bombardment and selection (regeneration control—to check for regeneration efficiency). Based on the optimization of the key variables individually and in combination, stable transgenics were developed in castor (6) and sorghum using PDS-1000/He and PIG methods, respectively, which could be successfully used for other crops limited by difficulties in tissue culture based regeneration.

2. Materials

Autoclaved ultra pure water is used for preparation of stock solutions and media. All reagents should be tissue culture or molecular biology grade. Follow the local waste disposal regulations.

2.1. Production of Target Tissues for Particle Bombardment

1. MS medium: Murashige and Skoog (8) basal medium (macro- and microelements, vitamins and inositol) with 30 g/L sucrose, 0.8% agar. Adjust pH of the medium to 5.8 with NaOH before adding the gelling agent and autoclaving. All media can be prepared from readymade tissue culture media powders (Himedia, India).

   Target tissues for castor transformation

2. Culture medium: MS medium with 0.5 mg/L thidiazuron (TDZ).

3. Seeds of castor (cv. DCS-9): Decoat mature seeds and surface sterilize with 0.1% (w/v) $HgCl_2$ for 8 min and rinse thoroughly four times in sterile distilled water. Dissect whole embryos with papery cotyledons carefully (see Note 1). Culture excised embryo axes on culture medium (MS medium with 0.5 mg/L TDZ) for 5–7 days in dark at 26±2°C.
4. Selection medium: MS medium with 0.5 mg/L benzylamino purine (BAP) and hygromycin (20, 40, 60 mg/L).
5. Shoot proliferation medium: MS medium with 0.5 mg/L BAP.
6. Shoot elongation medium: MS medium with 0.2 mg/L BAP.
7. Rooting medium: Half-strength MS medium with 1 mg/L naphtheleneacetic acid (NAA).
8. Osmoticum medium: Shoot proliferation medium with 0.2 M sorbitol and 0.2 M mannitol (see Notes 2 and 3).

Target tissues for sorghum transformation

9. Immature embryos (12–20 days after pollination) and young inflorescences (1–2 cm long).
10. Callus induction medium: MS medium with 2.5 mg/L 2, 4-dichlorophenoxyacetic acid (2, 4-D).
11. Medium for multiple shoot bud induction from shoot apices: MS medium with 0.5 mg/L TDZ and 2 mg/L BAP.
12. Selection medium: To the above callus or shoot induction medium, add 50 mg/L kanamycin (Kan) or 50 mg/L hygromycin (Hyg) or 3 mg/L phosphinothricin (PPT) depending on the selection marker gene used.
13. Shoot regeneration medium: MS medium with 2 mg/L BAP and 0.25 mg/L indolebutyric acid (IBA).
14. Rooting medium: Half-strength MS medium with 1 mg/L NAA.
15. Osmoticum medium: Callus induction or multiple shoot bud induction medium with 0.25 M sorbitol and 0.25 M mannitol.

2.2. Solutions for Preparation of Cartridges

1. 0.1 M Spermidine (free base, tissue culture grade): Solid spermidine is hygroscopic. Hence, weigh 0.145 g rapidly and dissolve it in 10 mL of cold sterile pure water in laminar flow. Conduct all operations in ice and in laminar flow hood. Filter sterilize through 0.22 μ membrane into a sterile container. Aliquot 100 μL volumes in 1.5 mL micro centrifuge tubes. Store at −20°C. Discard individual tubes after first use (see Note 4).
2. 2.5 M $CaCl_2 \cdot 2H_2O$ (DNA tends to precipitate in the presence of $CaCl_2$):

Dissolve 1.83 g of $CaCl_2 \cdot 2H_2O$ in water and make up to 5 mL with pure water. Autoclave the solution at 120°C and 15.0 lb pressure for 20 min. Store at 4°C till use (see Note 5).

3. 70% and 100% Ethanol (HPLC grade).

2.3. Bombardment

1. Biorad PDS-1000/He device or PIG.
2. Helium gas cylinder; high pressure (2,400–2,600 psi), 99.995% or higher purity.
3. Vacuum pump: oil filled, pumping speed of 90–150 L/min.
4. Rupture disks (Biorad): 900, 1,100, 1,350 psi—sterilize by briefly dipping in 70% isopropanol just prior to insertion in the retaining cap.
5. Stopping screens (Biorad)—sterilize by autoclaving.
6. Opaque plastic box with lid for sterilization of accessories.
7. Gold/Tungsten particles: Microcarriers (gold—0.6 μm, 1.0 μm or 1.6 μm, tungsten—1.1 μm, 1.3 μm in diameter) (Biorad) (See Notes 6 and 7).
8. Macrocarriers—sterilize by dipping in 70% ethanol; macrocarrier holders—sterilize by autoclaving.
9. Baffles (PIG)—autoclave.
10. 13 mm Filter holder (Gelman Sciences)—autoclave.
11. Whatman No 1 filter paper.
12. Whatman No 4 filter disks.

2.4. Transient GUS Expression Analysis-Standard GUS Histochemical Assay

1. 0.1 M sodium phosphate buffer (pH 7.0): Mix 57.7 mL 1 M Na_2HPO_4 and 42.3 mL 1 M NaH_2PO_4 and make up to 1 L with water.
2. X-GlcA buffer: 50 mM sodium phosphate buffer pH 7.0, 10 mM EDTA, pH8.0, 0.3% (v/v) Triton X-100.
3. X-Gluc solution: Dissolve 5 mg of 5-bromo-4-chloro-3-indolyl β-D glucuronide (X-gluc) sodium salt in 50 μL DMSO and add to 10 mL X-GlcA buffer. Store at 4°C in the dark (brown bottle). The stock is good indefinitely if stored at −20°C.
4. ELISA plates.
5. Stereomicroscope.

2.5. Selection Medium—Stock Solutions for Selection Marker Genes (see Note 8)

1. Kanamycin (stock 25 mg/L)—dissolve in water.
2. Hygromycin (stock 50 mg/L)—dissolve in water.
3. PPT (1.0 mg/L)—dissolve in water.

2.6. Plasmid DNA

1. Plasmid DNA vectors carrying gene of interest, cloned in *Escherichia coli*.

2. Qiagen plasmid isolation kit.
3. TE (Tris-EDTA):

1 M Tris (crystallized free base) Tris(hydroxymethyl) aminomethane, (FW 121.4 g/mol)

Weigh 60.57 g in 0.5 L distilled water and adjust the pH to 7.5 using HCl.

0.5 M EDTA -Diaminoethane tetraacetic acid, (FW 372.2 g/mol) Weigh 18.6 g, dissolve in 100 mL distilled water, adjust the pH to 8.0 using NaOH. EDTA will not be soluble until pH reaches 8.0. We can add a pinch of NaOH in the beginning to easily dissolve it. Use vigorous stirring and moderate heat, if desired.

To make 1.0 L of TE buffer take 10 mL 1 M Tris-Cl pH 7.5 and 2 mL 500 mM EDTA pH 8.0, make it to 1.0 L with distilled water.

3. Methods

Preparation for bombardment should begin a week in advance. All the stock solutions as recommended are to be prepared once in 90 days for better results. All the steps should be carried out in a laminar flow hood to avoid microbial contamination.

3.1. Preparation of Plasmid DNA

1. Prepare purified plasmid DNA (without RNA and A_{260}/A_{280} = 1.8–2.0) using Qiagen plasmid isolation kits (see Note 9).
2. Prepare DNA stock solution of 1.0 or 0.5 μg/μL in TE (Tris-EDTA pH 8.0) buffer.

3.2. Sterilization of Gold/Tungsten Particles (Microcarriers)

1. Weigh 30 mg gold/tungsten particles in a 1.5 mL eppendorf tube.
2. Keep the gold/tungsten containing tube in a water bath at 95°C for 2 h with intermittent inversion and tapping (see Note 10).
3. Add 500 μL of 70% ethanol and vortex vigorously for 1–2 min (see Note 11).
4. Allow the particles to soak in 70% ethanol for 15 min and pellet the particles by spinning in a microfuge for 1 min.
5. Take out the supernatant carefully without removing any microcarriers and wash in sterile ultra pure water three times: Add 500 μL of sterile water, vortex vigorously for 1 min, allow the particles to settle for 1 min, pellet the microparticles by briefly spinning in a microfuge. Remove the supernatant and discard.
6. Finally suspend the pellet in 500 μL of sterile ultra pure water (see Note 12).

7. Vortex thoroughly and aliquot 50 μL stocks into 1.5 mL eppendorf tubes. While aliquoting it is important to continuously vortex the tube from the bottom to get uniform suspension of particles.
8. This brings the microparticle concentration to 3.0 mg/mL.
9. Store at 4°C for use (can be used up to 2 weeks after preparing without losing significant effect on transient expression). Tungsten aliquots can be stored at −20°C to prevent oxidation while gold aliquots can be stored at 4°C.

3.3. Coating of DNA on Gold/Tungsten Particles

Keep all the stocks including DNA ready in ice box. The entire procedure is carried out under sterile conditions in laminar flow hood.

The following procedure is sufficient for six bombardments and it is ideal to prepare DNA-coated carriers in separate batches for 6 shots each.

1. Vortex the tube containing the microcarriers continuously and gently to resuspend and disrupt agglomerated particles and obtain uniform suspension of particles (see Note 13). Take 50 μL of gold/tungsten particles (ca. 3 mg of microcarriers for each DNA preparation) in a 1.5 mL eppendorf tube (see Note 14).
2. Vortex gently and continuously, add in order and along the walls 5 μg plasmid DNA (5 μL of 1 μg/μL stock), 50 μL of $CaCl_2$ (2.5 M; autoclaved) and 20 μL freshly made spermidine (0.1 M)—vortex briefly and tap gently with finger for 2–3 min (see Notes 15 and 16).
3. This gives a DNA concentration of ca. 2 μg/mg of microcarriers (see Note 17).
4. Allow the microcarriers to settle for 5 min on ice.
5. Pellet microcarriers in a microfuge (3,000–5,000 rpm for 1–2 s). Save the supernatant (see Note 18).
6. To the pellet add 140 μL of chilled 70% ethanol (HPLC grade) without disturbing the pellet. Tap with finger for thorough mixing. Pellet it again. Remove the liquid.
7. To the pellet add 140 μL of chilled 100% ethanol (HPLC grade) without disturbing the pellet. Tap with finger. Pellet it again.
8. Remove the liquid. To the pellet add 48 μL of 100% chilled ethanol.
9. Particles must be gently pelleted always (3,000–5,000 rpm for 1–2 s) as high speed enhances particle agglomeration.

10. Gently resuspend the pellet by tapping the side of the tube several times to obtain uniform suspension. DNA-coated microcarriers should be used as soon as possible and within 2 h of preparation.
11. Remove 8 μL aliquots of microcarriers immediately (ca. 500 μg) and transfer them to the center of a macrocarrier. Spread microcarriers over the central 1 cm of the macrocarrier using the pipette tip. Allow it to dry immediately in a small desiccating chamber (a small petriplate containing calcium chloride powder) (see Note 19).
12. Wait until macrocarriers dry (10–20 min) in a sterile environment at RT (a laminar flow hood) (see Note 20).
13. A thin film of gold/tungsten particles can be seen (see Note 21).
14. If fewer bombardments are needed, prepare enough microcarriers for three bombardments by reducing all volumes by one-half.
15. In case tungsten particles begin to clump, the batch should be discarded.

3.4. Transformation of Embryo Axes of Castor by Biorad PDS-1000/He

3.4.1. Production of Target Tissues for Particle Bombardment

1. Use 5–7-day-old germinating embryos for bombardment.
2. Arrange about 50–60 decotyledonated embryo axes on Whatman No 1 filter paper in a circle of 2.5 cm in the center of a 9.0 cm petriplate with osmoticum (see Note 22).
3. Pre-plasmolyse the explants for 2 h prior to bombardment (see Notes 23 and 24).

3.4.2. Preparation of Biorad PDS-1000/He Device

1. The above preparation of DNA-coated particles is sufficient for six petriplates of target tissues using Biorad PDS-1000/He device. Wear safety glasses and latex gloves.
2. Verify that Helium tank has 200 psi in excess of desired rupture disk pressure for bombardment.
3. Sterilize the shelves by wiping with 70% ethanol followed by drying for an hour before bombardment.
4. Sterilize macrocarriers and rupture disks required for the day by dipping for few seconds in 70% isopropanol or absolute alcohol. Dry and use them immediately (see Note 25).
5. Dip all the other metal accessories for half an hour in 70% ethanol and dry before use.
6. Place macrocarriers with DNA in the holders with cap plug and set them in the macrocarrier holders. The macrocarriers must be kept clean and free of grease and should be handled with sterile forceps.

3.4.3. Bombardment with the Biorad PDS-1000/He Device

Follow the manufacturer's instructions for the gun.

1. Do all the following steps in the same order for each shot to avoid confusion and leaving out something.
2. Place a rupture disk into the rupture disk holder using sterile forceps. Ensure that the disk is properly placed in the holder by tapping the holder a few times gently. Screw the holder tightly into the vacuum chamber (see Note 26).
3. Place a stopping screen onto the diaphragm in the flying disc holder assembly. On top of this, place upside down the macrocarrier holder containing a macrocarrier with DNA on it. Place the stage in the second slot from the top in the vacuum chamber.
4. Insert the petriplate with target tissues in the chamber. The sample platform should be positioned in the chamber in the fifth slot from the top (second from below).
5. Place about 50–60 embryo axes on a sterile filter disc (Whatman #4 circles) placed on MS media plate with osmoticum.
6. Take care to see that the meristematic region of the explant faces the particle jet (see Notes 27 and 28; Fig. 1).
7. Shoot the DNA at 900, 1,100, and 1,350 psi as per manufacturer's instructions.
8. Bombardments are done under a vacuum of 27 in. of Hg (see Note 29).
9. The distance from rupture disk to macrocarrier was 25 mm.
10. The macrocarrier flight distance was set at 10 mm.
11. Bombard the tissue once (see Note 30).
12. Take the petriplate out of the chamber and set to one side with the lids replaced. Repeat until all samples have been processed.

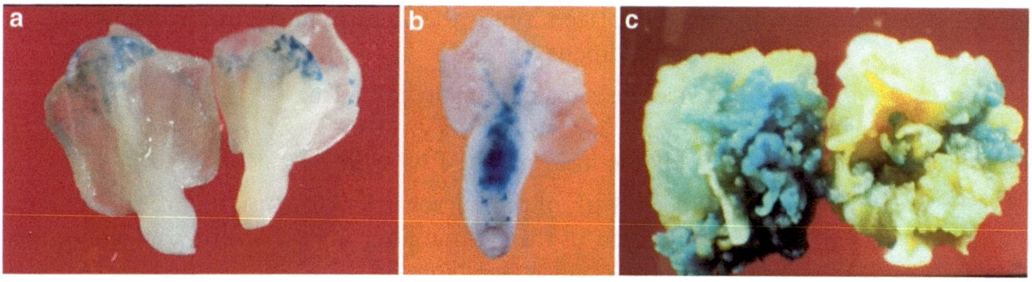

Fig. 1. Particle gun bombardment of embryo axis derived from mature seeds of castor showing the importance of placement of embryos. (**a**) Whole embryos showing GUS expression in the cotyledons, (**b**) decotyledonated embryos placed upright with swollen embryo axis towards the particle jet showing GUS expression in the meristematic region, (**c**) GUS expression in shoots during selection on hygromycin.

After all samples have been shot, seal the petriplates with parafilm and transfer them to the culture room (see Notes 31 and 32).

3.4.4. Post-bombardment Treatment of Castor

1. Following bombardment, incubate the explants at 25°C in dark overnight on osmoticum medium (shoot proliferation medium + 0.2 M each sorbitol and mannitol) (see Note 33).
2. Transfer the explants to shoot proliferation medium (MS medium with 0.5 mg/L BAP) for 15 days.
3. Determine transient GUS expression 24, 48, and 72 h following bombardment.

3.5. Transformation of Embryogenic Callus Cultures of Sorghum by Particle Inflow Gun

3.5.1. Production of Target Tissues for Particle Bombardment

1. Select the embryogenic calli derived from immature embryos and young inflorescences and subculture them to fresh callus induction medium (Subheading 2.1, step 10) 4 d prior to bombardment (see Note 34). Alternatively, isolate immature embryos (12–20 d after pollination) and culture them in callus induction medium for 3–4 days prior to bombardment for direct transformation.
2. While subculturing, calli should be dissected out into 2–4 mm small pieces.
3. Induce multiple shoot buds from shoot apical meristems (20–25 days old) by culture in callus induction medium prior to bombardment.
4. A day before bombardment, check all the subcultured small pieces of calli and select robust calli for use as target tissues.
5. Transfer all the target tissues to the corresponding osmotic medium 4 h prior to bombardment (see Note 24).

3.5.2. Preparation of Particle Inflow gun

1. Sterilize the shelves by wiping with 70% ethanol followed by drying an hour before bombardment.
2. Follow steps 1–7 in Subheading 3.3.
3. Carefully remove 100 μL of the supernatant.
4. Aliquot 4 μL from the remaining mixture into the syringe filter holder placed onto the centre grid. Immediately use for bombardment.
5. Place the remaining mixture on ice.

3.5.3. Bombardment with the Particle Inflow gun

1. Spray the inside of the PIG with 70% alcohol.
2. Pipette 4 μL of the microcarrier suspension onto the center of the screen of the syringe filter unit. Place the remaining mixture back on ice (see Note 35).
3. Reassemble and screw the syringe filter unit (finger tight is enough) into the Luer-lok needle adaptor.

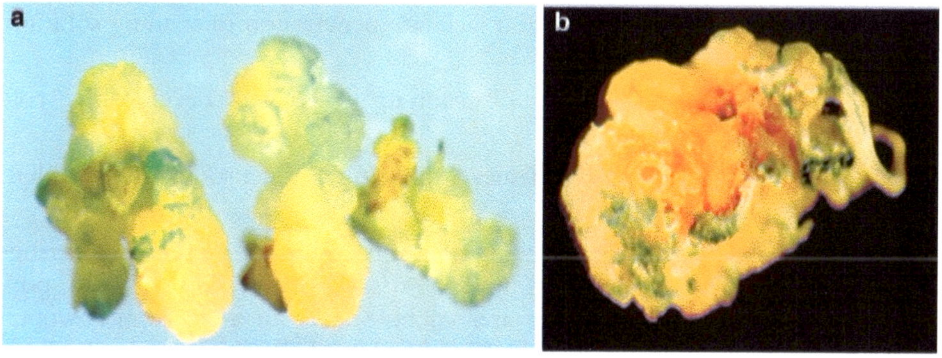

Fig. 2. Particle inflow gun (PIG) transformation of sorghum. (**a**) *gus* expression in calli derived from immature inflorescence, (**b**) *gus* expression in multiple shoot buds in sorghum.

4. Take the first petriplate with pretreated calli, remove the lid and then place a baffle over the top and in the center. Place this on the 11th shelf and directly underneath the filter holder. In case of calli, 12–15 calli are placed per plate.
5. Close the perspex door and latch the handles.
6. Open the vacuum valve and bombard the tissues when the vacuum reaches 90 Kpa.
7. All other operations post-bombardment are similar to that followed for PDS-1000/He method.

3.5.4. Post-bombardment Treatment of Sorghum

1. The following day (12–15 h) after shooting, transfer the bombarded material to corresponding osmotic free medium (multiple shoot buds to the medium described in Subheading 2.1 step 11 and calli to 2.1 step 10). Place individual calli/target tissues separately from each other.
2. Determine transient GUS expression 24, 48, and 72 h following bombardment (Fig. 2).
3. Incubate the targeted tissues for one week before transferring to selection medium.
4. Calculate transformation efficiency on the basis of at least three independent experiments with each having an adequate number of replicates and explants/plate. It is determined as the number of surviving shoots after the final cycle of selection over the total number of explants bombarded.

3.6. Analysis of GUS Expression

1. Transfer some of the shooted calli, embryo axes, and shoot apices to small wells in ELISA plates in sterile fumehood.
2. Cover the tissue with standard X-GlcA solution. Calli should not be dipped in the solution. Vacuum infiltrate for 5 min to

remove air and facilitate penetration of staining solution (see Note 36).

3. Incubate at 37°C for up to 16 h to develop blue color of product.
4. Cells and tissues expressing β-glucuronidase turn blue (see Note 37).
5. Remove staining solution and wash with several changes of 70% ethanol until the tissue clears. Count the blue spots and measure the area of blue spots in a stereomicroscope after approximately 12 h. Chlorophyll clearing can also be done with acetone or 10% lactic acid.
6. Calculate comparative evaluation of different bombardment parameters (see Note 38).

3.7. Selection and Regeneration of Castor

1. Select putative transformed shoots after 3 cycles of selection on selection medium for 15 days each with 20 mg/L hygromycin (first selection cycle), 40 mg/L hygromycin (second selection cycle) and 60 mg/L hygromycin (third selection cycle) (see Note 39).
2. Transfer the shoots recovered following selection to shoot proliferation medium for shoot proliferation and maintain for 15 days.
3. Subculture shoots to shoot elongation medium for shoot elongation and maintain for 15 days.
4. Transfer elongated shoots to rooting medium for rooting.

3.8. Selection and Regeneration of Sorghum

1. After one week of resting period, transfer the calli to the corresponding selection medium.
2. Selection agent depends on the plant selection marker gene in the vector.
3. For *npt II* gene 50 mg/L kanamycin, for *hpt* gene 50 mg/L hygromycin, and for *bar* gene 3.0 mg/L PPT are added to the medium.
4. Displace calli every third day in the medium for effective selection pressure.
5. Subculture calli regularly after every 15 d to fresh selection medium (see Note 40).
6. After completion of 3 cycles of selection, transfer calli to shoot regeneration medium.
7. Transfer elongated shoots to rooting medium.
8. Acclimatize rooted plantlets in soil.

4. Notes

1. Regular availability of immature embryos is not always possible. Embryo axes from mature seeds proliferate on medium with high cytokinin and represent useful targets for routine bombardment experiments.
2. Biolistic transformation procedures cause physical trauma and osmotic pretreatment of target tissues results in plasmolysis of cells and thereby reduces damage by preventing extrusion of the protoplasm from bombarded cells and thus, improving the cell viability.
3. Instead of mannitol and sorbitol, 10–12% sucrose or maltose can also be used to maintain the osmolarity.
4. After adjusting the concentration to 0.1 M, spermidine can be stored in small aliquots in a deep freezer (−80°C). At the start of bombardment experiments, when the solution is being used, the tubes are kept in a −20°C freezer and used within 2 weeks from thereon. Spermidine containing tube is used only once. Hence, the stock volume can be reduced to 5 mL, if possible.
5. 2.5 M $CaCl_2$ could be replaced with 1 M $Ca(NO_3)_2$ (9).
6. Choice of microcarriers: Tungsten particles are less expensive than gold but are more heterogenous in size and shape and are known to be toxic to some cell types. Gold particles do not degrade DNA but tend to agglomerate irreversibly in aqueous solutions and hence, need to be prepared fresh on the day of bombardment.
7. Regardless of the microcarrier type, there is a reduction in transformation frequency with increase in particle size. Larger particle size probably destroys the cells upon bombardment.
8. Some plants show high level of inbuilt resistance to antibiotics. Cereals, in general, have high levels of intrinsic tolerance to antibiotics such as kanamycin, geneticin, paromomycin and hence, hygromycin and bialophos/ppt are used in most studies.
9. DNA should be very pure (free of RNA or protein) to prevent clumping of microprojectiles.
10. Heating of microcarriers reduces particle agglomeration and significantly enhances the transformation efficiency.
11. Ethanol keeps the particles sterile and also maintains them in their dispersed condition. It is important to use the driest ethanol possible, and hence, use freshly opened bottle of ethanol each time for preparation of DNA-coated microcarriers and also the sterilization of rupture disks and stopping screens. Otherwise make aliquots of 1 mL and put them in −20°C

freezer. When using 70% ethanol, prepare it afresh just before the experiment. Any residual moisture will result in uneven particle delivery during shooting, or in extreme cases will prevent dislodging of the particles carrying DNA by the helium gas after firing.

12. In general, 50% sterile glycerol is recommended for storing the microcarriers. Storing microcarriers (gold/tungsten) particles in pure water rather than 50% glycerol was found to reduce particle agglomeration and obtain uniform film of spread on macrocarriers. For particle preparation, 50% glycerol could be substituted with 40% PEG as precipitation of DNA is complete besides easy resuspension of particles in PEG as compared to glycerol (10).
13. Care should be taken to make the precipitation reaction mixture as homogenous and reproducible as possible as it is the most important source of variation affecting the transformation efficiency.
14. Testing of microcarrier concentrations at 1.5, 3.0, and 6.0 mg per bombardment mix indicated that lower concentrations of particles are optimal with a threefold reduction in transient GUS expression with 6.0 mg of microcarriers.
15. Coating of microcarriers with DNA should be done as quickly as possible. In case of a delay between addition of any one of the components, the mixture should be kept on ice.
16. If cobombardment is performed, then equimolar concentrations of the two plasmids are mixed for co-transformation.
17. In general, 0.5–8.0 μg of DNA quantity could be used. Increasing the quantity of DNA above a certain concentration (2.0 μg per mg tungsten) did not result in the appearance of larger number of expression units. Relatively high concentrations of DNA resulted in severe aggregation of the microprojectiles and the resulting aggregates are not readily dispersed by vortexing. Larger aggregates of microprojectiles apparently are not effective for DNA delivery.
18. Supernatant in the first step of coating gold particles should be run for 10 min in the agarose gels. After ethidium bromide staining DNA should not be visible. This ensures that the entire DNA is coated on metal particles.
19. Macrocarriers deposited with microcarriers can be dried in individual "dessicators" (petriplates with $CaCl_2$) immediately upon application, and should be used within 2 h of drying.
20. After spreading all the coated gold/tungsten particles onto macrocarriers 10 μL of TE buffer can be added to the tube in which microcarriers were coated with DNA and run on agarose gels. After ethidium bromide staining DNA should be visible. This ensures that DNA is coated on gold particles.

21. When the DNA-coated microcarriers are placed on the macrocarrier, it should spread out into a fine circular layer ca. 1 cm in diameter of evenly dispersed gold/tungsten, followed by complete drying of ethanol. Improper spread indicates the presence of water in the gold/tungsten sample. The problem can be fixed by washing the gold/tungsten particles a few more times in absolute ethanol making sure that the ethanol is from freshly opened bottle. If the gold/tungsten on the disc looks clumpy, it indicates that the particles were not sufficiently dispersed during the ethanol washes. The sample can be pipetted up and down several times before dispensing on the disc.
22. Explants placed on filter paper as solid support is essential to ensure that the tissue is not dislodged from the plate during bombardment.
23. Unlike callus cultures, meristems have a streamlined topology and are outlined by relatively thick cell walls that contain higher turgor stress of meristematic cells. Culturing of explants on media with increased osmoticum prior to bombardment releases the interior turgor of the meristems and consequently reduces the pressure shock wave in the cells and the deleterious effects of particle penetration leading to increased transient gene expression.
24. Calli/explants could be cultured on osmotic medium for 0, 4, 12, 24, and 48 h prior to bombardment and analyzed for transient expression to determine the optimal time for osmotic adjustment.
25. Each rupture disk and macrocarrier should be dried separately from each other, else they will stick to each other. We sterilize them and store them for a week in petriplates lined with paper towels. Do not soak rupture disks in 70% isopropanol for more than a few seconds as it may delaminate the disks.
26. Rupture disk failure is a common problem and it is overcome by tightening the rupture disk retaining cap onto the gas acceleration tube as much as possible.
27. Placement of embryos for bombardment was found to be critical in castor. Preculture of embryo axis on cytokinin medium resulted in swelling of the embryonic axis region. Placing the explants upright with the swollen embryo axis portions closer to the particle accelerator enabled bombardment and particle penetration only in the embryonic meristematic region which is in an active state of division due to the influence of TDZ—a highly potent cytokinin (Fig. 1).
28. Shoot meristems are small and highly organized targets for biolistic transformation. These are organized in two to three germ layers and germline cells are presumed to originate from the cells of the subepidermal layer (11). Gene transfer targeted

to these cells pave way for development of transgenic floral organs which have the potential to yield transgenic progeny. Therefore, meristem transformation requires a system that allows precise targeting and controlled delivery and equal distribution of dispersed particles at a high density.

29. In general, the optimum helium pressure range for any crop species is narrow. Because, microprojectile bombardment will result in cell damage, it is advisable to use more gentle conditions for bombardment than those which are found optimum for transient expression for obtaining maximum stable expression.
30. Generally, 1–3 shots are done with 2–4 h duration between bombardments. Repeated bombardments allow better coverage of the target area and compensate for misfires from faulty and poorly set rupture disks and thereby increase the frequency of transformation of the target tissues. However, increased number of bombardments could result in greater tissue damage, impaired cell proliferation and regeneration. In case of castor embryo axes, increased number of bombardments resulted in reduced frequency of transient and stable expression events (6).
31. For crop species where bombardment parameters are standardized, minor changes in few parameters could always give the optimal results. However, in crops like castor where information is rather limited, it is essential to determine through transient assays the effects of different variables individually as well as in a fractional factorial experiment design to study their interactions in order to define the ideal conditions for bombardment.
32. In our laboratory, at least two researchers work simultaneously to process and transfer the target material to osmoticum medium, coating of DNA onto the microcarriers and performing bombardments. This enables completion of bombardment of 4–5 batches comprising of six plates each and thus, accounting for a total of 1,200–1,400 embryo axes each day.
33. Post-bombardment osmotic treatments favor a high frequency of transient expression besides a higher recovery of stable events (12).
34. Age of the calli/target tissues plays a crucial role and capacity of regeneration into morphologically normal green plants diminishes with prolonged period in tissue culture. As a consequence, the transformation frequency declines drastically with increased time in culture. In embryogenic calli, the time period from the latest subculture to bombardment is critical as the peak transient GUS expression corresponds to the peak mitotic index which varies significantly for the crop and the target tissue.

Thus, regular subculture of callus is necessary for stimulating callus growth and shoot differentiation.

35. In PIG method, it is important to use fresh and sterile filter holders and baffles each time a new plasmid mixture is used to avoid cross-contamination between the DNA mixtures. Unlike in PDS-1000/He system where most of the consumables are disposable, in PIG, filter holders and baffles are reusable. Hence, the used filter holders and baffles are soaked for a day in a mixture of water and detergent followed by rinsing and autoclaving to prevent cross-contamination.
36. Since X-GlcA is expensive, we economize its use by putting only the meristematic regions (epicotyls) or plant parts of interest rather than the whole explant into X-GlcA staining solution and reusing it one more time.
37. One has to be cautious with false positives caused by endogenous GUS-like activity which most often is less intense and less discrete than real GUS activity. Castor microspores show endogenous GUS activity which is as intense as that of the transformed cells.
38. GUS foci: For determining the efficiency of transient expression, blue spots (cells or clumps of cells) are counted under a stereo microscope. Clumps of blue cells are scored as a single spot. Average number of blue spots per plate is collected from at least six plates and transformation efficiency is calculated as the mean number of blue spots per plate. Since we had targeted the swollen embryonic axis region, frequency of transient GUS expression was calculated as the number of explants showing intense GUS expression to the total number of explants stained following bombardment and is expressed as percentage. Counting GUS spots manually under microscope is a tedious task. The transient GUS expression in plant transformation experiments could be monitored using image analysis. The image can be captured and subjected to analysis with software that counts the number of blue-stained GUS positive spots as well as the area and percent of the total area of each spot.
39. Time scale for production of transgenic plants ready to be acclimatized varies with the type of selection agent and also the concentration of selection agent. After determining the lethal dose, a supra lethal dose is used for stringent selection of putative transformants.
40. During selection, the calli turn brown and dark. Though the entire callus turns dark it may be continued for subculture and regeneration because the transformed tissues respond slowly.

References

1. Sanford JC et al (1987) Delivery of substances into cells and tissues using a particle bombardment process. Particul Sci Technol 5:27–37
2. Birch RG, Bower R (1994) Principles of gene transfer using particle bombardment. In: Yang N, Christou P (eds) Particle bombardment technology for gene transfer. Oxford University Press, New York, pp 3–38
3. Vasil V, Vasil IK (1981) Somatic embryogenesis and plant regeneration from tissue cultures of *Pennisetum americanum* and *P. americanum* x *P. purpureum* hybrid. Am J Bot 68: 864–872
4. McCabe DE et al (1988) Stable transformation of soybean (*Glycine max* L.) by particle acceleration. Bio/Technol 6:923–926
5. Brar GS et al (1994) Recovery of transgenic peanut (*Arachis hypogaea* L.) plants from elite cultivars utilizing ACCELL technology. Plant J 5:745–753
6. Sailaja M, Tarakeswari M, Sujatha M (2008) Stable genetic transformation of castor (*Ricinus communis* L.) via particle gun-mediated gene transfer using embryo axes from mature seeds. Plant Cell Rep 27:1509–1519
7. Sujatha M et al (2009) Expression of the *cry1EC* gene in castor (*Ricinus communis* L.) confers field resistance to tobacco caterpillar (*Spodoptera litura* Fabr) and castor semilooper (*Achoea janata* L.). Plant Cell Rep 28:935–946
8. Murashige T, Skoog F (1962) A revised medium for rapid growth and bioassays with tobacco tissue cultures. Physiol Plant 15:473–497
9. Walter C et al (1994) A biolistic approach for the transfer and expression of a *gusA* reporter gene in embryogenic cultures of *Pinus radiata*. Plant Cell Rep 14:69–74
10. Lis JT (1980) Fractionation of DNA fragments by polyethylene glycol induced precipitation. Methods Enzymol 65:347–353
11. Medford J (1992) Vegetative apical meristems. Plant Cell 4:1029–1039
12. Vain P, McMullen MD, Finer JJ (1993) Osmotic treatment enhances particle bombardment-mediated transient and stable transformation of maize. Plant Cell Rep 12:84–88

Chapter 4

Biolistic DNA Delivery to Leaf Tissue of Plants with the Non-vacuum Gene Gun (HandyGun)

Anssi L. Vuorinen, Arto Nieminen, Victor Gaba, Sidona Sikorskaite, and Jari P.T. Valkonen

Abstract

Non-vacuum gene guns such as HandyGun are flexible tools for bombardment of targets of varying size. Construction of HandyGun is simpler and cheaper than vacuum gene guns and will be described here. The conditions for maximal transient transformation efficiency of plant cells with plasmid DNA using HandyGun will be provided.

Key words: HandyGun, Infectious virus clone, Sodium acetate precipitation

1. Introduction

HandyGun is a further improved version of the previous HandGun (1, 2) which in turn was essentially built according to the vacuum gene gun of Gray et al. (3) by simply excluding the vacuum chamber. The costs of parts needed for construction and maintenance of HandyGun are low. Most parts are commonly available.

Please note, however, that installation of the timer which uses strong alternating current of 230 V must be done by a qualified electrician.

Other improvements include the DNA precipitation method (4) which takes less than 30 min. The ease by which HandyGun is operated allows a high speed in the workflow in larger experiments and 100 bombardments per hour can be achieved routinely.

Consumable costs consist mainly of plasmid DNA isolation, gold particles, and helium gas.

Stephan Sudowe and Angelika B. Reske-Kunz (eds.), *Biolistic DNA Delivery: Methods and Protocols*,
Methods in Molecular Biology, vol. 940, DOI 10.1007/978-1-62703-110-3_4, © Springer Science+Business Media, LLC 2013

Our experiences in using HandyGun are so far limited to bombardment of leaf tissue of plants with plasmids containing infectious clones of plant viruses or with viral RNA (infectious in vitro transcripts) ((4), unpublished data). Hence, there seems to be a lot of scope in broadening the application of HandyGun to many other target organisms including vertebrates, invertebrates, and microbes such as filamentous fungi.

2. Materials

2.1. The Components of HandyGun

The parts needed for building HandyGun are listed below and exemplified in Fig. 1. They can be replaced with other similar parts. The parts comply with the British Piping Standard (BSPP) and the European 230 VAC electric voltage. If necessary, parts that comply with the NPT standard and 110 VAC can also be obtained.

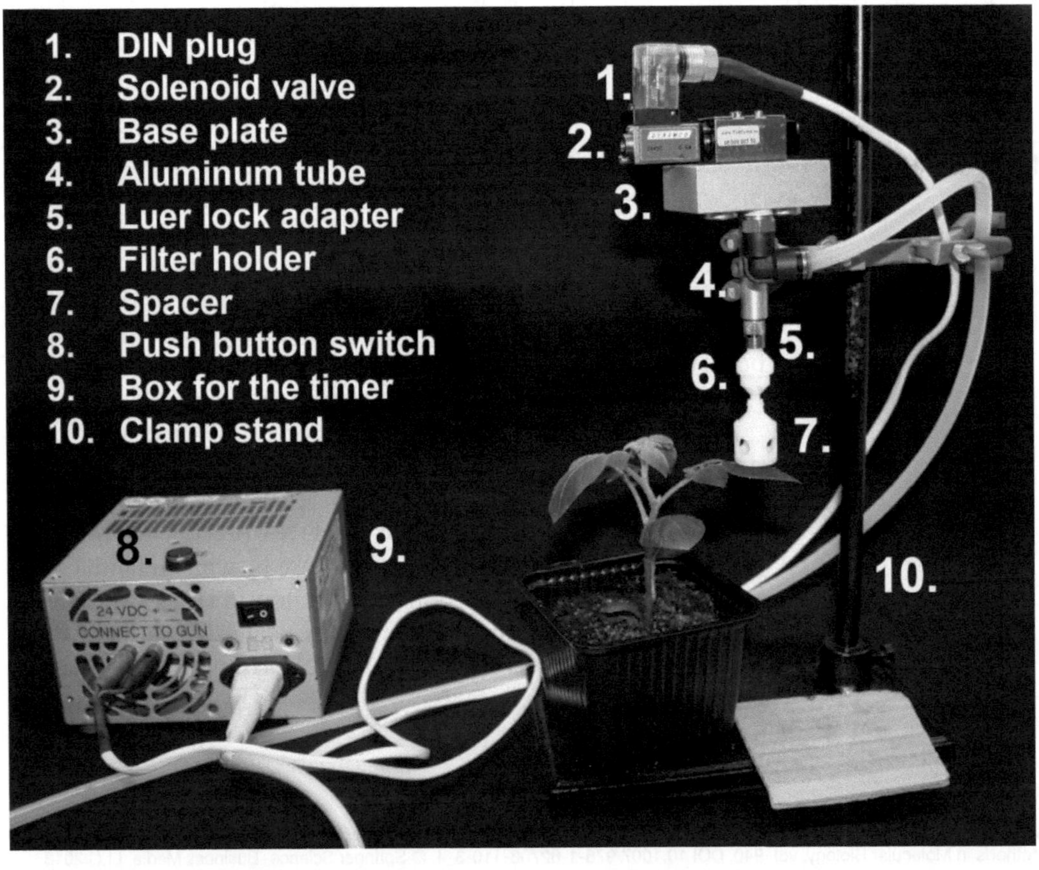

Fig. 1. The HandyGun assembly.

1. DIN plug to fit to the solenoid valve. Materials are commonly available. We used Standard DIN, transparent, 24 V, Hirschmann Automation and Control GmbH, Neckartenzlingen, Germany: EN 175 301-803-A.
2. Solenoid valve, magnetic (Dash 3, 24 VDC, 0.6 W, Dynamco, Commerce, GA, USA: cat. no. D3533KL0). The valve needs to be fast enough for a 100 ms burst.
3. Baseplate for the solenoid valve (Dynamco: cat. no. B03B2B) and three 1/4-in. plugs to fill the extra apertures (see Note 1).
4. Aluminum tube lathe-machined from a cylinder with a male thread at one end to fit the baseplate, and at the other end machined with a female fitting to accept the Luer-Lock adapter. Length 50 mm excluding the threads. Inner diameter 5 mm (drilled), outer diameter 12 mm (self-manufactured). The length is not critical, but the part should not be long enough to cause gas flow turbulence.
5. Luer-Lock adapter, male (Cole-Parmer, Vernon Hills, IL, USA: cat. no. 31507-73).
6. Swinney filter holder, 13 mm plastic or 13 mm steel (Pall, East Hills, NY, USA: cat. no. 4317 or 4042).
7. Spacer made from the top of a 13 mm plastic Swinney filter holder and a plastic tube glued on it (self-manufactured). The plastic tube was lathed to fit the diameter of the top part of the filter holder and four holes were drilled in the sides to allow the propellant gas to exit.
8. Push button switch attached to the timer (commonly available, we used the following: Type LUMOTAST 75 RAFI GmbH, Berg bei Ravensburg, Germany). The switch needs to be of good quality, because some switches we tested made the valve release several times on one contact.
9. Suitable box for the timer. For example, a computer power unit box (e.g., FSP Group Inc, Taoyuan City, Taiwan, R.O.C.: cat. no. FSP250-60GTA, partly self-manufactured). Only the main switch of the power unit was kept and all other parts were removed.
10. Clamp stand to hold the HandyGun.
11. Timer, 0–100 ms, 230 VAC, set to 100 ms (Megatron Electronics and Controls, Haifa, Israel: cat. no. MSST-700-CPT) (see Note 1).
12. Two glass fuses for the timer (5 × 20 mm IEC127 II, 100 mA, Camden Electronics Ltd, St. Albans, UK).
13. A cylinder of instrument helium (50 L, compressed at 200 bar, AGA, Espoo, Finland).
14. A suitable valve for the gas cylinder (Unicontrol 500 HT [helium/nitrogen], AGA: cat. no. 213 007 280/309254).

15. A pipe tubing elbow to connect the baseplate to the gas tube (CL Compact, Camozzi, Brescia, Italy: cat. No. 7522-8-1/4).
16. Suitable plastic tubing for connecting the gas bottle to the baseplate.
17. Two conductor electric wires for connecting the timer to the push button and to the solenoid valve.
18. Plastic pipe for the gas.
19. (Optional) two banana connectors for connecting the 24 VDC wire from the timer box to the solenoid valve.

2.2. Other Necessary Materials

1. A set of micropipettes for volumes 1–1,000 μL and sterile tips.
2. A microfuge.
3. 1.5 mL microcentrifuge tubes.

2.3. Solutions for Nucleic Acid Precipitation

1. 3 M sodium acetate (NaAc), pH 7.0, store at 4°C.
2. 0.05 mg/mL PVP in 99.5% ethanol, store at –20°C.
3. Gold microcarriers (BioRad, Hercules, CA, USA), diameter 1.0 μm; stock 10 mg/mL in 99.5% ethanol, store at –20°C.
4. 250 ng/μL Plasmid DNA containing the infectious virus clone.
5. 99.5% ethanol.

2.4. Plant of Interest

1. Leaf tissue of any plant of interest.

3. Methods

3.1. Assembly of the Gun

1. Attach the solenoid valve to the baseplate (Fig. 2).
2. Screw the aluminum tube to the gas exit port and attach to a clamp stand (Fig. 2).
3. Use the three 1/4-in. plugs to close the additional ports.
4. Screw the Luer-Lock adapter to the other end of the aluminum tube (Fig. 2).
5. Attach the pipe tubing elbow to the gas entry port.
6. Tighten the plastic tubing to the pipe tubing elbow at one end, and to the valve of the gas cylinder at the other (Fig. 2).
7. Attach the valve to the gas cylinder.
8. Connect a two-conductor electric wire to the glass fuses and onwards to the main switch of the computer power unit. Then connect the wire from the switch to the entry ports 1 and 2 of the timer (230 VAC). This step must be done by a qualified electrician (Fig. 3).
9. Connect the timer to the push button through ports 11 and 12 with another two-conductor wire (24 VDC) (Fig. 3).

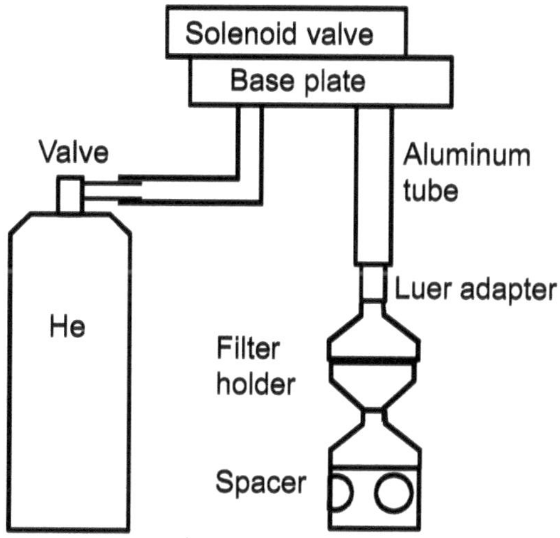

Fig. 2. Schematic presentation of the assembly of the pneumatic parts of HandyGun.

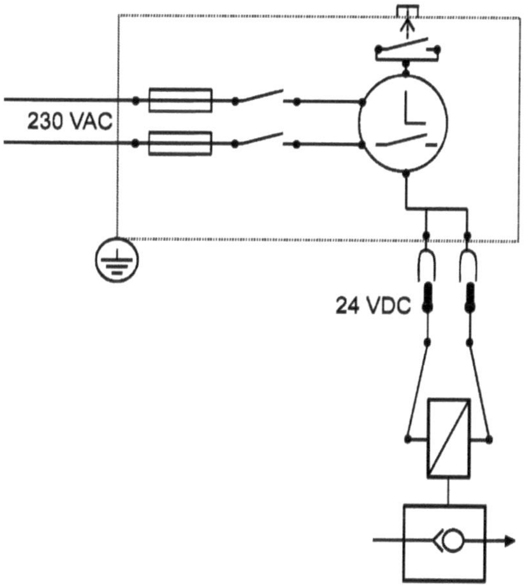

Fig. 3. Circuit diagram of the electric connections in HandyGun.

10. Place the timer inside the box and either use existing holes on the box for the wires or drill additional ones. Connect one more two-conductor wire (24 VDC) to the ports 6 (−) and 9 (+) of the timer. It may be convenient to attach this wire to female banana connectors on the side of the timer box, so that the timer can be easily disconnected from the rest of the gun.
11. Connect the 24 VDC wire from the timer to the solenoid valve using the DIN plug.

3.2. Preparation of DNA-Coated Microprojectiles for 50 Bombardments (See Notes 2 and 3)

1. Take 500 μL of well-mixed gold ethanol stock (10 mg/mL) to a new tube.
2. Add 200 μL of plasmid DNA (250 ng/μL), mix well.
3. Add 20 μL of 3 M NaAc, pH 7.0, mix well.
4. Allow to precipitate 10 min at room temperature.
5. Remove and discard the supernatant without disturbing the pellet.
6. Resuspend the pellet in the remaining supernatant and wash three times with 1 mL 99.5% ethanol. Spin 5 s in a microfuge between each wash.
7. Resuspend the pellet to 250 μL PVP solution (0.05 mg/mL in ethanol).

3.3. Bombardment

1. Adjust the gas pressure to 3 bar using the valve at the gas cylinder. Shoot several times without microprojectiles to ensure that the pressure remains stable.
2. Open a filter holder and dispense 5 μL of well-mixed microprojectiles on to the center of the grid (see Notes 3 and 4).
3. Close the filter holder and attach the spacer to its pointed end. Then attach the filter holder to the Luer-Lock adapter.
4. Place a plant under the HandyGun. Position a plant leaf under the spacer and hold it in place by supporting it from beneath with a piece of cardboard.
5. Push the button switch to shoot.
6. Spray the leaf with tap water after shooting to alleviate damage and enhance wound-healing (Fig. 4).
7. Clean and sterilize the filter holder after use (see Note 5).

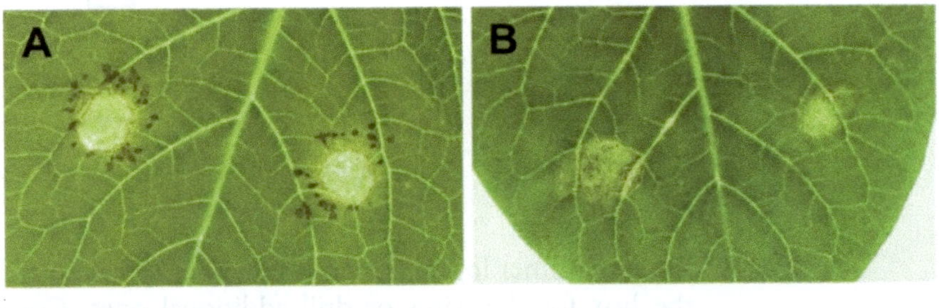

Fig. 4. Potato leaves bombarded with (**a**) microprojectiles coated with plasmid DNA for expression of infectious full-length RNA transcripts of *Potato virus A* (PVA) or (**b**) bombarded with uncoated microprojectiles as a control. The center of the bombarded area suffers from mechanical damage caused by high pressure and wounding by the microprojectiles. The initial infection sites indicating the numbers of successful transient transformation events in the bombarded leaf tissue are evident in (**a**) due to expression of a hypersensitive resistance response (necrotic local lesions) to PVA in the potato genotype "A6" used. Leaves were photographed 5 days post-inoculation.

4. Notes

1. The setup described here is functional at 230 VAC. For 110 VAC the timer needs to be replaced. BSPP are used.
2. It is necessary to optimize the bombardment parameters for new plant materials. The amount of DNA and gold are important parameters in terms of the bombardment efficiency. It may also be necessary to adjust the pressure. We found that 2 bar is a pressure too low for the optimal operation of the magnetic valve. On the other hand, the maximum pressure allowed for the valve (6 bar) breaks the leaves of tobacco and potato plants (4).
3. Ethanol evaporates quickly and the total volume of the microprojectile suspension may decrease relatively fast as the tube needs to be opened frequently.
4. For experiments with a large number of bombardments, it is advantageous to have several filter holders available. The bombardment mixture can be dispensed to all of them at once and the top part of one filter holder can stay permanently attached to the Luer-Lock adapter, while the lower parts are switched between bombardments.
5. Before reusing the filter holder for a different construct, make sure that the remainders of DNA are thoroughly removed. This can be achieved by, e.g., soaking the holder in soap water over night, followed by rinsing and autoclaving.

Acknowledgments

The work was supported financially by the Academy of Finland (grants 118766 and 1134759 to J.P.T.V.).

References

1. Gal-On A et al (1997) Simple hand-held devices for the efficient infection of plants with viral-encoding constructs by particle bombardment. J Virol Methods 64:103–110
2. Gaba V, Gal-On A (2005) Inoculation of plants using bombardment. In: Simon A, Kowalik T, Quarles J (eds) Current protocols in microbiology. Wiley Interscience, New York, pp 16B.3.1–16B.3.14
3. Gray DJ et al (1994) Simplified construction and performance of a device for particle bombardment. Plant Cell Tissue Organ Cult 37:179–184
4. Sikorskaite S et al (2010) HandyGun: an improved custom-designed, non-vacuum gene gun suitable for virus inoculation. J Virol Methods 165:320–324

4 Notes

Chapter 5

HandGun-Mediated Inoculation of Plants with Viral Pathogens for Mechanistic Studies

Victor Gaba, Moshe Lapidot, and Amit Gal-On

Abstract

Particle bombardment is an efficient method for virus inoculation of intact plants. This technique enables inoculation with full-length infectious clone cDNA, PCR products, virus from sap or virus preparation, and in vitro viral transcripts. The inoculation of some phloem-limited RNA and circular DNA viruses is also possible. The technique of bombardment without the use of vacuum permits the inoculation of soft-leaved plants that do not usually survive bombardment inoculation, the investigation of viral recombination in planta, promoter analysis, monitoring virus movement using an infectious clone bearing a reporter gene and the inoculation of large numbers of plants. The inoculation of whitefly-borne circular DNA begomoviruses is now possible due to direct genome amplification by Rolling Circle Amplification (RCA), followed by bombardment using a device that does not require a vacuum for operation. Here we describe the inoculation of intact plants with (a) RNA virus infective clones and (b) begomoviruses after direct genome amplification by RCA, using a handheld bombardment device.

Key words: Particle bombardment, Virus inoculation, Viral infective clones, Seedling inoculation, RNA virus, DNA virus, Rolling circle amplification, Potyviruses, Begomoviruses

1. Introduction

Particle bombardment has been used for viral inoculation of plants since the ground-breaking work of Klein et al. (1) where, following the bombardment of onion scale leaves with *Tobacco mosaic virus* RNA, virions accumulated in epidermal cells. Subsequently Gal-On et al. (2) demonstrated that a cDNA clone of *Zucchini yellow mosaic virus*, an RNA virus, can infect a plant when bombarded under the control of a strong promoter.

Particle bombardment is an excellent means for inoculation of whole plants with full-length cDNA of plant viruses or RNA virus

Stephan Sudowe and Angelika B. Reske-Kunz (eds.), *Biolistic DNA Delivery: Methods and Protocols*,
Methods in Molecular Biology, vol. 940, DOI 10.1007/978-1-62703-110-3_5, © Springer Science+Business Media, LLC 2013

transcripts, and complements classical virus inoculation methods. Particle bombardment with infective cDNA avoids the issue of 5′ capping of viral transcripts (2). *Agrobacterium tumefaciens* infiltration can also be used for cDNA virus inoculation, but requires lengthier cloning procedures and introduces another pathogen into the plant.

Particle bombardment of an infectious RNA virus requires production of a full-length infectious viral cDNA clone downstream from a promoter, e.g., the *Cauliflower mosaic virus* 35S promoter (2, 3) or the *Strawberry vein banding virus* promoter (4), with a terminator or viral poly(A) signal (2). Alternately, the clone can be linearized at the 3′ end vitiating the need for a termination signal.

The bombardment equipment required in this chapter is the HandGun (3), which is very similar to the HandyGun (5, 6). There are several bombardment devices built with vacuum chambers (BioRad PBS (1), Particle Inflow Gun ("PIG") (7), "Plastic PIG" (8)), which are more effective as micron-sized particles propagate faster in partial vacuum. Such enhanced equipment is probably only essential for a minority of cloned RNA viruses. Particle bombardment by the HandGun device is 10^5-fold more efficient for virus inoculation than mechanical inoculation of a viral plasmid promoter-driven clone (3). In turn, particle bombardment using a vacuum-device is tenfold more effective again (2, 3).

Uses for particle bombardment include inoculation with capped and uncapped transcripts (2), PCR cDNA products (9), full-length infectious cDNA clones, virus from plant sap (3), RNA virus preparation (3), phloem-limited RNA viruses (10), and total RNA from a diseased plant (Gal-On and Gaba, unpublished). Additionally, particle bombardment permits analysis of virus movement following infective clones bearing reporter genes (11), swift promoter analysis (4), inoculation of plants with soft, thin leaves, such as *Nicotiana bentamiana* or *Chenopodium quinoa* (3), examination of viral recombination in intact plants (12), and inoculation of a large number of plants (13). Additionally, such a method permits infection with an accurate inoculum quantity.

Recently we developed a technique for the bombardment inoculation of intact plants with phloem-limited, whitefly-borne, circular DNA begomoviruses. Previously such inoculations (14) would require cloning of the begomoviral DNA and bombardment with a vacuum-type PIG (7). We describe a method where Rolling Circle Amplification (RCA) of total DNA from infected plants permits the inoculation of several begomoviruses using non-vacuum equipment similar to the HandyGun (15). Moreover, infectious DNA was obtained from freeze-dried or desiccated plant material, plant leaf squashes on Whatman™ FTA cards (for the field collection of nucleic acid samples) and viruliferous whiteflies, using RCA. Plant material collected 25 years ago similarly yielded infectious begomoviral DNA (15).

Fig. 1. Typical bronzing damage to squash cotyledon following particle bombardment with tungsten particles.

The virus genome bombarded can originate from a variety of sources. Usually, we start from cloned virus under the control of a strong plant promoter. RCA products can be from viral clones or from total DNA extracts of infected plants. Bombardment of a new plant species/virus combination should be calibrated (5, 6). Bombardment with tungsten particles is necrotic to *N. benthamiana*, which can survive gold (Au) particle bombardment. Nevertheless, many species (e.g., cucurbits, tobacco, potato, tomato) can survive tungsten particle bombardment inoculation.

Commonly, leaf damage due to bombardment with tungsten particles is observed 1–2 days after bombardment (Fig. 1), but does not limit infection. Visible symptoms should be scored several days after inoculation (Fig. 2). Molecular techniques to detect systemic infection should be employed 10–14 days after inoculation (depending on virus). Previously inactive full-length clone constructs can become infective due to efficient bombardment inoculation (2). Nevertheless, symptom expression due to recombinant RNA may take longer or be weaker, or even be asymptomatic, requiring molecular investigation. Scoring of results should be as "number infected per number inoculated." Always include non-inoculated and mock-inoculated controls, and a positive control when checking mutant clones.

In this paper we describe two techniques: bombardment with a cloned RNA virus under control of a strong plant promoter and inoculation with an RCA-amplified begomovirus genome.

2. Materials

Make up all solutions with double distilled water; autoclave all components that can be sterilized.

Fig. 2. Symtomatic squash plant inoculated with an infectious cDNA clone of *Zucchini yellow mosaic virus*.

2.1. HandGun as Described (3)

1. The bombardment equipment is described by Gal-On et al (3) and an improved version in this volume by Vuorinen et al (6). The use of this equipment is described here.

2.2. Materials Required for Bombardment with Tungsten Particles

1. Helium gas cylinder (99.999% pure) with gas regulator; available from many suppliers. Secure the gas cylinder according to safety instructions.
2. Tungsten (W), M17 particles; size 1.6 μm (Bio-Rad, USA). Take care with toxic W (see Note 1).
3. 13 mm Plastic Swinney Filter Holder, part number 4317, Pall Corporation, USA (see Note 2).
4. 50% glycerol (sterile) (see Note 3).
5. Low power sonicator bath.
6. Minicentrifuge.
7. Vortex.
8. Sterile double distilled water.
9. Disposable gloves.
10. 70% ethanol (EtOH) (analytical).
11. cDNA of cloned virus in double distilled water (see Note 4), or RCA-specific source material.

12. Host plants (we use cotyledon stage and/or first-true leaf stage).
13. 1.25 M calcium nitrate, $Ca(NO_3)_2.4H_2O$ pH 10.5. For cDNA inoculation (see Note 5).
14. 1.25 M calcium nitrate, $Ca(NO_3)_2.4H_2O$ pH 7–8. For inoculation of transcripts, virus, viral RNA, and sap.

2.3. Additional Materials Required for Particle Bombardment with Au Particles, and/or RCA Technique

1. Gold powder, spherical, APS 1.5–3.0 μm 99.96+% (metals basis), Johnson Matthey Alfa Aesar Company. Similar material is available from Biorad Inc., USA (see Note 6).
2. 2.5 M $CaCl_2$ newly prepared, for bombardment with Au powder (see Note 7).
3. Spermidine (see Note 8).
4. TempliPhi Amplification kit (GE Healthcare, Chalfont St. Giles, UK) for RCA.
5. Eppendorf company microfuge tubes (1.5 mL) and tips (200 μL and 1 mL) (see Note 9).

3. Methods

3.1. Preparations for Bombardment with Tungsten Particles

3.1.1. Preparation of Tungsten Particle Stock

1. Wash ca. 50 mg tungsten (W) particles for 1 h in 1 mL EtOH in a 1.5 mL microfuge tube.
2. Vortex, then spin down briefly.
3. Wash three times with 1 mL aliquots of sterile double distilled water.
4. Resuspend washed tungsten particles in 50% glycerol, to a final concentration of 50 mg W/mL 50% glycerol.
5. Optionally, sonicate the tube of tungsten in 50% glycerol in an ultrasonic bath (2–3 min) to separate the particles.
6. Store at −20°C until use.

3.1.2. Preparation of Tungsten Mixture with cDNA

1. Prepare bombardment mixture for five plants. Take aliquots of prepared tungsten (or gold) particle stock.

   Use 1:1:1 proportions of these components; the total amount can vary according to need. For example, we use:

   (a) 5 μL $Ca(NO_3)_2$ (1.25 M, pH 10.5). For bombardment with RNA, use $Ca(NO_3)_2$ (1.25 M, pH 7–8).

   (b) 5 μL tungsten or gold stock.
   (c) 5 μL DNA. Add DNA last (50–250 ng DNA approximately is required for the bombardment of five plants).
2. Mix at room temperature, leave for 5–10 min. Mix by holding between thumb and forefinger of the left hand, and tapping with the right forefinger at beginning, middle and end of the 5–10 min period.
3. Use 2.5 μL per shot, for 5–6 shots (see Note 10).
4. Mix tube contents by pipetting up and down before loading on the grid of the filter holder.

3.2. Preparations Required for Bombardment with Gold Particles and RCA Products

3.2.1. Preparation of Gold Particles (Alex Lipsky's Method)

1. Use genuine Eppendorf tubes and tips to reduce particle clumping when DNA is added.
2. Wash ca. 50 mg Au particles for a few minutes in 1 mL fresh 70% EtOH.
3. Wash ×3 with 1 mL aliquots of sterile 50% glycerol, centrifuge between each wash. Prepare 50% glycerol in a glass, NOT a plastic vessel.
4. Vortex, then spin down briefly.
5. Resuspend in 50% glycerol (sterile): 50 mg Au/mL 50% glycerol.
6. Sonicate tube for several seconds.
7. Make the final solution such that you can easily dispense aliquots with 4 mg gold per Eppendorf tube, and store at –20°C, i.e., 100 mg Au/1,000 μL 50% glycerol = 40 μL per tube containing 4 mg Au.

3.2.2. Rolling Circle Amplification

Circular DNA is amplified by RCA using the TempliPhi Amplification kit:

1. Add 1 μL of plant total DNA Dellaporta preparation (16) to 5 μL sample buffer.
2. Heat to 95°C for 3 min to denature the DNA.
3. Snap-chill on ice.
4. Add 5 μL of kit reaction buffer plus 0.2 mL of kit enzyme mix (containing Phi29 DNA polymerase and random hexamers in 50% glycerol).
5. Incubate for 18 h at 30°C.
6. Inactivate the enzyme at 65°C for 10 min.

3.2.3. Preparation of Au-RCA Mixture for Begomovirus (or for Hard-to-Inoculate Cloned Virus Samples)

1. Vortex defrosted Au stock tube for 5 min.
2. Add 1–5 μL 1 μg/μL DNA.
3. Vortex.
4. Add 10 μL 2.5 M $CaCl_2$.
5. Vortex.
6. Add 4 μL 1 M spermidine.
7. Vortex 3 min.
8. Leave 1 min at room temperature.
9. Spin up to 1650 to 2500×*g*. Stop when the centrifuge reaches this speed.
10. Remove supernatant.
11. Wash Au/DNA mixture with 1 mL fresh 70% EtOH—put 70% EtOH gently and *DO NOT* resuspend Au.
12. Similarly wash with 1 mL EtOH.
13. Resuspend the pellet with 20 μL EtOH; use 4 μL per shot for bombardment.

3.3. Bombardment of Cloned Virus or RCA Products

1. Set gas pressure at the secondary regulator or cylinder. Generally, we use 2.5–3.5 bars pressure, depending on the target species.
2. Fire bombardment apparatus twice to fill tube with gas and check the pressure and fittings. Do not exceed 6 bars gas pressure with this apparatus, as the solenoid valve is not rated for a higher pressure.
3. Pipette up and down before each "shot" to blend the metal-DNA mixture well (essential).
4. Pipette 2.5 μL W mixture (4 μL Au mixture) onto the center of the grid of the open filter holder. The grid will hold the liquid (see Note 10).
5. Screw the filter holder closed. Attach the filter holder to the Luer-Lok adapter at the end of the bombardment apparatus.
6. Adjust the clamp stand so that the end of the filter holder is about 2–3 cm away from the target leaf or cotyledon. If using the HandyGun, touch the spacer to the target leaf.
7. Support the cotyledon or leaf to be bombarded with a gloved hand.
8. Press button to fire.
9. Fire the bombardment apparatus again at a second cotyledon or leaf to discharge entirely (W), or at the same target when using Au.
10. Repeat steps 1–9 until no bombardment mixture remains. Discard the microfuge tube.
11. Make a new bombardment mixture (above).

12. Use a clean, autoclaved, filter holder for each different DNA or mixture of DNAs.
13. At the conclusion of a bombardment experiment, close the gas valve on the cylinder.
14. Keep plants isolated following inoculation.
15. In case of low efficiency, failure, and for problem solving (see Note 11).

4. Notes

1. We have used this tungsten particle size, but additional sizes have been used successfully by others. We have not found problems with W stored frozen for long periods, but it has been noted that some W batches lose effectiveness during storage.
2. A similar metal filter holder is also available, but the plastic model lasts longer. Filter holders are washed by flushing with distilled water using a syringe several times, opening and allowing to air-dry, finally autoclaving. A harmless residue of W or Au often remains on the net after cleaning.
3. In case of low infection efficiency it is possible to prepare the 50% glycerol in a glass, not a plastic vessel, which reduces particle "clumping."
4. Such a plasmid can be prepared from an miniprep derived from a single colony grown overnight in 2 mL volume in Luria Broth using, i.e., Wizard *Plus* SV Minipreps kit, Promega, USA, following manufacturer's instructions. Note that DNA loses effectiveness when stored frozen, and it is preferable not to use stocks older than a month.
5. The pH can vary initially from 8 to 10.5. Autoclave, aliquot, and freeze.
6. Au is used for increased bombardment efficiency and with the RCA reaction. Au tends to stick to plasticware easily when in aqueous solution (but not in 50% glycerol or ethanol), and the use of Eppendorf original plastic ware is recommended. Siliconized plastic ware can also be used.
7. Prepare 10 mL 2.5 M $CaCl_2$ just before use; dispose of after use.
8. Dilute the spermidine to 1 M with double distilled water, aliquot in volumes of about 20 μL, freeze. Remove a spermidine aliquot when necessary and defrost at room temperature. Discard aliquot after use: use a new aliquot each day.
9. Use original Eppendorf plastic ware specifically with gold powder, or to reduce clumping due to use of DNA-containing protein (due to poor preparation quality), or when infection rate is low.

10. The bombardment mixture volume should be kept small (2–5 μL). Such a liquid volume does not interfere with the bombardment process.
11. If the cDNA is not infectious it is most likely that the cDNA quality is poor, contaminated, or old: prepare fresh cDNA. If a transcript is not infectious, prepare all reagents freshly (including W and Au), and sterilize all equipment. Alternately, cDNA might not be infectious due to the presence of inhibitory material in the preparation: dilute the cDNA preparation. If this does not help, efficiency can be gained by optimization of gas pressure for inoculation, the use of Au in place of W (as in Subheading 3.1.1) (and preparation of the bombardment mixture with spermidine as described), the use of authentic Eppendorf company tips and tubes, and sterilizing the 50% glycerol in a glass vessel. A tenfold improvement in inoculation efficiency can be gained by the use of bombardment technology using vacuum (3). Finally, the clone should be checked (perhaps sequenced). The promoter and 5′ terminus should be contiguous without intervening sequence (2).

    To demonstrate that cDNA is authentically uninfectious, bombard many plants ca. 50–100. Alternately, tissue specificity might be a cause, so bombard other organs. Heavy damage to leaves following bombardment can be due to W clumping: dilute cDNA, reduce gas pressure, or increase distance between apparatus and leaf.

Acknowledgment

Contribution from the Agricultural Research Organization, The Volcani Center, Bet Dagan, Israel, No. 503/11. The authors' work was supported by research grants from BARD (The United States-Israel Binational Agricultural Research and Development Fund), USAID-MERC, and the Chief Scientist of the Israeli Ministry of Agriculture.

References

1. Klein TM et al (1987) High-velocity microprojectiles for delivering nucleic acids into living cells. Nature 327:70–73
2. Gal-On A et al (1995) Particle bombardment drastically increases the infectivity of cloned cDNA of zucchini yellow mosaic potyvirus. J Gen Virol 76:3223–3227
3. Gal-On A et al (1997) Simple handheld devices for the efficient infection of plants with viral encoding constructs by particle bombardment. J Virol Methods 64:103–110
4. Wang Y et al (2000) Identification of a novel plant virus promoter using a potyvirus infectious clone. Virus Genes 20:11–17
5. Sikorskaite S et al (2010) HandyGun: an improved custom-designed, non-vacuum gene gun suitable for virus inoculation. J Virol Methods 165:320–324

6. Vuorinen AL et al (2012) Biolistic DNA delivery to leaf tissue of plants with the non-vacuum gene gun (HandyGun). In: Sudowe S, Reske-Kunz AB (eds) Biolistic DNA delivery: methods and protocols, Methods in molecular biology. Humana, Totowa
7. Finer JJ et al (1992) Development of the particle inflow gun for DNA delivery to plant cells. Plant Cell Rep 11:323–328
8. Gray DJ et al (1994) Simplified construction and performance of a device for particle bombardment. Plant Cell Tissue Organ Cult 37:179–184
9. Fakhfakh H et al (1996) Cell-free cloning and biolistic inoculation of an infectious cDNA of potato virus Y. J Gen Virol 77:519–523
10. Yang G et al (1997) A cDNA clone from a defective RNA of citrus tristeza virus is infective in the presence of the helper virus. J Gen Virol 78:1765–1769
11. Kimalov B et al (2004) Maintenance of coat protein amino terminal net charge is essential for zucchini yellow mosaic virus systemic infectivity. J Gen Virol 85:3421–3430
12. Gal-On A, Meiri E, Raccah B et al (1998) Recombination of engineered defective RNA species produces infective potyvirus *in planta*. J Virol 72:5268–5270
13. Shiboleth YM et al (2001) A new approach for weed control in a cucurbit field employing an attenuated potyvirus-vector for herbicide resistance. J Biotechnol 92:37–46
14. Lapidot M et al (2007) Biolistic inoculation of plants with *Tomato Yellow Leaf Curl Virus* DNA. J Virol Methods 144:143–148
15. Guenoune-Gelbart D et al (2010) Inoculation of plants with begomoviruses by particle bombardment without cloning: using rolling-circle amplification of total DNA from infected plants. J Virol Methods 168:87–93
16. Dellaporta SL, Wood J, Hicks JB (1983) Maize DNA miniprep. Plant Mol Biol Rep 1:19–21

Chapter 6

Biolistics-Based Gene Silencing in Plants Using a Modified Particle Inflow Gun

Kevin M. Davies, Simon C. Deroles, Murray R. Boase, Don A. Hunter, and Kathy E. Schwinn

Abstract

RNA interference (RNAi) is one of the most commonly used techniques for examining the function of genes of interest. In this chapter we present two examples of RNAi that use the particle inflow gun for delivery of the DNA constructs. In one example transient RNAi is used to show the function of an anthocyanin regulatory gene in flower petals. In the second example stably transformed cell cultures are produced with an RNAi construct that results in a change in the anthocyanin hydroxylation pattern.

Key words: RNA interference, Biolistics, Particle bombardment, Transient transformation, Cell culture

1. Introduction

At least three natural pathways of RNA silencing, collectively known as RNA interference (RNAi) have been described in plants. These are (1) small/short interfering RNA (siRNA) silencing, (2) the silencing of endogenous messenger RNAs by microRNAs (miRNAs), and (3) RNA-triggered DNA methylation and suppression of transcription (1). These pathways are thought to be of ancient origin and have diverse roles in genome control, including regulation of gene expression, defense against viruses and damage caused by transposons, and the formation of heterochromatin (2). All the pathways involve Dicer-like RNAse III endonuclease enzymes, which cleave double stranded RNA (dsRNA) into siRNAs (21–24 nucleotides (nts)), including miRNAs (21–22 nts) (1). These pathways also have in common RNA-induced silencing complexes (RISCs) that depend on the

Stephan Sudowe and Angelika B. Reske-Kunz (eds.), *Biolistic DNA Delivery: Methods and Protocols*,
Methods in Molecular Biology, vol. 940, DOI 10.1007/978-1-62703-110-3_6, © Springer Science+Business Media, LLC 2013

highly specialized small-RNA-binding Argonaute proteins (3). The RISCs produce single-stranded short-guide RNAs from siRNAs to direct function to a specific nucleic acid target. This in turn leads to homology-dependent gene silencing, for example by forming a duplex region with target mRNA that leads to transcript degradation (1–4).

RNAi as a response to transgene introduction and the subsequent generation of dsRNA was first described in plants for antisense RNA experiments with carrot cell lines (5). Napoli et al. (6) and van der Krol et al. (7) unexpectedly obtained similar results using a sense transgene against the flavonoid biosynthetic gene chalcone synthase (CHS) in petunia, the methodology of which at the time was termed co-suppression, sense suppression, or post transcriptional gene silencing. The subsequent elucidation of the mechanisms behind these sequence-dependent, gene-silencing phenomena has allowed the rational design of efficient RNAi suppression vectors for transient and stable genetic modification approaches, for both stable trait engineering and high throughput gene function screens (reviewed in refs. (8, 9)). Innovations include high throughput hairpin vectors, vectors using an inverted repeat of the transcript terminator sequence rather than the target gene sequence, tandem knockdowns, and tissue-specific and gene-specific RNAi-based gene silencing (4, 9–12).

Most RNAi systems employed in gene knockdown studies in plants have used virus-induced gene silencing (VIGS), with hosts including *Nicotiana benthamiana*, barley, arabidopsis, tomato, pea, and soybean (13, 14). VIGS circumvents the need for plant transformation and selection, but a suitable virus vector must be identified for each target plant species. Local biosecurity regulations may further limit the range of virus vectors that can be used. In some cases the virus DNA is delivered using biolistics (15).

The biolistic process, also known as particle bombardment, microprojectile bombardment, or particle gun transformation, has several advantages for both transient and stable gene-silencing studies when compared with biological vector-based processes, such as *Agrobacterium* or viruses. The physical nature of the particle bombardment process means it does not suffer from genotype range limitations of biology-based gene transfer systems. Furthermore, the plant response to pathogen infection from *Agrobacterium* or a virus is avoided. Biolistics also eliminates the need to employ treatments to kill or reduce *Agrobacterium* populations immediately after co-cultivation of explants. In transient transformation studies, multiple vectors can be delivered simultaneously, which is much more difficult when using *Agrobacterium*. Furthermore, DNA construct preparation is often simplified, as T-DNA vectors suitable for *Agrobacterium* binary transformation are not needed. Indeed, biolistics using only expression cassettes (promoter, transgene, and gene terminator), without extraneous plasmid backbone, has been used for delivery of multiple transgenes (16). For these reasons, biolistics has become a preferred method for transient gene over-expression (17) and silencing studies (18). The disadvantages of biolistics compared with VIGS and *Agrobacterium* for transient transformation are the typically small amounts of transformed tissue generated, although the RNAi signal does propagate out of the single transformed cell, and the variability that can occur between each transformation event. The previous disadvantages of biolistics for stable transformation, principally the variable nature of the DNA insertion event, have been mostly overcome through advances in the design of the vectors and the protocols used. This includes the use of minimal cassettes and reduced amounts of DNA in the bombardments (19).

In this chapter, we give two example protocols for gene silencing in plants using a particle inflow gun (20); (1) Transient RNAi using an inverted repeat of the sequence of the *Antirrhinum majus* (*antirrhinum*) gene *Rosea1* and (2) generation of potato cell cultures stably transformed with an RNAi vector against a *flavonoid 3′5′-hydroxylase* (*F3′5′H*) gene. *Rosea1* is an R2R3MYB transcription factor that activates anthocyanin pigment biosynthesis in the flowers. *F3′5′H* encodes an enzyme that adds hydroxyl groups to the B-ring of the anthocyanin precursors, resulting in the production of delphinidin-based rather than pelargonidin- or cyanidin-based anthocyanins, and an associated shift in the apparent color from pink/red to purple/blue.

2. Materials

Prepare and store all reagents at room temperature (unless indicated otherwise). All solutions and materials (plasticware and glassware) should be sterilized by autoclaving (unless indicated otherwise). The particle bombardment procedure is conducted within a laminar flow cabinet.

2.1. Plant Material

1. *A. majus* plants: *A. majus* (*antirrhinum*) wild-type line JI522, obtained as seed from Professor Cathie Martin of the John Innes Centre (Norwich, UK), grown in a glasshouse under ambient environmental light.
2. Plant cell lines: Because of the unlimited host range of biolistic transformation, gene silencing could be utilized in any plant cell culture for which a transformation protocol exists. In this example, we perform gene silencing on a purple-colored potato cell line to modify anthocyanin hydroxylation patterns.

2.2. In Vitro Culture of Antirrhinum Floral Buds

1. Sterilizing solution: add 1–2 drops of Tween20 per 100 mL of 10% (v/v) bleach. A commercial bleach solution (4.2% (v/v) sodium hypochlorite) is used. Prepare the sterilizing solution fresh for each experiment.
2. Sterile water.
3. Culture plates for the buds: 7.5% (w/v) agar plates of half-strength Murashige and Skoog (½ MS) medium.
4. Culture room: 25°C with a 16 h photoperiod under artificial lights (20–50 μmol/m^2/s light from Osram 36W grolux fluorescent tubes).

2.3. Preparation of Potato Cell Lines for Bombardment

1. Culture medium: Murashige and Skoog (MS) medium plus 1 μg/mL 2,4-D.
2. Conical flask.
3. Shaking platform.
4. 5 mL Gilson Pipetteman with a widened tip orifice.
5. Tubs: sterile clear plastic disposable tubs with clear snap-on lids (Propak NZ, 290mL K-resin).
6. Vacuum filter apparatus (e.g., Whatman 3-piece filter funnel) and Whatman No. 1 filter disc (50 mm diameter).
7. Culture room: 25°C in the dark.

2.4. Preparation of DNA-Coated Gold Particles

1. DNA constructs: Constructs using inverted repeats (IRs) to produce hairpin RNA are more effective in inducing RNAi than using antisense or co-suppression constructs (10), and there is flexibility in how an IR construct is designed (see Note 1). In the examples given here, pDAH2 (see Note 2) was used to make constructs with inverted repeat transgenes for hairpin RNA. The pDAH2-based vector pKES17 utilized the last 323 bp of the *Rosea1* open reading frame, and pJCH008 utilized a 203 bp sequence corresponding to positions 954–1,134 bp of the published potato F3′5′H cDNA (HQ860267). The 35S:GFP construct is pPN93, which is pRT99:GUS (22) modified by replacement of the GUS gene with the Green Fluorescent Protein (GFP) (courtesy of Dr Simon Coupe, formerly of Plant & Food Research).
2. Gold particles: Weigh approximately 100 mg of gold particles (1.0 μm diameter, Bio-Rad Laboratories, Australia) into a 1.5 mL microfuge tube and wash briefly in 1 mL isopropanol using a microfuge. Then wash three times with sterile water. On the last wash add a volume of water so that the concentration of gold is 100 mg/mL and aliquot the gold particles at 50 μL portions into microfuge tubes (5 mg gold per aliquot). Frequent mixing is necessary during this process to maintain an even suspension. The use of low-retention microfuge tubes is

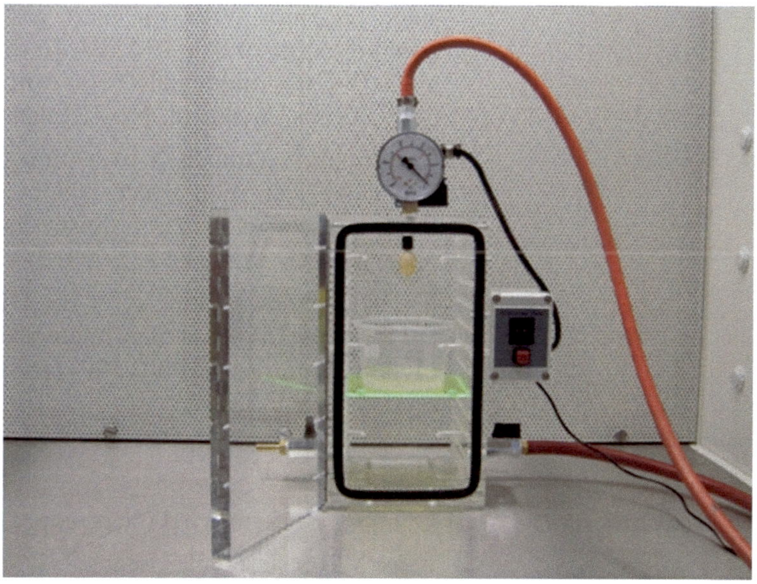

Fig. 1. Helium-driven particle inflow gun used in this study.

an advantage for better mixing and recovery of the gold particles after the DNA binding procedure. Store at 4°C until used.

3. 0.1 M Spermidine in distilled water, store at –20°C.
4. 2.5 M $CaCl_2$ in distilled water, store at –20°C.
5. Vortex machine.

2.5. Particle Inflow Gun

1. The particle bombardment described here uses a helium-driven particle inflow gun based on Vain et al. (21). However, the design has been modified (Fig. 1) by one of the authors (S. Deroles). Dr. Deroles also constructed the gun (see Note 3). A high speed solenoid valve has been added and is paired with a trigger mechanism tuned to the reaction time of the valve (3 ms). This significantly improves the accuracy of the shot time to ±1 ms for accurate valve opening times down to 8 ms. The high level of accuracy in particle delivery allows for a wide range of shot conditions to be trialed for the optimization of protocols for different plant tissues.
2. Set up the gun within a laminar flow cabinet and connect to a compressed helium supply and a vacuum pump.
3. The helium supply can be from either an in-house supply (as in our laboratories) or from a free-standing cylinder. In both cases a low-pressure inert gas regulator is required to deliver helium pressures between 200 and 600 kPa. It will need to be a twin

stage regulator rated for inert gasses and have an output pressure range of 0–1,400 kPa. In the examples here we used a BOC Gases model HPT500 with a 0–1,400 kPa output (BOC Gases, Auckland, New Zealand).

4. The vacuum pump is a twin-vane type with a flow rate of 5–6 cubic feet/min, which is approximately 150 L/min or 9 m^3/h e.g., a Telstar 2F-9 (Telstar Vacuum Solutions, Madrid, Spain) or Edwards RV12 (Edwards Vacuum, Crawley, UK).
5. The filter holders which carry the DNA-coated gold particles are Swinnex 13 in line filter holders (Millipore, Billerica, USA, catalogue number SX0001300) (see Note 4).
6. 80% (v/v) Ethanol.
7. 2–5% (v/v) Decon 90.

2.6. Microscopy

1. Olympus BH2 inverted microscope.
2. Leica M205 stereomicroscope.
3. Leica DC 500 digital camera to record images.

3. Methods

3.1. In Vitro Culture of Antirrhinum Floral Buds

1. Remove whole buds (3–10 mm in length) from the plants.
2. Remove the sepals and immerse the buds for 10 min in sterilizing solution, then rinse them three times with sterile water.
3. Place the buds on agar plates of ½MS. Maintain them on ½ MS both during and after the particle bombardment.
4. After bombardment incubate the buds in the culture room at 25°C with a 16 h photoperiod under artificial lights (20–50 μmol/m^2 s light from Osram 36W grolux fluorescent tubes).

3.2. Preparation of Potato Cell Lines for Bombardment

1. Maintain potato cell cultures as a liquid suspension (100 mL in a 250 mL conical flask) on a shaking platform (90 rpm) at 25°C in the dark and subculture fortnightly.
2. For bombardment, withdraw 5 mL of a 4–7-day-old culture using a 5 mL Gilson Pipetteman with a widened tip orifice and spread on a Whatman No. 1 filter disc (50 mm diameter) mounted on a vacuum filter apparatus. Spread the cells in a circle approximately 30 mm in diameter, to match the target diameter of the inflow gun at 13 cm shot distance. The cells should be loaded onto the filter disc under a mild vacuum to remove excess liquid.

3. Transfer the disc to a media tub containing MS media plus 1 μg/mL 2,4-D for bombardment.
4. After bombardment incubate the tubs at 25°C in the dark.

3.3. Preparation of DNA-Coated Gold Particles

Preparation of the gold-DNA may use variations around the method presented here (see Note 5).

1. Add the construct DNA (usually 2–5 μg, see Note 6) to the gold particle aliquot (50 μL), add water to bring the volume to 60 μL total, and then mix briefly.
2. More than one DNA construct can be included in the bombardment, although the total volume should be kept to 60 μL. If a 35S:GFP control construct is being used then 2 μg of DNA in the 60 μL total volume is usually adequate.
3. Place the microfuge tube onto a vortex machine set at a low speed so that gentle mixing of the solution is maintained.
4. Using two pipettes, simultaneously add 20 μL of 0.1 M spermidine and 50 μL of 2.5 M $CaCl_2$. Then cap the tube and vortex thoroughly at high speed for 3 min.
5. Pellet the gold using a brief burst in a microfuge and remove 90 μL of the supernatant. Then resuspend the gold particles in the remaining liquid by either pipetting and/or "flicking" the base of the microfuge tube.
6. Use a 5 μL aliquot for each bombardment "shot." Before removing each 5 μL aliquot disperse the gold evenly into suspension by pipetting. Each gold-DNA preparation typically allows for six shots.

3.4. Particle Bombardment

1. Clean the microfuge racks and laminar flow cabinet using 80% (v/v) ethanol. Clean the gun chamber by wiping with a cloth dipped in diluted (2–5% v/v) Decon90.
2. Place the Swinnex filter unit in a rack ready for pipetting the gold-DNA into it.
3. The bombardment conditions (see Note 7) are a solenoid valve opening time (SVO) of 30 μs, a helium pressure setting of 300 kPa for *antirrhinum* and 400 kPa for potato cell cultures, a shooting distance of 13 cm, and a partial vacuum of approximately -96 kPa (−14 psi).
4. Place the agar plate (petals) or tub (cell cultures) containing the target material into the inflow gun, ensuring the target material is within the blast zone (see Note 8).
5. Aliquot 5 μL of gold-DNA onto the Swinex filter grid support by pipetting into the Swinnex filter holder through the top opening and then screw the filter holder onto the helium port in the top of the gun chamber.

6. Open the valve to the vacuum pump until the pressure inside the chamber is reduced to the desired level, then close the valve to the pump. Press the solenoid trigger to conduct the gold-DNA shot.
7. Open the second chamber valve (that does not lead to the pump) to release the vacuum. Make sure this is done at a rate that does not cause excessive disturbance within the chamber.
8. The same tissue may be repeatedly bombarded. For the *antirrhinum* buds, between two and six bombardments are carried out on each bud (and multiple buds are bombarded simultaneously).
9. The buds should be transferred to fresh agar plates, with the bases of the buds pushed slightly into the agar to give an upright position. Transfer the plant material (buds or cell lines) to the culture room.

3.5. Observation of Phenotype in Antirrhinum

1. If 35S:GFP is included as a control this will typically be visible by 24 h after bombardment. A short wavelength blue light source and dissecting microscope is used to observe the GFP.
2. *Antirrhinum* buds at 3–10 mm are relatively un-pigmented, and they pigment as they subsequently develop in culture. As the pigmentation forms, notable differences should be apparent in RNAi buds compared with control buds bombarded with 35S:GFP alone. Check the buds after 3–4 days for the first signs of an induced phenotype. At approximately 10 days, buds have expanded to open flowers, allowing easier observation of phenotype (and sampling of tissue for additional analysis, if desired). The zones of pigment inhibition often extend from the bombarded abaxial (outer) epidermis to the adaxial (inner) epidermis (Fig. 2).

3.6. Observation of Phenotype in Potato

1. GFP may be observed as for *antirrhinum*.
2. If the target phenotype is a visible change (e.g., color) then phenotypes resulting from transient expression will be visible 1–3 days after shooting in a rapidly growing cell line. For more subtle changes stably transformed cell lines must be selected for observation of the phenotype. Selection in cell culture can be via a selective marker gene e.g., 35S-nptII. Alternatively, selection can use visible markers e.g., GFP. In the experiment described here, a visible color change occurs due to the inhibition of the production of the purple-colored 3′5′-hydroxylated anthocyanins (Fig. 3), and this can be used for selection of the transformed cells. Such visible selection is useful when performing intragenic/cisgenic transformations that cannot use transgenic selective marker genes.

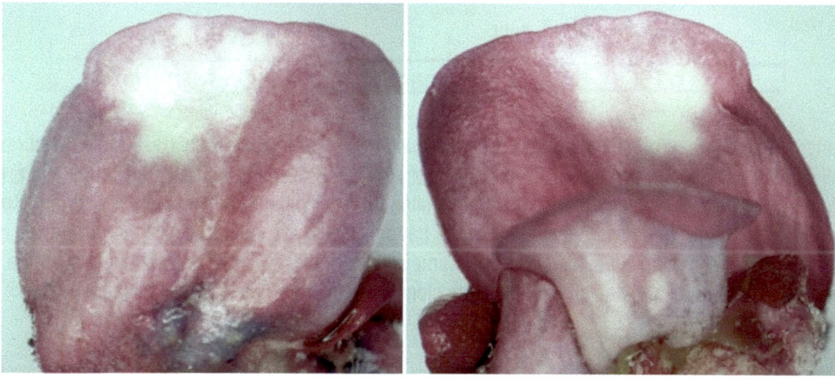

Fig. 2. Inhibition of *Rosea1* activity in *antirrhinum* flower buds using transient RNA interference (RNAi). Petals of buds cultured in vitro are shown 12–17 days after the biolistic introduction of the plasmid for *Rosea1* RNAi inhibition. The adaxial (*left*) and abaxial (*right*) epidermis of the same region of one petal are shown. The same pattern of inhibition on both surfaces demonstrates that the silencing signal was transmitted from the bombarded abaxial epidermis to the adaxial epidermis.

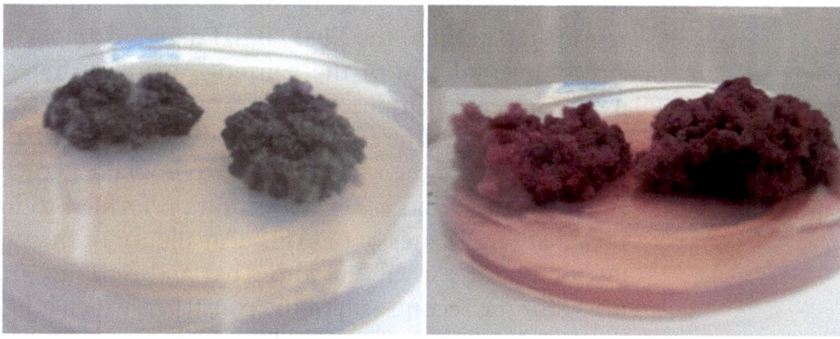

Fig. 3. Inhibition of *F3′5′H* activity in stably transformed potato cell lines using RNAi. The wild-type colored line is shown on the *left* and the *F3′5′H RNAi* line on the *right*.

4. Notes

1. Varying sizes of the IRs, the region of the gene sequence used in the IRs (5′ UTR, 3′ UTR, or coding sequence), the size of the spacer, the use of an intron as the spacer and other vector design details have been investigated (e.g., (10, 23)), and there is no single way to make an effective vector.
2. pDAH2 was the vector used for initiating RNAi in the biolistically shot tissues (Fig. 4). The vector contains an isocaudamer-based multiple cloning site (Xba1-BglII-127 bp spacer-BamHI-Nhe) oriented to enable convenient construction of the hairpin sequence between a double 35SCaMV promoter and NOS terminator sequences. The advantage of the

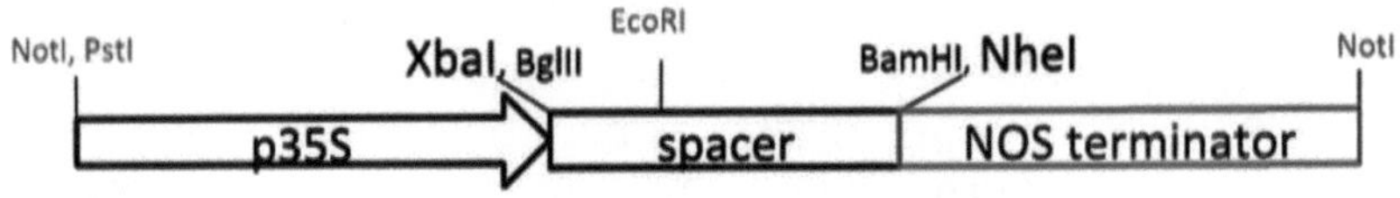

Fig. 4. Schematic of the region of pDAH2 used for driving production of hairpin constructs (HDR region). Abbreviations are: p35S, CaMV35S promoter; Spacer, synthetic 127 bp region of DNA; NOS terminator, transcript termination region of the *nopaline synthase* gene of *A. tumefaciens*. The recognition sites for the isocaudamer restriction enzymes XbaI and NheI (*large font*) and BglII and BamHI (*small font*) that are used for hairpin cloning are indicated. The hairpin-driving region (HDR) was constructed in pGEM5Zf, which contains the gene for ampicillin resistance. The HDR is modular to enable interchanging of promoter and terminators, and has sites to facilitate cloning into pGREENII and pART27 binary vectors.

isocaudamer-based multiple cloning site is that the same digested PCR product can be used for both sense and anti-sense arm ligations. The target sequence is amplified using primers that add XbaI (5′ end) and BglII (3′ end) restriction sites to the amplicon. The digested amplicon is then ligated into XbaI/BglII digested pDAH2 and transformed into a standard *E. coli* strain such as NovaBlue. pDAH2 is then isolated from a positive clone, digested with BamHI and NheI and ligated again to the XbaI/BglII digested amplicons to create the second arm of the hairpin. A similar strategy can be used to create hairpins of two target genes in tandem. For example, the first target is amplified to contain XbaI (5′) and EcoRI (3′) sites the second target to contain EcoRI (5′) and BglII (3′) sites and a 3-way ligation performed to insert into the XbaI/BglII sites of pDAH2. An aliquot of the successfully ligated vector is then cut with XbaI and BglII to provide the insert for the second arm of the construct, which is ligated into the BamHI/NheI site of an aliquot containing the first arm.

3. For detailed specifications of the particle inflow gun contact Dr. Deroles (KiwiScientific, Levin, NZ. Email: deroles@ihug.co.nz).
4. The Swinnex 13 filters that carry the DNA-coated gold particles can be used for more than one shot, but should be changed between different DNA samples. They can also be reused for subsequent experiments. Once the experiment is complete they should be cleaned by a 30 min treatment in Decon90 in a sonicator bath followed by autoclaving under standard conditions.
5. Several variant methods for the mixing of the gold and DNA have been published. We have found the method presented here to be effective, although we have not conducted quantitative comparisons to show whether it is more effective than other published methods. Common variations include: the amount of gold or DNA used, how the spermidine and $CaCl_2$ are added, and whether the gold is centrifuged down or simply allowed to settle out.

6. Although 5 μg of DNA is commonly used, 1–10 μg can also be effective, depending on the specific experiment and plant material. When 5 μg is used this results in approximately 1 μg being used in each bombardment.
7. We have found that the inflow gun and method included in this report are effective for petals of *antirrhinum*, petunia, pea, sandersonia, and viburnum; leaves and petals of sunflower; and cell cultures of potato, carrot, sweet potato, petunia, and tobacco. For each species and type of tissue variations in the SVO time and helium pressure may be needed. These are determined by reiterative trials using the GFP reporter gene.
8. It is helpful to draw the general blast zone of the inflow gun at different shooting distances (e.g., 13 cm) onto the Perspex shelf used to support the shot material. This can be determined by conducting a few shots onto a piece of white filter paper cut to the same size as the shelf.

Acknowledgments

We thank Dr Simon Coupe for use of pPN93.

References

1. Axtell MJ, Bowman JL (2008) Evolution of plant microRNAs and their targets. Trends Plant Sci 13:343–349
2. Baulcome D (2004) RNA silencing in plants. Nature 431:356–363
3. Ender C, Meist G (2010) Argonaute proteins at a glance. J Cell Sci 123:1819–1823
4. Eamens AL, Waterhouse PM (2011) Vectors and methods for hairpin RNA and artificial microRNA-mediated gene silencing in plants. Methods Mol Biol 701:179–97
5. Ecker JR, Davis RW (1986) Inhibition of gene expression in plant cells by expression of antisense RNA. Proc Natl Acad Sci USA 83: 5372–5376
6. Napoli C, Lemieux C, Jorgensen R (1990) Introduction of a chimeric chalcone synthase gene into *Petunia* results in reversible co-suppression of homologous genes in *trans*. Plant Cell 2:279–289
7. van der Krol AR et al (1990) Flavonoid genes in petunia: addition of a limited number of additional copies may lead to a suppression of gene activity. Plant Cell 2:291–299
8. Mansoor S et al (2006) Engineering novel traits in plants through RNA interference. Trends Plant Sci 11:559–565
9. Frizzi A, Huang S (2010) Tapping RNA silencing pathways for plant biotechnology. Plant Biotechnol J 8:655–677
10. Brummell DA et al (2003) Inverted repeat of a heterologous 3'-untranslated region for high-efficiency, high-throughput gene silencing. Plant J 33:793–800
11. Watson JM et al (2005) RNA silencing platforms in plants. FEBS Lett 579:5982–5987
12. Higuchi M et al (2009) Simple construction of plant RNAi vectors using long oligonucleotides. J Plant Res 122:477–482
13. Bernacki S et al (2010) Virus-induced gene silencing as a reverse genetics tool to study gene function. Methods Mol Biol 655:27–45
14. Llave C (2010) Virus-derived small interfering RNAs at the core of plant-virus interactions. Trends Plant Sci 15:701–707
15. Zhang C et al (2009) Development and use of an efficient DNA-based viral gene silencing vector for soybean. Mol Plant Microbe Interact 22:123–131
16. Francois IEJA, Broekaert WF, Cammue BPA (2002) Different approaches for multi-trans-gene-stacking in plants. Plant Sci 163:281–295
17. Taylor NJ, Fauquet CM (2002) Microprojectile bombardment as a tool in plant science and

agricultural biotechnology. DNA Cell Biol 21:963–977

18. Shang Y et al (2007) Methods for transient assay of gene function in floral tissues. Plant Methods 3:1–12
19. Lowe BA et al (2009) Enhanced single copy integration events in corn via particle bombardment using low quantities of DNA. Transgenic Res 18:831–840
20. Finer J et al (1992) Development of a particle inflow gun for DNA delivery to plant cells. Plant Cell Rep 11:323–328
21. Vain P et al (1993) Development of the particle inflow gun. Plant Cell Tissue Organ Cult 33:237–246
22. Töpfer R, Schell J, Steinbiss H (1988) Versatile vectors for transient gene expression and direct gene transfer in plants. Nucleic Acids Res 16:8725
23. Wesley SV et al (2001) Construct design for efficient, effective and high-throughput gene silencing in plants. Plant J 27:581–590

Part II

Biolistic DNA Delivery in Nematodes

Chapter 7

Biolistic Transformation of *Caenorhabditis elegans*

Meltem Isik and Eugene Berezikov

Abstract

The ability to generate transgenic animals to study gene expression and function is a powerful and important part of the *Caenorhabditis elegans* genetic toolbox. Transgenic animals can be created by introducing exogenous DNA into the worm germline either by microinjection or by microparticle bombardment (biolistic transformation). In this chapter we describe a simple and robust protocol to generate transgenic *C. elegans* animals by biolistic transformation with gold particles using the Bio-Rad PDS-1000/He system with Hepta adapter and *unc-119* selection marker. We also point out the steps that need special attention to achieve successful transformations.

Key words: *Caenorhabditis elegans*, Biolistic transformation, Bio-Rad Biolistic PDS-1000/He system, Hepta adapter, *unc-119*(*ed3*)

1. Introduction

Caenorhabditis elegans is a small, free living round worm, which feeds on bacteria and fungi in soil. In the laboratory, it is grown by feeding on *Escherichia coli* that is spread onto a plate. Rapid generation time and short life span, easy and inexpensive maintenance, fully sequenced genome and array of RNAi resources and mutant animals make *C. elegans* a popular biological model. Generation of transgenic animals is a powerful way to study and manipulate gene expression in vivo, and hence an important part of worm research. Transgenic animals can be created either by microinjection of exogenous DNA into the worm germline or through the use of biolistic transformation.

Microinjection of DNA into the gonadal syncytium is commonly used to transfer DNA into *C. elegans* (1, 2). Although successful in forming transgenic lines, the technique requires some

Stephan Sudowe and Angelika B. Reske-Kunz (eds.), *Biolistic DNA Delivery: Methods and Protocols*,
Methods in Molecular Biology, vol. 940, DOI 10.1007/978-1-62703-110-3_7, © Springer Science+Business Media, LLC 2013

time and effort to become proficient and requires specialized equipment. Additionally, the transgenic animals can show either mosaic expression or changes during continued passage, since they carry the transgene as an extrachromosomal array (2).

Another means of generating transgenic worms is the use of microparticle bombardment. In this approach, a population of hermaphrodites, rather than individuals, is bombarded with DNA-coated beads, and transformed individuals are selected from the subsequent generations. This method relies upon gold particles forming a complex with DNA in the presence of $CaCl_2$. DNA is protected from nuclease degradation in vivo by using cationic polyamines, such as spermidine.

The advantages of the biolistic transformation approach include ease-of-application (minimal specialized training and practice are required from a researcher), and the ability to obtain both extrachromosomal array as well as low-copy integrated transgenic lines (3–5). This method also provides a way to introduce homologous recombinants of genes (6) and dsRNA hairpin constructs (hpRNAi) to target tissues refractory to other RNAi delivery methods (7). Additionally, bombardments can be used to favor gene conversion events at transposon insertion sites (8).

The microparticle bombardment approach also has some limitations. First of all, the range of cobombardment marker genes is much more limited with bombardment in contrast to microinjection (4, 5, 9). This limitation is dictated by the need for a strong selection marker, since transformants are selected from a very large initial population of animals, rather than from individual hermaphrodites. Genetic interactions between the selectable marker and gene of interest can make it difficult to create healthy double mutants that can be transformed (10). Overexpression or misexpression of the selectable marker gene can have phenotypical consequences, making the functional gene expression studies difficult to interpret. Low-copy integrated lines that are obtained from bombardments may have problems associated with the site of integration and higher copy number. Expression of the gene of interest may not reflect that of the endogenous gene and integrated exogenous DNA may also interfere with the expression of other genes located at the site of integration.

Cotransformation markers that have been used successfully in microparticle bombardment experiments are selectable markers that rescue a mutant phenotype (e.g., *unc-119*, *pha-1*, *dpy-20*, *spe-26*), produce dominant mutant phenotype (e.g., *rol-6*), or express green fluorescent protein (GFP). Among these selection markers the *unc-119* rescue, developed by Praitis et al. (5), is the most commonly used, and the transformation protocol described here is geared towards the *unc-119* selection strategy. The strength of *unc-119* rescue selection is based on the fact that *unc-119(ed3)* mutant animals cannot form dauers and therefore die out in the

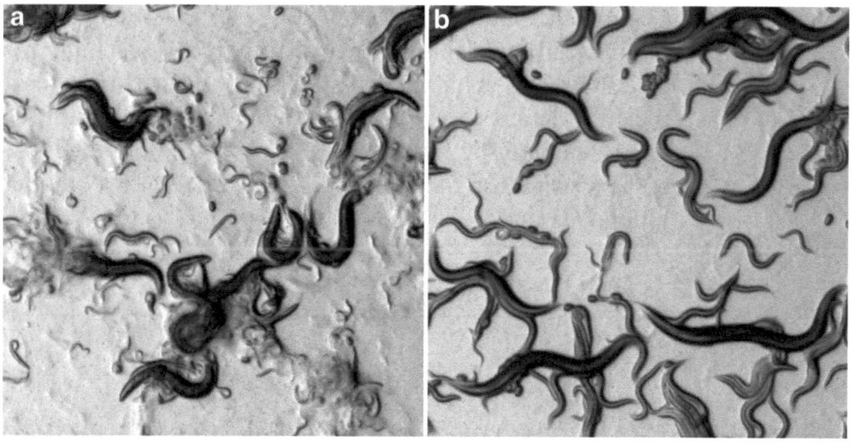

Fig. 1. (**a**) *unc-119(ed3)* worms before transformation. The worms are dumpy, cannot freely move throughout the plate and are unable to form dauers under starvation conditions. (**b**) *unc-119(ed3)* worms after transformation with *unc-119* selection marker. The worms can move as wild-type worms and they can form dauers under starvation conditions.

absence of food, whereas rescued (transformed) animals survive and can be easily identified at low magnification (Fig. 1).

unc-119(ed3) worms are difficult to grow on standard Nematode Growth Medium (NGM) plates since they hardly move and therefore starve on parts of a plate while other parts of the plate still contain food. Egg plates (11) appeared to solve this problem. The thick food layer on egg plates allows *unc-119* worms to crawl more easily and to take over all the plate. Usually five egg plates are sufficient to grow worms for one bombardment. Note that special care should be taken to prevent contaminations during preparation of egg plates.

The original DP38 strain (*unc-119*(*ed3*)) carries an unrelated dauer-formation constitutive (*daf-c*) mutation, which could affect the gene expression and phenotype of the lines obtained from bombardments (12). We use HT1593 strain, which is derived from DP38 by outcrossing seven additional times to remove the *daf* mutation present in DP38.

The cotransformation marker and gene of interest do not need to be in the same construct to produce integrated lines (5, 10, 13). However, bombardment of worms with multiple plasmids can create transgenic animals carrying some, but not all the plasmids (4, 14). Recently, a protocol for inserting the *unc-119* gene into the ampicillin resistance gene by homologous recombination has been described (14), and can be used to efficiently combine the gene of interest and the *unc-119* selection marker into a single plasmid.

There are several microparticle delivery systems that have been used for biolistic transformation of *C. elegans*: Bio-Rad Biolistic® PDS-1000/He Particle Delivery system, Bio-Rad Helios Gene Gun, and a "home-made" microparticle bombardment apparatus

(3–6). Although the PDS-1000/He Particle Delivery system requires substantial upfront investment, this machine provides the easiest way to achieve reproducible transformation results and is the most often used device for biolistic transformation of *C. elegans.* The biolistic transformation protocol described in this chapter is specific for the PDS-1000/He system in conjunction with Hepta adapter.

2. Materials

2.1. *C. elegans* Strains and Culture Conditions

1. HT1593 [*unc119(ed3)*] worms should be maintained on NGM/OP50 plates at 15–20°C.
2. NGM plates: plates with 9 cm diameter, NaCl, Bacto-agar, bacto-peptone, cholesterol (5 mg/mL in ethanol), dH_2O, 1 M $CaCl_2$, 1 M $MgSO_4$, 1 M KPO_4 (pH 6) (preparation described in Subheading 3.1).
3. *E. coli* OP50 strain.
4. LB: 10 g Bacto-tryptone, 5 g Bacto-Yeast Extract, 5 g NaCl, bring volume to 1 L with H_2O, autoclave.
5. Egg plates: 500 mL sterile bottle, 60°C waterbath, ten chicken eggs, LB and 60 NGM plates (preparation described in Subheading 3.2).
6. Egg Salts Buffer: 6.9 g NaCl, 3.6 g KCl, 1 L H_2O. Autoclave.
7. 50 mL Flasks for growing OP50 *E. coli.*

2.2. Preparation of DNA for Transformation

1. Plasmid DNA required for transformation can be isolated by a plasmid purification kit or a standard alkaline lysis protocol. 5 μg of DNA is necessary per bombardment. If the cotransformation/selection marker and the construct of interest are not in the same plasmid, 5 μg of each plasmid should be added.
2. Restriction endonculease of choice for plasmid linearization outside the selection marker and target regions.

2.3. Microparticle Bombardment Components

1. Bio-Rad Biolistic® PDS-1000/He Hepta system (includes Hepta adapter and five macrocarrier holders).
2. Rupture discs (1,350 psi) Bio-Rad.
3. Macrocarriers Bio-Rad.
4. Stopping screens Bio-Rad.
5. Hepta adapter Bio-Rad.
6. Gold bead microcarriers (ChemPur 0.3–3 μm #009150 or Bio-Rad 1 μm beads).

7. 0.1 M Spermidine (Sigma S-4139 free base, tissue culture grade; filter sterilized; store frozen at −20°C). 2.5 M $CaCl_2$ (filter sterilized).
8. 2.5 M $CaCl_2$ (filter sterilized).
9. Dehydrated ethanol (70% and 96%). Do not use old ethanol that will have absorbed a lot of water from the air.
10. 1.5 and 2 mL tubes.
11. Vortex.
12. Deionized water.
13. Isopropanol.
14. 50% Glycerol (filter sterilized).

3. Methods

3.1. Preparation of NGM Plates

1. Mix 3 g NaCl, 21 g bacto-agar, 7.5 g bacto-peptone (see Note 1) and make up the volume to 1 L with dH_2O. Autoclave.
2. Cool to 55 °C, and add (using sterile technique and swirling) 1 mL cholesterol (5 mg/mL in ethanol), 1 mL 1 M $CaCl_2$, 1 mL 1 M $MgSO_4$, and 25 mL 1 M KH_2PO_4 (pH 6.0).
3. Pour 9 cm plates about half-full, and flame the agar surface to remove air bubbles (or worms will burrow). Let it dry at least one day prior to seeding with OP50 *E. coli*.

3.2. Preparation of Egg Plates

Day 1

1. Prepare 400 mL LB, ten eggs, sterile bottle (500 mL) and 60 NGM plates.
2. Grow 40 mL of OP50 *E. coli* culture overnight.

Day 2

3. Separate yolks of ten chicken eggs into 500 mL sterile bottle. Shake well until the yolks form a homogenous solution. Bring volume to 400 mL with LB medium. Shake well.
4. Incubate at 60°C for 1 h to deactivate the enzymes present in the yolk. Cool to room temperature.
5. Add 40 mL of overnight OP50 *E. coli*. Shake well. Distribute 5–8 mL of mixture per NGM plate. Allow to settle overnight.

Day 3

6. Gently pour off remaining liquid from plates. Allow to dry one more day.
7. Wrap plates and store at 4°C (see Note 2).

3.3. Growing unc-119(ed3) Strain for Transformation

1. Prepare several clean master plates of *unc-119(ed3)* worms by bleaching the worms and seeding the eggs on NGM plates with OP50. Once the plates are almost starving (when the amount of worms is highest) transfer pieces from the plates on egg plates (see Note 3).
2. Place egg plates (lid on top) inside a clean box. Parafilm can be used to avoid contamination (see Note 4).

3.4. Preparation of Gold Particles

1. Weigh 60 mg of gold particles (0.3–3 μm, ChemPur) into a 2 mL tube.
2. Add 2 mL 70% EtOH, vortex 5 min, allow the particles to settle for 15 min, spin 3–5 s, and discard supernatant.
3. Add 2 mL deionized water, vortex for 1 min, allow the particles to settle for 1 min, spin 3–5 s and discard supernatant. Repeat this washing step for two more times.
4. Resuspend the gold particles in 1 mL of 50% sterile glycerol. The final concentration of gold is 60 mg/mL and is sufficient for ten bombardments. Prepared suspension can be stored for 1–2 month at 4°C or room temperature.

3.5. Coating of Gold Particles with DNA

1. Linearize 5 μg of plasmid DNA by digesting the plasmid with a restriction enzyme that does not cut inside the transgene of interest and the *unc-119* gene. Cleaning of DNA after restriction digestion is not necessary. Final reaction volume should be 50 μL.
2. Vortex gold beads (60 mg/mL) for 5 min to disperse clumps (see Note 5), take 100 μL into 1.5 mL tube, add 50 μL of linearized DNA and vortex 1 min at a moderate speed (see Note 6).
3. Add 150 μL 2.5 M $CaCl_2$, vortex at a moderate speed for 1 min.
4. Add 60 μL of 0.1 M spermidine (see Note 7), vortex 3–5 min, settle 1 min, spin for 3 s, remove supernatant.
5. Add 300 μL 70% EtOH, vortex briefly, settle, spin for 3 s, remove supernatant.
6. Add 500 μL absolute EtOH, vortex briefly, settle, spin for 3 s, remove supernatant.
7. Add 140 μL absolute EtOH, vortex briefly and proceed to loading DNA-coated gold particles onto macrocarriers (see Note 8). By this point plate with worms should be nearly ready for bombardment.

3.6. Preparation of Worms

1. Put one 9 cm NGM plate to 37°C for 1 h. The plate should be dry enough to quickly absorb all liquid from a pellet of worms. Place the dried plate on ice to cool down.

2. Wash worms with egg salts buffer from 5 to 6 egg plates into a 50 mL tube (see Note 9).
3. Allow worms to settle for 5–10 min. Remove supernatant containing younger animals and most of the bacteria.
4. Resuspend worms in 30 mL of egg salts buffer. Allow worms to settle, remove supernatant. Repeat the procedure several times. The goal is to obtain 2 mL sediment of adult stage worms clean from large pieces of debris and early larval stages.
5. Remove the supernatant and transfer 2 mL of worms to the prechilled dry NGM plate. Allow the worm liquid to distribute uniformly on the plate. Keep the plate on ice (see Note 10). Once the liquid is absorbed by agar, a uniform layer of worms would form. While plate with worms is drying, proceed to loading of macrocarriers with DNA-coated gold particles (see Note 11).

3.7. Bombardment

1. Read the manufacturer's instructions to set up the Bio-Rad Biolistic PDS-1000/He particle delivery system with the Hepta adapter (15).
2. Macrocarrier holder, Hepta adapter, and stopping screens should be sterilized in EtOH or autoclaved regularly. Clean the bombardment chamber with 70% ethanol before each use.
3. Rinse seven macrocarriers in absolute ethanol and allow them to dry on a tissue paper.
4. Place macrocarriers into Hepta adapter holder using the seating tool. A pipette tip can be used to ensure that the macrocarriers are fastened into the holder.
5. Spread 20 μL of DNA-coated gold beads onto each macrocarrier and let dry (see Note 12).
6. Assemble stopping screen and macrocarrier holder together.
7. Soak a 1,350 psi rupture disc in isopropanol for 3–5 s, place in the retaining cap of the Hepta adapter, and tighten the adapter onto the helium gas acceleration tube.
8. Place the assembled macrocarrier holder into PDS-1000 chamber on the third shelf from the top.
9. Place the plate holder into PDS-1000 chamber on the fourth shelf from the top and place the open NGM plate with worms in the center of the plate holder.
10. Close the chamber and adjust vacuum to 27 in Hg pressure, press "fire" button and hold until disc ruptures.
11. Release vacuum, remove the plate, and shut down the bombardment machine according to manufacturer's instructions.

3.8. Post-bombardment Care of Worms

1. Allow worms to recover for 30 min.
2. Wash the worms off the plates with 11 mL of Egg salts buffer and distribute to 20 9 cm NGM plates seeded with OP50.
3. Incubate plates at appropriate temperature depending on the transgene used (see Note 13).
4. Allow worms to grow and starve for 10–14 days at 20°C (see Note 14).
5. Stable transformants with wild-type phenotype are easily identified on starved plates. Single several worms from each plate for further analysis (see Note 15).

4. Notes

1. Rich 2% NGM plates supports long-term bacterial growth and less burrowing of worms into the agar.
2. It takes several days for egg plates to dry sufficiently. Do not turn egg plates upside down until they are dry, since this can lead to leakage and contamination at the sides of the plates. We generally keep them lid up even after seeding the plates with worms.
3. Egg plates can be seeded by using bleached eggs or chunking from a clean master plate. Since *unc-119(ed3)* is a slow growing strain, it is handy to maintain master plates. Never seed egg plates by washing or chunking worms from previous egg plate!—it would inevitably lead to contamination.
4. Using parafilm to cover the seeded egg plates can sometimes (depending on the preparation of egg plates) lead to sickness or death of worms because of hypoxic conditions. The state of the worms is important for successful transformations, therefore it is better to keep seeded egg plates in clean boxes rather than using parafilm.
5. It is essential that the beads do not clump. If beads clump, use additional vortexing, pipeting, or sonication until all large clumps are broken up.
6. Do not vortex at high speeds to avoid shearing of DNA.
7. Since spermidine is a labile molecule, it is important to store it in small aliquots at −20°C, and thaw the solution right before performing the bombardment. Protamine, which is a powder and more stable at room temperature, can also be used instead of spermidine to deliver foreign DNA (16).
8. Since the beads tend to form clumps at this point, transfer the microparticles onto the macrocarriers as soon as possible.
9. The best time for harvesting worms is when the food in the egg plate is almost finished. Young adults transform most efficiently.

10. If the plate is not kept on ice, worms would start to crawl and form clumps.
11. The timing of the procedure is important for the end result. The order that should be followed during the procedure is:
    (a) Prepare gold particles and linearize DNA.
    (b) Put an NGM plate to 37°C for dehydration.
    (c) Harvest worms with egg salts buffer.
    (d) Start coating gold particles with DNA and continue washing the worms in the meantime.
    (e) After addition of spermidine, start sterilizing the macrocarriers.
    (f) Put NGM plate on ice.
    (g) Put worm solution on prechilled NGM agar.
    (h) Place macrocarriers into the macrocarrier holder and load coated gold particles.
    (i) After the gold particles are dried and the worm solution forms a viscous layer on top of the NGM plate, bombard.
12. Use a pipette tip to ensure that the macrocarrier is fastened into the holder. The DNA suspension should be added after macrocarrier is fastened to the macrocarrier holder. Trying to fasten the macrocarrier after DNA suspension is dry may lead to the loss of gold particles.
13. Some transgenes may be temperature sensitive, and may affect the phenotypes of worms at permissive temperatures. For instance, twk-18 transgene expresses a dominant, temperature sensitive K+channel in the body wall muscles. Worms move normally at 15°C; however when shifted to 25°C, the worms carrying the extrachromosomal array are paralyzed (17).
14. Usually transformants can be identified already in the F1 progeny within 5 days after bombardment but most of them would be transient. Therefore, it is necessary to wait until ***unc-119(ed3)*** worms die and wild-type moving worms outgrow the plate. This also makes the selection easier.
15. In some cases it is possible to identify two or more independent transformants per plate. Therefore, it is necessary to single several worms per plate.

References

1. Stinchcomb DT et al (1985) Extrachromosomal DNA transformation of *Caenorhabditis elegans*. Mol Cell Biol 5:3484–3496
2. Fire A (1986) Integrative transformation of *C.elegans*. EMBO J 5:2673–2680
3. Jackstadt P et al (1999) Transformation of nematodes via ballistic DNA transfer. Mol Biochem Parasitol 103:261–266
4. Wilm T et al (1999) Ballistic transformation of *Caenorhabditis elegans*. Gene 229:31–35

5. Praitis V et al (2001) Creation of low-copy integrated transgenic lines in *Caenorhabditis elegans*. Genetics 157:1217–1226
6. Berezikov E, Bargmann C, Plasterk R (2004) Homologous gene targeting in *Caenorhabditis elegans* by biolistic transformation. Nucleic Acids Res 32:e40
7. Johnson NM, Behm CA, Trowell SC (2005) Heritable and inducible gene knockdown in *C. elegans* using Wormgate and the ORFeome. Gene 10:26–34
8. Barrett PL, Fleming JT, Gobel V (2004) Targeted gene alteration in *Caenorhabditis elegans* by gene conversion. Nat Genet 36:1231–1237
9. Evans TC (ed) (2006) Transformation and microinjection. In: WormBook (ed) The C. elegans research community, WormBook, doi/10.1895/wormbook.1.108.1, http://www.wormbook.org
10. Praitis V (2006) Creation of transgenic lines using microparticle bombardment methods. Methods Mol Biol 351:93–107
11. Krausea M (1995) Techniques for analyzing transcription and translation. Methods Cell Biol 48:513–529
12. Hochbaum D, Ferguson AA, Fisher AL (2010) Generation of transgenic *C. elegans* by biolistic transformation. J Vis Exp. doi: 10.3791/2090
13. Askjaer P et al (2002) Ran GTPase cycle and importins alpha and beta are essential for spindle formation and nuclear envelope assembly in living *Caenorhabditis elegans* embryos. Mol Biol Cell 13:4355–4370
14. Ferguson AA, Fisher AL (2009) Retrofitting ampicillin resistant vectors by recombination for use in generating *C. elegans* transgenic animals by bombardment. Plasmid 62:140–145
15. Bio-Rad (1997) Bio-Rad biolistic® PDS-1000/He particle delivery system instruction manual. Bio-Rad Laboratories, Hercules, CA
16. Sivamani E, DeLong RK, Qu R (2009) Protamine-mediated DNA coating remarkably improves bombardment transformation efficiency in plant cells. Plant Cell Rep 28:213–221
17. Kunkel MT et al (2000) Mutants of a temperature-sensitive 2-P domain potassium channel. J Neurosci 20:7517–7524

Chapter 8

Improved Vectors for Selection of Transgenic *Caenorhabditis elegans*

Annabel A. Ferguson, Liquan Cai, Luv Kashyap, and Alfred L. Fisher

Abstract

The generation of transgenic animals is an essential part of research in *Caenorhabditis elegans*. One technique for the generation of these animals is biolistic bombardment involving the use of DNA-coated microparticles. To facilitate the identification of transgenic animals within a background of non-transformed animals, the *unc-119* gene is often used as a visible marker as the *unc-119* mutants are small and move poorly and the larger size and smoother movement of rescued animals make them clearly visible. While transgenic animals can be identified from co-bombardment with a transgene of interest and a separate *unc-119* rescue plasmid, placing the *unc-119* in *cis* on the transgene increases confidence that the resulting transgenic animals contain and express both the marker and the transgene. However, placing the *unc-119* marker on the backbone of a plasmid or larger DNA construct, such as a fosmid or BAC, can be technically difficult using standard molecular biology techniques. Here we describe methods to circumvent these limitations and use either homologous recombination or Cre-LoxP mediated recombination in *Escherichia coli* to insert the *unc-119* marker on to a variety of vector backbones.

Key words: *C. elegans*, Transgenic animals, *unc-119*, Microparticle bombardment, Recombination, Cre recombinase, Plasmid, Biotechnology

1. Introduction

In *Caenorhabditis elegans* research, transgenic animals are routinely generated for the purpose of tagging proteins, visualizing gene expression, over-expressing a gene, or otherwise modifying the worm genome (1, 2). As a result, the generation of transgenic animals is an essential aspect of *C. elegans* research, and effective and facile methods to reliably generate animals with stable transgenic arrays or integrated transgenes are necessary for lab

Stephan Sudowe and Angelika B. Reske-Kunz (eds.), *Biolistic DNA Delivery: Methods and Protocols*, Methods in Molecular Biology, vol. 940, DOI 10.1007/978-1-62703-110-3_8,

© Springer Science+Business Media, LLC 2013

productivity. Two commonly employed methods for introducing foreign DNA into the worm genome are microinjection, and biolistic bombardment (2, 3). Both of these techniques require the use of a visible marker to identify animals which carry the transgene. However, these methods differ in the stability, raw number, and proportion of transgenic lines obtained, as well as the ease of carrying out the protocol (1, 2). Biolistic bombardment is the best method to use to obtain multiple lines with transgenes that are either integrated or extrachromosomal arrays from a single bombardment. This is also an attractive method because it has a much smaller learning curve and success is less dependent on the skill of the operator.

One challenge to bombardment is that the identification of transgenic worms requires visually screening a large quantity of animals. For this technique, a visible marker that is detectable at lower magnifications is optimal. To date, the best marker for biolistic bombardment is the *unc-119* rescue gene (3). The *unc-119* mutants are smaller, have uncoordinated locomotion, are deficient in their ability to form dauers, and have lower fitness in lab conditions compared to wild type (4). Rescue of the *unc-119* mutation with a transgene produces worms that have normal mobility and size which makes them easy to distinguish from non-transformed animals. One way to use *unc-119* rescue as a selectable marker is to bombard two separate plasmids at the same time—one containing the *unc-119* rescue gene and the other containing the gene of interest. There exists an expansive plasmid collection generated by the lab of Dr. Andrew Fire (available from Addgene, Inc., Cambridge, MA), and a *C. elegans* fosmid library created by the lab of Dr. Donald Moerman (available at Source BioScience, Nottingham, UK) that are ready to be used to create transgenic worms in a co-bombardment fashion. However, a disadvantage to this strategy is that the lack of physical linkage between the selective marker and the transgene of interest can lead to some selected transgenic lines containing the marker but not the gene of interest, and vice versa (5). This is especially a problem when there is no way of easily confirming the presence of the transgene of interest (i.e., by GFP expression, western blot, or PCR genotyping). Physically linking the *unc-119* marker and the transgene of interest by putting them on the same plasmid ensures that this will not happen. Generating constructs such as these with traditional molecular biological techniques poses challenges such as identification of available restriction sites, which may make this approach not worthwhile.

The following vectors and protocols overcome the obstacles of traditional molecular cloning through the use of homologous or Cre-LoxP recombination, and offer facile methods to produce DNA constructs that improve the selection of transgenic

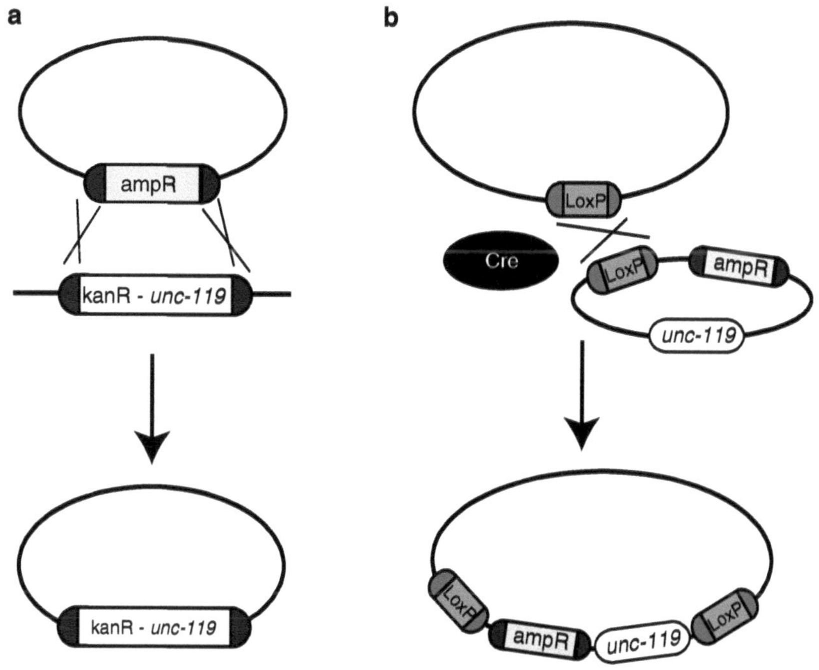

Fig. 1. Schematic overview of vector modifications. (**a**) For plasmid vectors with an ampicillin resistance gene, the *unc-119* marker can be added via homologous recombination between the ampicillin resistance gene and a kanamycin resistance—*unc-119* cassette which is flanked by regions of homology to the ampicillin resistance gene. This results in the cassette replacing the ampicillin resistance gene. (**b**) For vectors such as fosmids containing a LoxP site, Cre-LoxP recombination is used to insert a replication-defective plasmid containing the *unc-119* marker and the ampicillin resistance gene on to the fosmid backbone.

animals (Fig. 1). In the first protocol, the *unc-119* rescue gene in the form of either *C. elegans* cDNA or *Caenorhabditis briggsae* genomic DNA may be inserted into any plasmid with an ampicillin resistance gene by recombination (Fig. 1a) (5). Among the plasmids that can be modified by this protocol are the large collection of plasmids generated by the lab of Dr. Andrew Fire, which expedite common tasks such as generating a GFP reporter or expressing a cDNA with a tissue-specific promoter. This collection is available from Addgene Inc. (Cambridge, MA). The second protocol provides a means to modify existing fosmids or BACs containing large portions of genomic DNA or modified genomic DNA, by inserting the *C. elegans* genomic *unc-119* rescue gene via Cre-LoxP-mediated recombination (Fig. 1b) (6). Using these methods we are able to routinely bombard with vectors from the Fire lab that have been retrofitted to contain the *unc-119* rescue gene, and also to bombard with fosmids containing stretches of modified *C. elegans* genome, including the gene of interest and most of the regulatory elements, and the *unc-119* rescue gene, to obtain transgenic animals (7, 8).

2. Materials

2.1. Retrofitting Ampicillin-Resistant Plasmids by Inserting the unc-119 Marker

2.1.1. Strains

1. DH5α (*fhuA2* Δ*(argF-lacZ)U169 phoA glnV44* Φ*80* Δ*(lacZ) M15 gyrA96 recA1 relA1 endA1 thi-1 hsdR17*) *Escherichia coli* bacteria transformed with the pKD78 plasmid (9) (see Note 1).
2. DH5α *E. coli* bacteria (Invitrogen Corp., Carlsbad, CA).

2.1.2. Plasmids

1. p*unc-119*c plasmid containing the R6K replication origin, kanamycin resistance, and *unc-119* cDNA (Addgene Inc., Cambridge, MA) (5) (see Notes 2 and 3).
2. p*unc-119*cR plasmid which is similar to p*unc-119*c except that the mCherry gene is fused to the C-terminus of the *unc-119* cDNA (Addgene Inc., Cambridge, MA) (5) (see Notes 2 and 3).
3. p*unc-119*cbr which is similar to p*unc-119*c except that the *C. briggsae unc-119* gDNA is used (Addgene Inc., Cambridge, MA) (see Notes 2 and 3).
4. Fire lab vectors or other plasmids containing the ampicillin resistance gene on the plasmid backbone.

2.1.3. Equipment

1. Electroporator, such as the Eppendorf 2510 electroporator (Eppendorf AG, Hamburg, Germany), with 0.1 cm gap cuvettes.
2. Microbial incubator.
3. Shaking incubator.
4. Refrigerated centrifuge with rotors.
5. Microcentrifuge.
6. Pipettors.
7. 1.5 mL Microfuge tubes.
8. 14 mL Round bottom snap-cap culture tubes.
9. 750 mL Centrifuge bottles.
10. Apparatus for agarose gel electrophoresis.
11. Spectrophotometer.
12. –80°C Freezer.
13. 4°C Refrigerator.
14. Dry ice.
15. 2 L Erlenmeyer flask.

2.1.4. Reagents and Media

1. 34 mg/mL Chloramphenicol stock solution (1,000×): Dissolve 0.34 g. chloramphenicol powder in 10 mL absolute ethanol. Store at −20°C.
2. 10% Arabinose stock solution: Dissolve 1 g. arabinose in 10 mL milliQ water. Filter sterilize. Store at −20°C.
3. 30 mg/mL Kanamycin stock solution (1,000×): Dissolve 0.3 g kanamycin in 10 mL milliQ water. Filter sterilize. Store at −20°C.
4. LB broth: Dissolve 10 g. tryptone, 5 g. NaCl, and 5 g. yeast extract in 1 L. milliQ water. Autoclave for 20 min (see Note 4).
5. LB agar: Dissolve 10 g. tryptone, 5 g. NaCl, 5 g. yeast extract, and 15 g. agar in 1 L. milliQ water. Autoclave for 20 min (see Note 5).
6. LB broth with 34 μg/mL chloramphenicol: Using sterile technique, add 5 μL chloramphenicol stock solution to 5 mL of LB broth which has been autoclaved and allowed to cool to room temperature.
7. LB agar with 34 μg/mL chloramphenicol: Using sterile technique, add 1 mL chloramphenicol stock solution to 1 L. LB agar which has been autoclaved and allowed to cool in a water bath to 60°C. Mix by swirling and pour agar into 10 cm. Petri dishes. Allow plates to cool and harden at room temperature before use.
8. LB broth with 30 μg/mL kanamycin: Using sterile technique, add 500 μL kanamycin stock solution to 500 mL of LB broth which has been autoclaved and allowed to cool to room temperature.
9. LB agar with 30 μg/mL kanamycin: Using sterile technique, add 1 mL kanamycin stock solution to 1 L. LB agar which has been autoclaved and allowed to cool in a water bath to 60°C. Mix by swirling and pour agar into 10 cm. Petri dishes. Allow plates to cool and harden at room temperature before use.
10. SOB with 34 μg/mL chloramphenicol and 0.015% Arabinose: Dissolve 10 g. tryptone, 2.5 g. yeast extract, 0.25 g. NaCl, 1.4 g. $MgSO_4$, and 0.092 g. KCl in 500 mL milliQ water in a 2 L. flask, autoclave, and let the media cool to RT. Using sterile technique, add 500 μL chloramphenicol stock solution and 750 μL arabinose stock solution (see Note 6).
11. 10% Glycerol solution: Measure 100 mL glycerol, and add water to a volume of 1 L. Autoclave, and chill on ice or in a cold room to 4°C.
12. BamHI, ScaI, and XhoI restriction endonucleases and buffers.

13. Zymo DNA Clean & Concentrator kit (Zymo Research Corp., Irvine, CA).
14. Qiagen QIAprep®Spin Miniprep Kit (Qiagen GmbH, Hilden, Germany) or similar mini-prep reagents.
15. 0.8% Agarose gel.

2.2. Inserting unc-119 into Fosmids by Cre-LoxP Recombination

2.2.1. Strains

1. SW106 (*mcrA* Δ(*mrr-hsdRMS-mcrBC*) Δ*lacX74 deoR endA1 araD139* Δ(*ara, leu*) *7697 rpsL recA1 nupG* φ80d*lacZ*Δ*M15* [λ*c*1857 (*cro-bioA*)<>*Tet*] (*cro-bioA*)<>*araC*-P_{BAD} *Cre* Δ*galK*) strain (10) (see Note 7).
2. EPI300 (*F mcrA* Δ*(mrr-hsdRMS-mcrBC)* φ*80dlacZ*Δ*M15* Δ*lacX74 recA1 endA1 araD139* Δ*(ara, leu)7697 galU galK* λ$^-$ *rpsL nupG trfA tonA*) strain (see Note 8).

2.2.2. Plasmids, Fosmids, and Oligos

1. Fosmids containing *C. elegans* genomic DNA in the pCC1FOS vector (see Notes 9 and 10).
2. pLoxP *unc-119* plasmid contains the *unc-119* gene, ampicillin resistance gene, and an R6K origin of replication (Addgene Inc., Cambridge, MA) (6) (see Note 2).
3. Oligos for colony PCR:

   unc-119 F :CAAATCCGTGACCTCGACAC

   unc-119 R :CACAGTTGTTTCTCGAATTTGG

2.2.3. Equipment

1. Electroporator such as the Eppendorf 2510 electroporator (Eppendorf AG, Hamburg, Germany), with 0.1 cm gap cuvettes.
2. Microbial incubator.
3. Shaking incubator.
4. Refrigerated centrifuge with rotors.
5. Microcentrifuge.
6. Pipettors.
7. 1.5 mL Microfuge tubes.
8. 14 mL Round bottom snap-cap culture tubes.
9. 250 mL Centrifuge bottles.
10. PCR machine.
11. Apparatus for agarose gel electrophoresis.
12. Spectrophotometer.
13. –80°C Freezer.

14. 4°C Refrigerator.
15. Dry ice.
16. 2 L and 250 mL Erlenmeyer flask.

2.2.4. Reagents and Media

1. 12.5 mg/mL Chloramphenicol stock solution (1,000×): Dissolve 0.125 g chloramphenicol in 10 mL absolute ethanol. Store at −20°C.
2. LB broth: Dissolve 10 g. tryptone, 5 g. NaCl, and 5 g. yeast extract in 1 L. milliQ water. Autoclave for 20 min (see Note 4).
3. LB agar: Dissolve 10 g. tryptone, 5 g. NaCl, 5 g. yeast extract, and 15 g. agar in 1 L. milliQ water. Autoclave for 20 min (see Note 5).
4. 50 mg/mL Ampicillin stock solution (1,000×): Dissolve 0.5 g ampicillin in 10 mL milliQ water. Filter sterilize. Store at −20°C.
5. LB broth with 12.5 μg/mL chloramphenicol: Using sterile technique, add 500 μL chloramphenicol stock solution to 500 mL of LB broth which has been autoclaved and allowed to cool to room temperature.
6. LB agar with 12.5 μg/mL chloramphenicol: Using sterile technique, add 1 mL chloramphenicol stock solution to 1 L. LB agar which has been autoclaved and allowed to cool in a water bath to 60°C. Mix by swirling and pour agar into 10 cm. Petri dishes. Allow plates to cool and harden at room temperature before use.
7. LB broth with 50 μg/mL ampicillin and 12.5 μg/mL chloramphenicol: Using sterile technique, add 500 μL chloramphenicol stock solution and 500 μL ampicillin stock solution to 500 mL of LB broth which has been autoclaved and allowed to cool to room temperature.
8. LB agar with 50 μg/mL ampicillin and 12.5 μg/mL chloramphenicol: Using sterile technique, add 1 mL chloramphenicol stock solution and 1 mL ampicillin stock solution to 1 L. LB agar which has been autoclaved and allowed to cool in a water bath to 60°C. Mix by swirling and pour agar into 10 cm. Petri dishes. Allow plates to cool and harden at room temperature before use.
9. Two sterile 2 L flasks and one sterile 250 mL flask.
10. 10% Glycerol solution: Measure 100 mL glycerol, and add water to a volume of 1 L. Autoclave, and chill on ice or in a cold room to 4°C.
11. 10% Arabinose stock solution: Dissolve 1 g. arabinose in 10 mL milliQ water. Filter sterilize. Store at −20°C.
12. LB + 0.1% arabinose (10 mL): Add 100 μL 10% arabinose stock solution to 10 mL sterile LB.

13. CopyControl™ Fosmid Autoinduction Solution (Epicentre Biotechnologies, Madison, WI).
14. FosmidMAX fosmid DNA purification kit (Epicentre Biotechnologies, Madison, WI).
15. GoTaq DNA polymerase master mix (Promega Corp., Madison, WI).
16. 0.8% Agarose gel.

3. Methods

3.1. Retrofitting Ampicillin-Resistant Plasmids by Inserting the unc-119 Marker

In this protocol, plasmid-based vectors carrying an ampicillin resistance gene can be modified to carry the *unc-119* marker on the vector backbone via homologous recombination in *E. coli* (Fig. 1a) (5). This procedure involves addition of an *unc-119* and kanamycin resistance gene cassette which is flanked by sequences homologous to the ends of the ampicillin resistance gene, and is then followed by destruction of the unmodified parental plasmid to obtain a pure population of modified plasmid. The destruction of the unmodified plasmid is necessary as most plasmids exist as multiple copies in *E. coli* so only a minority of the plasmids are modified during the recombination step (11). The final plasmid contains a kanamycin resistance gene and *unc-119* marker in place of the ampicillin resistance gene and is now kanamycin resistant.

3.1.1. Induction of Competent pKD78 Transformed DH5α Cells for Electroporation

1. Streak out pKD78 transformed DH5α on LB + 34 μg/mL chloramphenicol agar plates. Grow colonies at 30°C overnight in an incubator.
2. Inoculate 5 mL of LB + 34 μg/mL chloramphenicol with a colony of pKD78 transformed DH5α. Grow overnight at 30°C in a shaking incubator at 250 rpm.
3. Save 1 mL of the SOB + 34 μg/mL chloramphenicol and 0.015% arabinose to use as a blank for the spectrophotometer to determine the optical density of the bacteria.
4. Inoculate 500 μL of the pKD78 transformed DH5α overnight culture in the SOB + chloramphenicol + arabinose, and grow in a 30°C shaking incubator, until the bacteria reach an optical density of between 0.6 and 0.8. This will take between 2 and 4 h.
5. Immediately spin down the bacteria in a 750 mL centrifuge bottle at 6,000 × *g* for 15 min. at 4°C, and wash once with 500 mL 10% ice-cold glycerol (see Note 11).
6. Wash twice with 50 mL 10% ice-cold glycerol, leaving approximately 5 mL of glycerol solution after the final wash. Keep the bacteria on ice in between washes. Resuspend the pellet in the

remaining glycerol, make 100 μL aliquots in 1.5 mL microfuge tubes, and freeze on dry ice.

7. Store aliquots at –80°C. Long term storage for weeks to months in a –80°C freezer does not seem to reduce the efficiency of these cells.

3.1.2. Preparation of unc-119—Kanamycin Resistance Cassette

1. Cut 1 μg. p*unc-119*c, p*unc-119*cR, or p*unc-119*cbr with BamHI for 1 h at 37°C. If this protocol will be carried out to modify multiple plasmids, a larger quantity of plasmid DNA may be digested at this step, and the fragment stored at 4°C for later use.
2. Desalt and purify the digest with the Zymo DNA Clean & Concentrator kit and elute in 10 μL of water. Gel purification of this fragment does not improve the results.

3.1.3. Homologous Recombination of unc-119 Cassette with Target Vector

1. Thaw an aliquot of the induced pKD78 transformed DH5α cells on ice.
2. Mix 100 ng. of the *unc-119*—kanamycin resistance cassette with 50 ng. of the ampicillin-resistant recipient plasmid in a 1.5 mL microfuge tube.
3. Mix the DNA mixture with the competent cells, and immediately add this to a 0.1 cm gap electroporation cuvette. Electroporate the bacteria at 1,350 V.
4. Immediately add 1 mL LB broth to the bacteria, and transfer to a 14 mL snap-cap bacterial culture tube.
5. Place the tube in a shaking 37°C incubator, and allow the cells to recover for 2 h (see Note 12).
6. Plate 100 μL of bacteria on one LB + 30 μg/mL kanamycin agar plate, and the rest of the culture on three additional plates.
7. Incubate the plates at 37°C overnight. 50–500 colonies are expected following this step.
8. Inoculate four individual colonies in 5 mL LB + 30 μg/mL kanamycin in 14 mL snap-cap bacterial culture tubes. Grow overnight in a shaking incubator at 37°C.

3.1.4. Confirming the Presence of the unc-119 Cassette

1. Purify the plasmid DNA from the overnight culture with a QIAprep®Spin Miniprep Kit.
2. Digest 1/10 of the mini-prep DNA with XhoI for 1 h and run the digested DNA on a 0.8% agarose gel (see Notes 13 and 14).

3.1.5. Destruction of the Parent Plasmid by ScaI Digestion Followed by Retransformation

1. Digest 1/10 of the mini-prep DNA with ScaI (or another enzyme from Table 1) (see Note 15).
2. Desalt and purify the digest using the Zymo DNA Clean & Concentrator kit. Elute the DNA in 10 μL of water.

Table 1
Enzymes which cut within the ampicillin resistance gene and the number of sites present in the *unc-119*—kanamycin cassettes

Enzyme	*unc119c*	*unc119cR*	*unc119cbr*
ScaI	0	0	0
FspI	0	0	0
AhdI	1	1	0
BcgI	1	1	0
BmrI	0	0	0

3. Chemically transform DH5α with the digested DNA, and plate aliquots of the bacteria on LB + 30 μg/mL kanamycin. Incubate at 37°C overnight.
4. Select individual colonies to inoculate two 5 mL cultures in LB + 30 μg/mL kanamycin. Incubate at 37°C overnight in a shaking incubator.
5. Mini-prep and test digest the DNA using XhoI. Look for clones which only have the recombinant plasmid.
6. Generate a glycerol stock and use the bacterial culture for a midi or maxi-prep to obtain sufficient DNA for bombardment.

3.2. Inserting unc-119 into Fosmids by Cre-LoxP Recombination

In this protocol, fosmid- or BAC-based vectors carrying a LoxP site on the vector backbone can be modified to carry the *unc-119* marker via Cre-LoxP recombination in *E. coli* (Fig. 1b) (6). This procedure involves addition of a replication incompetent plasmid carrying the *unc-119* gene and the ampicillin resistance gene along with a LoxP site. The SW106 bacterial strain has a transgene which expresses Cre recombinase using the inducible arabinose promoter (10). A brief treatment with arabinose produces sufficient Cre to allow recombination between the LoxP sites on the vector and the fosmid/BAC backbone, and selection on LB containing ampicillin is then used to select for the recombinant plasmid.

3.2.1. Purify Fosmid DNA Containing the Gene of Interest

1. Streak out bacteria containing the fosmid of interest from a glycerol stock on a LB + 12.5 μg/mL chloramphenicol agar plate and incubate overnight at 37°C to obtain colonies.

2. Inoculate 5 mL LB + 12.5 μg/mL chloramphenicol with a single colony and grow overnight in a 37°C shaking incubator.
3. Purify the fosmid DNA from the overnight culture using the FosmidMAX DNA purification kit following the alternative protocol for purification of 1.5 mL of bacterial culture (see Note 16).

3.2.2. Generate Electrocompetent SW106 Bacteria

1. Streak SW106 on a LB agar plate and incubate overnight at 32°C to obtain colonies.
2. Inoculate 5 mL LB with a single colony and grow overnight in a 32°C shaking incubator.
3. Transfer 200 mL of sterile LB broth to a 2 L Erlenmeyer flask. Save 1 mL from the flask to use as a blank for measuring optical density.
4. Add 200 μL of the overnight culture to the LB in the 2 L Erlenmeyer flask and grow at 32°C to an OD_{600} of between 0.6 and 0.8. This should take ~4 h.
5. Transfer the bacteria to a 250 mL centrifuge bottle, and pellet the bacteria at 6,000×*g* for 15 min at 4°C. Wash once with 200 mL 10% ice-cold glycerol (see Note 11).
6. Wash twice with 50 mL 10% ice-cold glycerol, leaving approximately 2 mL of glycerol solution after the final wash. Keep the bacteria on ice in between washes. Resuspend the pellet in the remaining glycerol, make 100 μL aliquots in 1.5 mL microfuge tubes, and freeze on dry ice.
7. Transfer aliquots to a −80°C freezer for storage. The bacteria can be used for electroporation for >1 year.

3.2.3. Transform SW106 Bacteria with the Fosmid

1. Thaw competent SW106 bacteria on ice.
2. Mix 100 ng of purified fosmid DNA from the steps above with the bacteria, and electroporate at 1,350 V in a 0.1 cm gap cuvette.
3. Recover in 500 μL LB in a 32°C shaking incubator for 1 h.
4. Plate 100 μL and 200 μL on LB + 12.5 μg/mL chloramphenicol plates, and incubate overnight at 32°C.

3.2.4. Preparation of Electrocompetent Fosmid Transformed SW106 Cells

1. Inoculate 5 mL of LB + 12.5 μg/mL chloramphenicol with a single colony from the above transformation. Grow overnight in a 32°C shaking incubator.
2. Transfer 100 mL of sterile LB broth with 12.5 μg/mL chloramphenicol to a 2 L Erlenmeyer flask. Save 1 mL from the flask to use as a blank for measuring optical density.

3. Add 100 μL of the overnight culture to the LB in the 2 L Erlenmeyer flask and grow at 32°C to an OD_{600} of between 0.6 and 0.8.
4. Transfer the bacteria to a 250 mL centrifuge bottle, and pellet the bacteria at 6,000 × *g* for 15 min at 4°C. Wash once with 100 mL 10% ice-cold glycerol (see Note 11).
5. Wash twice with 50 mL 10% ice-cold glycerol. Keep the bacteria on ice in between washes. Resuspend the pellet in the remaining glycerol, make 100 μL aliquots in 1.5 mL microfuge tubes, and freeze on dry ice.
6. Transfer aliquots to a −80°C freezer for storage. The bacteria can be used for electroporation for at least several months.

3.2.5. Addition of the unc-119 Marker by Cre-LoxP Recombination

1. Thaw an aliquot of the competent fosmid transformed SW106 bacteria.
2. Mix 50 ng. pLoxP *unc-119* DNA with the SW106 bacteria (see Note 17).
3. Electroporate in a 0.1 cm gap cuvette at 1,350 V.
4. Add 500 μL LB + 0.1% arabinose to the cuvette, and put the bacteria in a 32°C shaking incubator for a 1 h recovery.
5. Plate 50 μL and 150 μL aliquots of the bacteria on LB + 50 μg/mL ampicillin and 12.5 μg/mL chloramphenicol agar plates. Incubate overnight at 32°C to obtain colonies.
6. Select 2–4 individual colonies to inoculate 5 mL of LB + 50 μg/mL ampicillin and 12.5 μg/mL chloramphenicol in 14 mL snap-cap culture tubes. Grow in a 32°C shaking incubator overnight (see Note 18).

3.2.6. Detecting the Presence of unc-119 by Colony PCR

1. Set up a PCR reaction using GoTaq DNA polymerase master mix including:

   0.5 μL Bacteria from overnight culture.

   1 μL 10 μM unc-119 F oligo.

   1 μL 10 μM unc-119 R oligo.

   7.5 μL Water.

   10 μL GoTaq master mix.
2. Perform PCR with an initial incubation of 5 min at 95°C to lyse the bacteria; 30 cycles of 95°C for 30 s, 50°C for 30 s, and 72°C for 30 s; and a final extension at 72°C for 5 min.
3. Run the PCR products on a 0.8% agarose gel. A single 248 bp band indicates the presence of the *unc-119* insert in the fosmid.

3.2.7. Transfer of the unc-119 Containing Fosmid to EPI300 Bacteria

1. Isolate fosmid DNA from 1.5 mL of an *unc-119* positive culture using the FosmidMAX DNA purification kit, as before.

2. Electroporate the electrocompetent EPI300 bacteria in 0.1 cm gap cuvettes at 1,350 V. Recover in 1 mL LB for 1 h with shaking at 37°C.
3. Plate 100 μL and 300 μL on LB + 50 μg/mL ampicillin and 12.5 μg/mL chloramphenicol agar plates. Incubate at 37°C overnight.
4. Pick a single colony and grow overnight in 5 mL LB broth + 50 μg/mL ampicillin and 12.5 μg/mL chloramphenicol in a 14 mL snap-cap culture tube. Use this culture to make a glycerol stock and seed a 40 mL culture of LB + 50 μg/mL ampicillin and 12.5 μg/mL chloramphenicol in a 250 mL flask for fosmid purification.
5. Induce the bacterial culture with the CopyControl induction solution to amplify the fosmid according to the manufacturer's instructions, and purify the fosmid using the FosmidMAX DNA purification kit. The purified DNA is usually sufficient for several bombardments.

4. Notes

1. The pKD78 plasmid encodes the lambda red recombination machinery under the control of the arabinose inducible araB promoter (9). The pKD78 plasmid is chloramphenicol resistant and uses the temperature-sensitive repA101 replicon so it must be grown at 30°C. The plasmid is available from the Coli Genetic Stock Center (CGSC) (http://cgsc.biology.yale.edu/index.php) in the BW25141 bacterial strain. BW25141 is not compatible with this technique as it is *pir+* and will allow the growth of the p*unc-119* plasmids. The pKD78 plasmid can be isolated via mini-prep and retransformed into DH5α for subsequent use.
2. All of these plasmids use the R6K replication origin which is only able to replicate in bacteria which express the *pir* gene, which is uncommon among routine lab bacterial strains. These are supplied in the EC100D *pir-116 (F- mcrA Δ(mrr-hsdRMS-mcrBC) ϕ80d lacZΔM15 ΔlacX74 recA1 endA1 araD139 Δ(ara, leu)7697 galU galK λ- rpsL nupG pir-116(DHFR))* bacteria (Epicentre Biotechnologies, Madison, WI) which express *pir* at high level and permit replication of the plasmid at medium copy number.
3. We have successfully obtained transgenic animals using either of three forms of the *unc-119* rescue gene (*C. elegans* cDNA, cDNA fused to mCherry, or *C. briggsae* gDNA). However, the *C. elegans* cDNA seems to require a higher copy number to

produce the rescue phenotype, therefore, if a lower copy number of the transgene is desired, use of the *C. briggsae* genomic DNA is recommended. The mCherry is particularly useful as it is visible using a fluorescent dissecting microscope and permits the transgene to be identified if it is crossed out of strains with the *unc-119* mutation (5).

4. LB broth mixes can be also purchased from many suppliers. We use LB mix from EMD4 Biosciences (San Diego, CA). LB broth containing chloramphenicol or kanamycin can be made by adding these antibiotics from the described stock solutions as a 1:1,000 dilution.
5. LB agar mixes can be also purchased from many suppliers. LB agar plates containing chloramphenicol or kanamycin can be made by cooling the molten agar to 60°C and then adding these antibiotics from the described stock solutions as a 1:1,000 dilution.
6. SOB media mix can also be purchased. We use SOB mix from EMD4 Biosciences (San Diego, CA).
7. The SW106 strain contains an arabinose inducible Cre transgene in addition to the heat inducible lambda red recombination machinery (10). To avoid inducing the recombination proteins, SW106 needs to be grown at 32°C or below. SW106 is available from the NIH (http://web.ncifcrf.gov/research/brb/recombineeringInformation.aspx).
8. The EPI300 strain lacks the Cre recombinase which prevents the *unc-119* marker from being lost via the reverse Cre-LoxP recombination event. EPI300 also permits induction of certain fosmids and BACs to high copy number to facilitate DNA purification. This strain is available from Epicentre Biotechnologies (Madison, WI).
9. Genomic DNA library fosmids in this vector have a LoxP site on the vector backbone which is used in the recombination reaction. The pCC1FOS vector can also be amplified in the EPI300 bacteria strain. Fosmids from the library generated by the lab of Dr. Donald Moerman are available from Source BioScience, Nottingham, UK. Other fosmid or BAC clones can also be used as long as they have a LoxP site and are not ampicillin resistant.
10. Prior to insertion of the *unc-119* marker, the fosmids can be modified by recombineering in the SW106 bacterial strain to add a TAP, GFP, or other tags or to introduce mutations following protocols described by our lab and others (6, 12–14).
11. Careful washing with the 10% glycerol is important to remove salts from the LB which could produce arcing during the electroporation. To facilitate resuspension of the pellet we use gentle vortexing on setting 3–4.

12. Incubation at 37°C inhibits growth of the pKD78 plasmid which will be subsequently lost from the DH5α bacteria. All subsequent growth steps can be carried out at 37°C. The prolonged outgrowth is needed as the recombination event usually only affects a single strand of the plasmid so the plasmid needs to replicate to give a double-stranded recombinant plasmid.
13. All of the *unc-119*—kanamycin resistance cassettes contain two XhoI sites so the insertion of the cassette into the target plasmid can be verified by restriction digest which yields a diagnostic band from the cassette and produces a change in the banding pattern compared to the parent plasmid. For p*unc-119*c cassette the diagnostic band is 2,329 bp in size; for p*unc-119*cR it is 3,171 bp, and for p*unc-119*cbr it is 2,306 bp. The XhoI digest allows plasmids lacking a map to be successfully retrofitted by looking for the diagnostic band and a change compared to the parent plasmid digested in parallel.
14. The digest usually gives a gel pattern that corresponds to a combination of the parent ampicillin-resistant plasmid mixed with the new recombinant plasmid. This occurs because the parent plasmids are present as multiple copies within the bacteria so most colonies consist of a mix of modified and unmodified plasmids (11). This is in contrast to BACs and fosmids which are present at only one copy per cell and are either modified or unmodified. We have found that enzymatic digestion of the mini-prep DNA followed by retransformation is the most effective means to obtain bacteria containing only the modified plasmid.
15. For the ScaI digestion to be successful, the parent plasmid must contain a ScaI site and the recombinant plasmid must NOT contain a ScaI site. ScaI sites are present in almost all ampicillin-resistant genes and appear to be uncommon in vectors and worm DNA. If a ScaI site is present elsewhere in the vector, an alternate enzyme from Table 1 can be selected for use. Alternatively, the parent plasmid may be removed by diluting the plasmid prep and re-transforming, however, this technique has been significantly less successful in our lab.
16. The FosmidMAX DNA prep kit has two protocols for the small-scale preparation of fosmids. The standard protocol adds the RiboShredder RNAse blend at step 18. This makes quantifying the yield using a spectrophotometer more difficult because the nucleotides from the digested RNA are still present. We instead use the alternate protocol which adds the Riboshredder mix at step 12 which allows the nucleotides to be removed by the ethanol precipitation in a later step.
17. The pLoxP *unc-119* DNA can be prepared by mini-prep of the EC100D *pir-116* bacteria carrying the plasmid. If this protocol

will be carried out more frequently, larger amounts of DNA can be obtained from a single maxi-prep, and the DNA can be stored at 4°C for later use.

18. We no longer routinely check the modified fosmid for the presence of the *unc-119* gene. The success rate of the protocol is very high in our lab, so we often move directly to the steps in Subheading 3.2.7 using an ampicillin-resistant colony.

References

1. Mello C, Fire A (1995) DNA transformation. Methods Cell Biol 48:451–482
2. Evans TC (2006) Transformation and microinjection. In: WormBook (ed) The *C. elegans* research community, WormBook. doi/10.1895/wormbook.1.108.1.
3. Praitis V et al (2001) Creation of low-copy integrated transgenic lines in Caenorhabditis elegans. Genetics 157:1217–1226
4. Maduro M, Pilgrim D (1995) Identification and cloning of unc-119, a gene expressed in the Caenorhabditis elegans nervous system. Genetics 141:977–988
5. Ferguson AA, Fisher AL (2009) Retrofitting ampicillin resistant vectors by recombination for use in generating C. elegans transgenic animals by bombardment. Plasmid 62:140–145
6. Zhang Y, Nash L, Fisher AL (2008) A simplified, robust, and streamlined procedure for the production of C. elegans transgenes via recombineering. BMC Dev Biol 8:119
7. Fisher AL et al (2008) The Caenorhabditis elegans K10C2.4 gene encodes a member of the fumarylacetoacetate hydrolase family: a Caenorhabditis elegans model of type I tyrosinemia. J Biol Chem 283:9127–9135
8. Ferguson AA, Springer MG, Fisher AL (2010) skn-1-Dependent and -independent regulation of aip-1 expression following metabolic stress in Caenorhabditis elegans. Mol Cell Biol 30:2651–2667
9. Datsenko KA, Wanner BL (2000) One-step inactivation of chromosomal genes in Escherichia coli K-12 using PCR products. Proc Natl Acad Sci U S A 97:6640–6645
10. Warming S et al (2005) Simple and highly efficient BAC recombineering using galK selection. Nucleic Acids Res 33:e36
11. Thomason LC et al (2007) Multicopy plasmid modification with phage lambda Red recombineering. Plasmid 58:148–158
12. Tursun B et al (2009) A toolkit and robust pipeline for the generation of fosmid-based reporter genes in C. elegans. PLoS One 4:e4625
13. Sarov M et al (2006) A recombineering pipeline for functional genomics applied to Caenorhabditis elegans. Nat Methods 3:839–844
14. Dolphin CT, Hope IA (2006) Caenorhabditis elegans reporter fusion genes generated by seamless modification of large genomic DNA clones. Nucleic Acids Res 34:e72

Chapter 9

Biolistic Transformation of *Brugia Malayi*

Tarig B. Higazi and Thomas R. Unnasch

Abstract

Biolistics has become a versatile tool for direct gene transfer to various cell and tissue types. Following its successful use on the parasitic nematode *Ascaris suum*, we developed and evaluated biolistics in the transfection of the model filarial parasite *Brugia malayi*. Biolistics was proven to be an efficient strategy for transfection of all life stages of the parasite and paved the way for studies on elements essential for promoter function and gene regulation of filarial parasites. Here we present a biolistics protocol for the transfection of *B. malayi* based on the Biolistics PDS 1000/He system and gold microcarriers.

Key words: Biolistics, Particle bombardment, Gold particles, Biolistics PDS 1000/He, Brugia, Transfection, Transformation, Luciferase, GFP

1. Introduction

Parasitic diseases caused by filarial helminths constitute a serious public health issue in tropical regions. The inability to genetically manipulate filarial nematodes was a major obstacle in the study of the regulation of gene expression in these parasites. Biolistics (also known as particle bombardment) has emerged as a versatile tool for direct gene transfer to various cell and tissue types. In nematodes, biolistics has been shown to be capable of transfecting intact *Caenorhabditis elegans* (1, 2) with the transfected animals being capable of transmitting the transgenes to their progeny. Biolistics has also been demonstrated to transiently transfect isolated embryos of *Ascaris suum* (3). Based on these successes, we developed and evaluated a biolistics-based transient transfection of *Brugia malayi* as a model filarial nematode (4). This transfection system has proven to be useful in defining the *cis* acting elements important

Stephan Sudowe and Angelika B. Reske-Kunz (eds.), *Biolistic DNA Delivery: Methods and Protocols*,
Methods in Molecular Biology, vol. 940, DOI 10.1007/978-1-62703-110-3_9, © Springer Science+Business Media, LLC 2013

for promoter structure and function and other aspects of gene regulation in this filarial parasite (4–14).

In our quest for the best gene transfer tool for *B. malayi* parasitic nematode, we tried several approaches e.g., microinjection, electroporation, and lipofection and found biolistics to be the most convenient for its productivity, consistency, and comparative ease. We tried both the gene gun and the Biolistics PDS 1000/He apparatus (BioRad, Hercules, CA, USA) for transfection of various *B. malayi* life stages and found the biolistics PDS 1000/He system to be superior to the gene gun in terms of reliability and consistency. Here we present a biolistics protocol based on the Biolistics PDS 1000/He system and gold microcarriers (BioRad, Hercules, CA, USA).

2. Materials

Unless otherwise noted, all reagents are prepared in sterile Millipore (double distilled) water. Reagents are all analytical and/or molecular biology grade.

2.1. Biolistics Apparatus and Supplies (BioRad, Hercules, CA)

1. The biolistic PDS-1000/He instrument.
2. Helium gas canister.
3. 0.6 μm gold microcarriers.
4. Macrocarriers.
5. Macrocarrier holder.
6. Rupture discs (1,100 and 2,200 psi).
7. Stopping screens.
8. Optimization kit (optional; see Note 1).
9. Vacuum pump or in-house vacuum facet.

2.2. Coating Gold Microcarriers

1. Expression vector with a reporter gene and insert of interest.
2. Vertical Microtube Holder for vortex (e.g., TurboMix attachment).
3. Endofree Plasmid Maxiprep kit (Qiagen Inc., Valencia, CA, USA).
4. 2.5 M Calcium Chloride ($CaCl_2$).
5. 0.1 M Free base, tissue culture grade Spermidine.
6. Absolute ethanol.
7. 70% Ethanol.
8. Sterile water.
9. 50% Glycerol.

10. 1.5 mL Eppendorf tubes.
11. Microcentrifuge.

2.3. Preparation of *Brugia malayi*

1. *B. malayi* adult males, adult females, microfilariae, and L3 larvae are obtained from NIAID/NIH Filariasis Research Reagent Resource Center (FR3), University of Georgia, Athens, GA, USA. In Europe, worms may be obtained from TRS services, Athens, GA.
2. CF-RPMI medium (RPMI 1640 tissue culture medium containing 25 mM HEPES, 20% fetal calf serum, 20 mM glucose, 24 mM sodium bicarbonate, 2.5 μg/mL amphotericin B, 10 U/mL penicillin, 10 U/mL streptomycin, and 40 μg/mL gentamicin).
3. Sterile scalpels.

2.4. Bombardment and Post-bombardment Treatment

1. Sterile Petri dishes (60 × 15 mm Dishes).
2. Isopropanol.
3. CF-RPMI medium (detailed above).
4. 37°C incubator equipped with 5% CO_2.

2.5. Analysis of Bombarded *B. malayi*: Reporter Gene Assays

2.5.1. Luciferase Assay

1. Luciferase Assay System (Promega, Madison, WI, USA) if using single firefly luciferase reporter vector. Dual-Luciferase Reporter Assay System or Dual-Glo Reporter Assay System (Promega, Madison, WI, USA) when using two luciferase reporter vectors (firefly and Renilla luciferase) (see Note 2).
2. Expression vector pGL3 basic and pRL-null (Promega, Madison, WI, USA).
3. Microglass homogenizers/tissue grinders.
4. Luminometer (e.g., FB12, Berthold Detection Systems, Bath, UK).
5. Protein content Assay kit (e.g., Bradford Protein Assay, BioRad, Hercules, CA, USA).

2.5.2. GFP Assay

1. Expression vector pPD95.75 (Addgene, Cambridge, MA, USA).
2. Fluorescent microscope outfitted with GFP channel/Green filter.
3. Methanol.
4. Glass microscopic slides, coverslips, mounting medium, and nail polish.

3. Methods

3.1. Preparation of Microcarriers

1. Add 1 mL 70% ethanol to 30 mg aliquots of 0.6 μm gold beads (microcarriers) (see Note 3) in a 1.5 mL microcentrifuge tube and vortex vigorously for 5 min on a vertical microfuge tube holder. Let the gold particles soak in the 70% ethanol for 15 min on the bench. Pellet the gold particles by spinning for 5 s in a microfuge. Carefully remove and discard the ethanol without disturbing the gold particles. Repeat this step three times.
2. Resuspend 30 mg gold particles in 1 mL sterile water and vortex vigorously for 1 min as above. Let the particles soak in water for 1 min, and then pellet them by spinning for 5 s in a microfuge. Carefully remove and discard the water. Repeat this step three times.
3. Add 500 μL 50% glycerol and resuspend the gold particles by pipetting for a final concentration of 60 mg/mL gold particles. Store gold particles at 4°C for coating with transgenic DNA (see Note 3).

3.2. Coating Microcarriers with Transgenic Plasmid DNA

1. Construct transgenic DNA in expression plasmid vectors containing reporter genes based on your study objectives (see Note 4).
2. Purify transgenic Plasmid DNA construct by cesium chloride gradient centrifugation or Qiagen EndoFree Plasmid Maxi Kit and resuspended in sterile water at 1 mg/mL (see Note 5).
3. Vortex 500 μL microcarrier solution (30 mg in 50% glycerol; see step 3 in Subheading 3.1) for 5 min on a platform vortex to resuspend gold particles. While holding the microfuge tube containing the particles on the vortex to maintain uniform suspension, aliquot 50 μL of the microcarrier solution into clean 1.5 mL microfuge tubes according to the number of bombardments planned. This corresponds to 3 mg of gold particles per tube.
4. Vortex the 1.5 mL microfuge tubes containing 3 mg gold particles in 50 μL 50% glycerol vigorously on a vertical microfuge holder (e.g., TurboMax attachment). While continuously vortexing, open the microfuge tube lids and add the following in strict order:

   5–8.4 μL purified transgenic plasmid DNA at 1 mg/mL (see Notes 6 and 7).

   50 μL 2.5 M $CaCl_2$ (1 M final concentration).

   20 μL 0.1 M Spermidine (0.015 M final concentration).

   Close the microfuge tube lids and continue agitation on the vertical vortex for three more minutes.

5. Allow coated gold beads to settle on the bench for 1 min then pellet them by spinning briefly for 2–3 s in microfuge. Remove the supernatant (see Note 8) and wash the coated gold beads with 140 μl 70% ethanol by adding the alcohol to tube and removing it immediately without disturbing the gold beads. Wash the coated beads once more with 140 μL absolute ethanol as before and resuspend the coated beads in 18 μL absolute ethanol. These coated gold beads will be sufficient for triplicate bombardments.

3.3. Preparation of B. malayi for Particle Bombardment

1. *B. malayi* is shipped in CF-RPMI-based transport media. Upon receipt, incubate worms in a 37°C water bath for 15–30 min then move to tissue culture hood and transfer into fresh CF-RPMI medium.
2. Prepare *B. malayi* developing embryos from aliquots of 50 adult females. Place female worms in tissue culture Petri dish and remove most of the culture medium to minimize their motility. Use a sterile scalpel to dice the worms into small pieces to release developing embryos from the worm uteri. Gradually add 1 mL of CF-RPMI to resuspend the embryos and transfer them to sterile 1.5 mL microfuge tube. Spin briefly in a microfuge and adjust medium volume to 300 μL. This procedure produces roughly 1×10^6 developing embryos that can be used for ten bombardments (roughly 1×10^5 embryos per bombardment).
3. Right before bombardment, place an aliquot of roughly 1×10^5 developed embryos in 30 μL CF-RPMI to the center of a sterile 60 × 15 mm petri dish and spread into a 20 mm diameter circle (penny size circle). Make sure to resuspend the embryos by gentle pipetting before removing an aliquot for bombardment. Slightly trim the tip of the pipette to ensure adequate transfer of the material.
4. Bombard adult female worms in groups of 1–5 worms per bombardment and bombard microfilariae and L3 larvae in groups of roughly 150 per bombardment according to the steps below.
5. Place aliquots of adult worms or mf/larvae in a penny size circle in the middle of sterile 60 × 15 mm petri dish. Chill adult worms on ice for 5 min right before bombardment to minimize their motility.

3.4. Bombarding Brugia malayi

1. Make sure the Biolistics unit PDS 1000/He unit is clean and connected to the helium and vacuum sources. Adjust helium pressure to 200 psi above the desired pressure, e.g., set the regulator to 1,300 psi when working with a 1,100 psi rupture disk for embryos. Choice of rupture discs depends on *B. malayi* life stage being transfected (see Note 9). Turn the unit on.

2. Immerse the rupture disk of the desired pressure (1,100 psi for developing embryos and larvae, 2,200 psi for adult females) in isopropanol. Let the disk dry on the bench for few seconds and insert on its retaining cap (see Note 10) and secure the cap in place. Insert macrocarriers into the macrocarrier holder using the plastic macrocarrier insertion tool.
3. Vortex microcarrier aliquots (see step 5 in Subheading 3.2) to resuspend the coated gold beads. While vortexing, remove 6 μL aliquots and place onto the middle of the macrocarrier. Make sure the beads are spread evenly in the center of the macrocarrier. Place macrocarriers for 5 min in a chamber containing $CaCl_2$ as a desiccant to dry the coated gold beads.
4. Place the macrocarrier holder and stopping screen on the microcarrier launch assembly (see Note 10) and insert into its level in the biolistics unit chamber. Then place *B. malayi* material on the target shelf and slide the shelf into the first level of the biolistics unit chamber.
5. Close the biolistics unit chamber door. Turn the vacuum on and hold it at 12.5 in. of mercury (optimal for all *B. malayi* life stages). Press and hold the illuminated Fire switch as the helium pressure build up to the desired level (e.g., 1,100 psi for developing embryos) and bombardment takes place. Release the Fire switch.
6. Set the vacuum switch to vent to release vacuum from chamber. Open the door and remove the petri dish containing bombarded *B. malayi* (see Note 11). Cover the petri dish and place it in a closed humid chamber for 5 min to allow recovery of the worms. Add 5 mL warm CF-RPMI (kept in a 37°C water bath) to each petri dish.
7. Incubate bombarded adult worms, larvae, or developing embryos at 37°C under an atmosphere of 5% CO_2 for 48 h (see Note 12). Inspect cultures daily and adjust the medium level if necessary. Then harvest transfected worms for reporter gene analysis.

3.5. Analysis of Transfected Parasites

3.5.1. Single and Dual Luciferase Assays

1. Transfer Biolistically transfected adult worms, larvae, or developing embryos into clean chilled glass homogenizers. Spin homogenizers for 5–10 s to pellet larvae or developing embryos and remove excess medium. For adult worms, remove medium with a pipette tip leaving worms on the bottom of the homogenizer. Insert matching homogenizer pestles, place homogenizers on a rack, and freeze at −80°C for at least 2 h.
2. Remove the rack containing the homogenizers from the freezer and homogenize the worms on the rack as they start to thaw.

Add 100 μL 1× cell culture lysis reagent contained in the Luciferase Assay Systems and continue homogenization. Spin the homogenizers briefly at 4°C. Remove 40 μL of the supernatant and use immediately to detect Luciferase activity. Remove additional aliquot to measure protein content of each homogenate.

3. For worms transfected with single constructs driving firefly luciferase expression, use the luciferase assay reagents in the Luciferase Assay System to measure luciferase activity in triplicate samples, following the manufacturer's instructions. Measure light units as counts per second in a luminometer Measure protein content in each homogenate using the Bradford method with reagents provided by BioRad. Results are expressed as net light units (gross counts per second minus the average of two negative control assays) per milligram of protein (see Note 6).
4. For worms transfected with a mixture of two luciferase reporter vectors (experimental constructs driving firefly luciferase expression in pGL3 vector and standard control construct driving Renilla luciferase expression in pRL-null vector), use reagents in the Dual-Luciferase Reporter Assay System or Dual-Glo Reporter Assay System to analyze both firefly and renilla luciferase in triplicate samples, following the manufacturer's instructions. Measure light units as counts per second in a luminometer. Results are expressed as the percentage of the firefly luciferase activity in each experimental construct (normalized to its internal renilla control) relative to that of the experimental standards (normalized to their internal renilla controls) bombarded on the same day (see Note 7).

3.5.2. GFP Reporter Gene Analysis

1. For worms intended for GFP localization assay, bombard with gold beads coated with *B. malayi* intergenic spacer domain of the 5S rRNA gene cluster driving GFP reporter gene in the vector pPD95.75 (see Note 13). Maintain bombarded worms in CF-RPMI at 37°C at 5% CO_2 for 48 h as before.
2. Fix transfected worms in 100% methanol at −20°C for 1 h and mount individual worms on a glass slide. Cover with glass coverslip and seal with nail polish.
3. Examine and capture GFP activity using appropriate microscope. We used a Leica DMIRBE inverted epifluorescence/Nomarski microscope outfitted with Leica TCS NT Laser Confocal optics. To image GFP, we excited with the 488 nm line of the argon laser with prism spectrophotometer set for a broad emission bandpass of 490–600 nm (see ref. (4) and Note 13).

4. Notes

1. The optimization kit contains all supplies for the Biolistics PDS 1000/He unit and is highly recommended for first time users.
2. Use of multiple reporter genes (e.g., Firefly and Renilla Luciferase vectors) will allow for determination of ratio and control for day-to-day fluctuations in light intensity readings (see Note 4 for more details).
3. Initial experiments show that 0.6 μm gold particles to be the optimal size for transfection of all *B. malayi* life stages. Dry gold particles from the manufacturer are aliquoted at 30 mg/1.5 microfuge tube and stored at room temperature. It is essential to keep record of the gold particles lot numbers for troubleshooting, if needed. Furthermore, we store washed, uncoated gold microcarriers up to 2 months at 4°C without adverse effects on the transfection efficiency.
4. We developed transient transfection system for *B. malayi* to study promoters and gene regulation in filarial nematodes (see ref. (5)). For this purpose, our transgenic DNA constructs included sequences encoding homologous putative promoter domains. We first used the intergenic spacer domain of the 5S rRNA gene cluster which was shown to control transcription of spliced leader pre-mRNA and to act as a promoter in *A. suum* transient transfection system (see refs. (3, 15)). We amplified and cloned this construct in frame with Firefly Luciferase reporter gene present in the expression vector pGL3 basic (see ref. (4)). Once the transfection system was established, we began studies on promoter analysis and the need arose to ensure day-to-day transfection efficiency and compare experiments done over time. For subsequent studies, gold bead microcarriers were coated with a 4:1 mix of transgenic constructs; one experimental promoter construct driving the firefly luciferase reporter gene in the pGL3 basic vector and one standard control construct driving the renilla luciferase reporter gene in the vector pRL-null (see ref. (5)).
5. We found more consistency and efficiency in reporter gene activity using transgenic constructs purified by Qiagen EndoFree Plasmid Maxi kit.
6. As mentioned in Note 4, this transfection system can be used to perform 1-day experiments in which a single luciferase expression vector can be used to evaluate experimental constructs and controls. In this case, 5 μL of 1 mg/1 mL purified transgenic DNA is used to coat 0.6 μm gold beads (5 μg DNA total) (see step 4 in Subheading 3.2). The reporter gene activity in bombarded worms is then expressed as specific activity (light units per mg of proteins of the extract) (see Fig. 1). Worms bombarded with gold beads coated with the expression

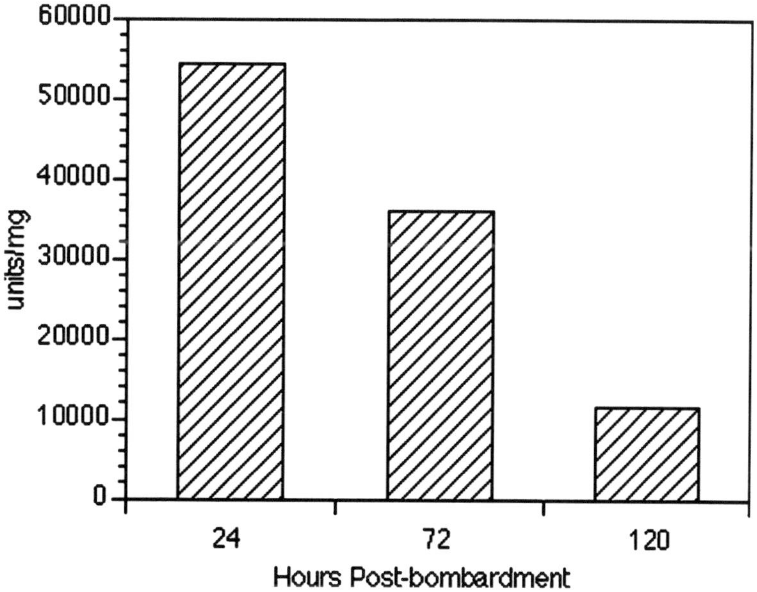

Fig. 1. Time course of luciferase expression in isolated embryos bombarded with pBm5S-luc: isolated embryos were bombarded with *B. malayi* intergenic spacer domain of the 5S rRNA gene cluster driving firefly luciferase reporter gene (pBm5S-luc) and maintained in culture for varying periods of time prior to being assayed for luciferase activity. Reprinted from Higazi et al. (4), Copyright (2002), with permission from Elsevier.

vector alone are used for background negative control to show that reporter gene activity is dependent on the presence of the putative experimental transgenic sequences (see Fig. 2).

7. This transfection system was used, for the most part, to compare and test multiple transgenic constructs over time (see ref. (5–14)). In this case normalizing the data for protein content does not control for differences in day-to-day transfection efficiency. To control for inter-sample differences in transfection efficiency, a dual luciferase assay containing an internal transfection control was devised (see ref. (5)). To accomplish this, a minimal *B. malayi* HSP70 promoter construct was cloned upstream of the renilla luciferase gene of the reporter vector pRL-null. Transfections were then carried out with 500 μg of 0.6 μm gold beads coated with a mixture consisting of 6.7 g of DNA containing experimental HSP70 promoter construct driving the expression of the firefly luciferase reporter and 1.7 g of plasmid DNA containing the HSP70 minimal promoter driving the expression of the renilla luciferase (4:1 proportion). To permit comparisons of data collected on different days, the dual reporter gene activity is expressed as the percentage of the firefly luciferase activity in each experimental construct (normalized to its internal renilla control) relative to

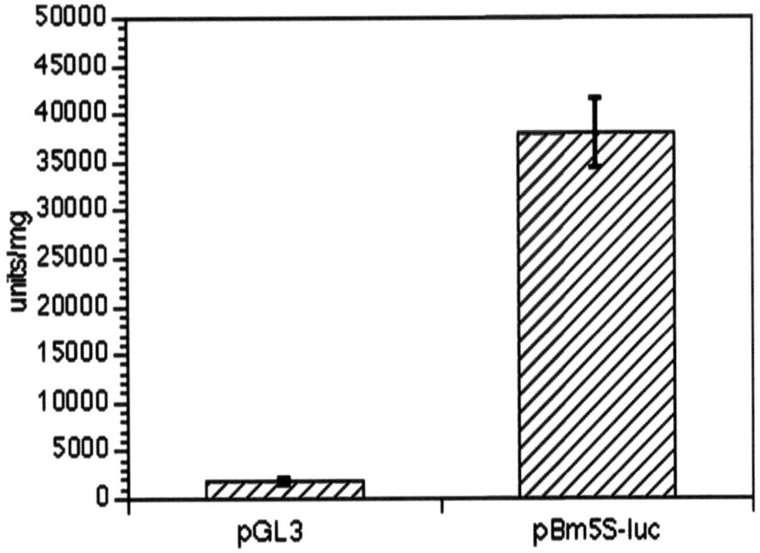

Fig. 2. Dependence of luciferase expression on promoter sequence in transiently transfected *B. malayi* embryos: parasites were bombarded with *B. malayi* intergenic spacer domain of the 5S rRNA gene cluster driving firefly luciferase reporter gene (pBm5S-luc) or pGL3 basic coated beads and maintained in culture for 48 h prior to being assayed for luciferase activity. Values shown represent means of duplicate determinations and error bars indicate the range of duplicate values. Reprinted from Higazi et al. (4), Copyright (2002), with permission from Elsevier.

that of the −659 to −1/luc standards (normalized to their internal renilla controls) bombarded on the same day (Fig. 3). Initial mixing experiments carried out to validate the dual luciferase assay, showed no difference in the specific activity (net luciferase light units/mg of protein) of *B. malayi* embryos transfected with the firefly constructs alone when compared to embryos transfected with a mixture of the renilla and firefly constructs (data not shown). These data confirmed that promoter interference was not occurring in worms transfected with both the firefly and renilla constructs. Furthermore, data obtained from both single and dual luciferase assays were found to be completely concordant (see ref. (5)).

8. In initial experiments, this supernatant can be saved to measure its DNA content and ensure the efficiency of the gold beads coating process.

9. In initial experiments we used the optimization kit to select optimal rupture disc pressures for various life stages of *B. malayi*. As expected, we found that the optimal pressure correlates with estimated thickness of the outer cuticle of the bombarded life stage. *B. malayi* developing embryos and L3 larvae were optimally transfected at 1,100 psi pressure. Optimal delivery of the gold beads into adult females required pressure

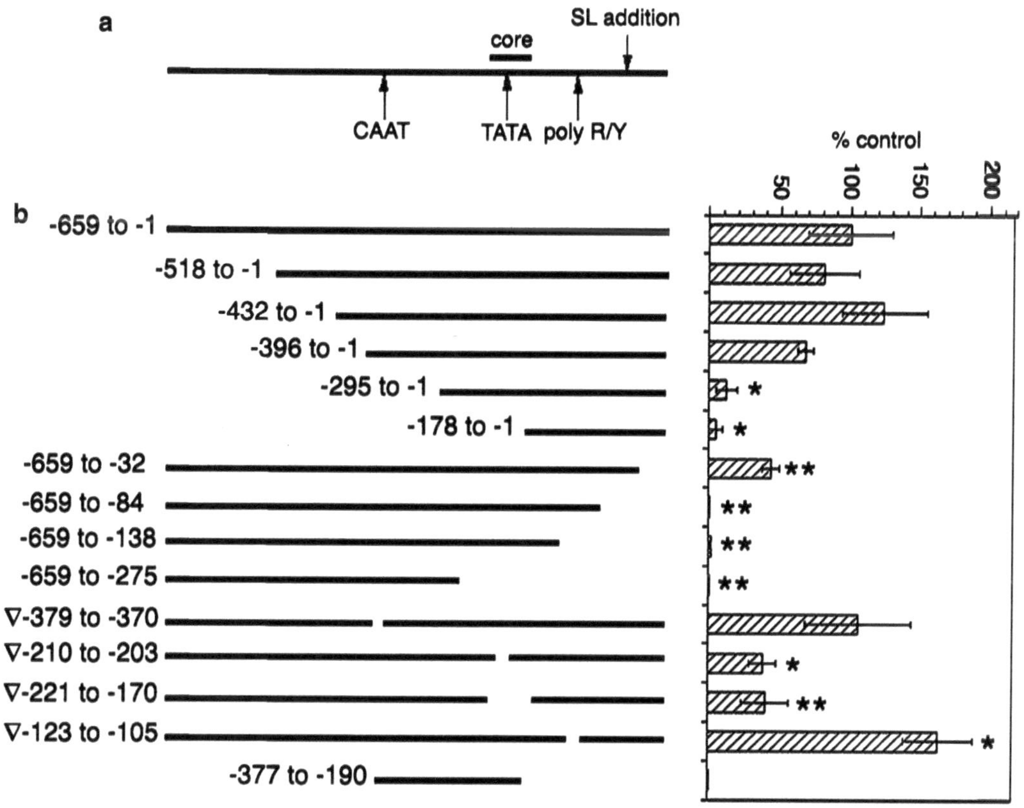

Fig. 3. Deletion analysis of *B. malayi* HSP70 upstream domain: the position of the putative CAAT box, TATA box, core promoter, poly R/Y stretch, and SL addition sites are indicated schematically in (**a**). (**b**) Contains a schematic representation of the deletion constructs tested, and the level of luciferase activity detected in embryos transfected with each construct. Luciferase activity was assayed as using the dual luciferase assay (see Subheading 3.5.1 and Note 7) (*asterisk*). Luciferase activity in embryos transfected with this construct was significantly different from that seen in embryos transfected with −659 to −1/luc ($0.05 \geq P > 0.01$) (*double asterisk*). Luciferase activity in embryos transfected with this construct was significantly different from that seen in embryos transfected with −659 to −1/luc ($P \leq 0.01$). Reprinted from Shu et al. (5), Copyright (2003), with permission from Elsevier.

of 2,200 psi. We also found that *B. malayi* developing embryos to be the most amenable to biolistics transfection with specific activities five and ten times those of L3 larvae and adult female worms respectively (data not shown).

10. It is essential to position the rupture disk firmly in it retaining cap. Failure to do this will result in leak of the helium gas and failure of the bombardment. It is also essential to use a new stopping screen for every couple of bombardments. We use a checklist to make sure all components are in place (e.g., rupture disc, macrocarrier with dried gold beads, stopping screen, etc.) before firing the system. Missing the stopping screen or using the same screen over three times are the most common mistakes which will result in loss of the worm material and

failure of the bombardment. In this case, it is necessary to clean up the instrument chamber before you proceed with other bombardments.

11. Inspection of the petri dish right after bombardment should show uniform distribution of the gold beads over the worm material in the center of the dish.
12. Luciferase activity was seen within 24 h of bombardment, and remained at relatively high levels for roughly 72 h post-bombardment (Fig. 1). The activity was found to decrease markedly after 72 h, although viability of the embryos cultured for 5 days remained above 70% and motile pretzel microfilaria continued to be visible within the eggshells of the cultured embryos, indicating the transient nature of the transfection. For this reason, we harvest bombarded worms 48 h post-bombardment to ensure optimal level of transfection.
13. Analysis of worms transfected with the luciferase reporter constructs required homogenization of the parasites; hence, it was impossible to locate tissues expressing the transgene. To locate activity of the transgene in transfected worms, *B. malayi* 5S promoter was cloned into a GFP reporter vector pPD95.75 and used to transfect intact adult females both by microinjection and biolistic bombardment (Fig. 4). Worms injected in the uterus expressed GFP in the individual intrauterine embryos in the vicinity of the injection site (Fig. 4a). In contrast, bombarded worms produced a punctate pattern of GFP fluorescence that localized to the hypodermal layer underlying the cuticle

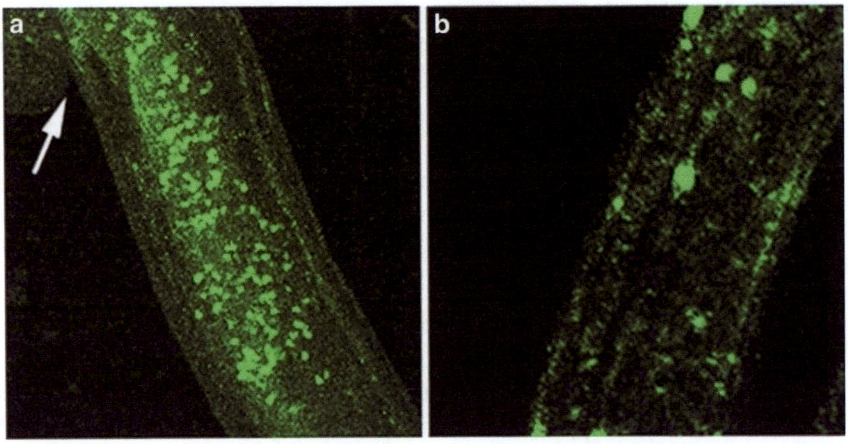

Fig. 4. GFP expression in biolistically transfected and microinjected adult parasites: Parasites were transfected with *B. malayi* intergenic spacer domain of the 5S rRNA gene cluster driving GFP reporter gene (pBm5S-GFP), maintained in culture for 48 h, and GFP expression was visualized as described (see Subheading 3.5.2). (**a**) Parasite microinjected with pBm5S-GFP. The *arrow* indicates the site of injection. (**b**) Parasite biolistically bombarded with beads coated with pBm5S-GFP. Modified from Higazi et al. (4), Copyright (2002), with permission from Elsevier.

(Fig. 4b). It is likely that the fluorescence in the biolistically transfected adult parasites results from penetration of individual gold particles into the hypodermis. Therefore the difference in the pattern of expression of the microinjected and biolistically transfected adult worm probably is the outcome of deposition of the exogenous DNA in different tissues of the adult worm by the two methods.

Acknowledgements

This work was supported by grants from US National Institutes of Health (project #R01-AI48562), the Edna McConnell Clark Foundation, and the Provost's office of the University of Alabama at Birmingham (to TRU).

References

1. Wilm T et al (1999) Ballistic transformation of *caenorhabditis elegans*. Gene 229:31–35
2. Jackstadt P et al (1999) Transformation of nematodes via ballistic DNA transfer. Mol Biochem Parasitol 103:261–266
3. Davis RE et al (1999) Transient expression of DNA and RNA in parasitic helminths by using particle bombardment. Proc Natl Acad Sci USA 96:8687–8692
4. Higazi TB et al (2002) *Brugia malayi*: transient transfection by microinjection and particle bombardment. Exp Parasitol 100(2):95–102
5. Shu L et al (2003) Analysis of the *Brugia malayi* HSP70 promoter using a homologous transient transfection system. Mol Biochem Parasitol 128(1):67–75
6. Higazi TB, Shu L, Unnasch TR (2004) Development and transfection of short-term primary cell cultures from *Brugia malayi*. Mol Biochem Parasitol 137(2):345–348
7. Higazi TB et al (2005) Identification of elements essential for transcription in *Brugia malayi* promoters. J Mol Biol 353:1–13
8. Liu C et al (2007) Sequences necessary for trans-splicing in transiently transfected *Brugia malayi*. Mol Biochem Parasitol 156:62–73
9. de Oliveira A, Katholi CR, Unnasch TR (2008) Characterization of the promoter of the *Brugia malayi* 12kDa small subunit ribosomal protein (RPS12) gene. Int J Parasitol 38:1111–1119
10. Liu C et al (2009) The splice leader addition domain represents an essential conserved motif for heterologous gene expression in *B. malayi*. Mol Biochem Parasitol 166:15–21
11. Liu C et al (2010) Functional analysis of putative operons in *Brugia malayi*. Int J Parasitol 40:63–71
12. Liu C, Chauhan C, Unnasch TR (2010) The role of local secondary structure in the function of the trans-splicing motif of *Brugia malayi*. Mol Biochem Parasitol 169:115–119
13. Tzertzinis G et al (2011) Molecular evidence for a functional ecdysone signaling system in *Brugia malayi*. PLoS Negl Trop Dis 4:e625
14. Bailey M et al (2011) The role of polymorphisms in the spliced leader addition domain in determining promoter activity in *Brugia malayi*. Mol Biochem Parasitol 176:37–41
15. Hannon GJ et al (1990) Trans splicing of nematode pre-messenger RNA in vitro. Cell 61:1247–1255

Part III

Biolistic DNA Delivery in Mammalian Cells In Vitro

Chapter 10

Biolistic Transfection of Human Embryonic Kidney (HEK) 293 Cells

Xiongwei Li, Masaki Uchida, H. Oya Alpar, and Peter Mertens

Abstract

Due to its excellent transfectability, the human embryonic kidney (HEK) 293 cell line is widely used as an in vitro model system for transfection experiments. Particle bombardment, or biolistics technology, provides a physical transfection approach that can deliver transgene materials efficiently into many different cell lines. Transfection of 293 cells by gene gun, allows examination of transgene expression in epithelial cells, as well as studies concerning a variety of questions in neurobiology. The present study of transfection of HEK 293 cells by biolistics technology uses the plasmids gWIZ-*luc* encoding luciferase and gWIZ-GFP encoding green fluorescence protein (GFP) as model transgenes. This system can be routinely used at varying bombarding conditions that can be adjusted according to experimental requirements and purpose, such as gene gun helium pressure, the sizes and the amount of the gold particles and the length of the spacer. The results obtained show that the Bio-Rad spacer for the gene gun should be optimized for travel distance and spreading of gold particles over a relatively small area, when used for biolistic transfection of cells dispersed in multi-well plate.

Key words: HEK 293 cells, Gene, Transfection, Biolistics, Gene gun, Gold particles, Confocal microscope, Plasmid DNA

1. Introduction

The "human embryonic kidney (HEK) 293 cell line" was generated over 30 years ago (1) following transformation of primary human embryonic kidney cells by exposure to sheared fragments of human adenovirus type 5 (Ad5) DNA. Originally, human embryonic kidney cells obtained from a healthy aborted fetus were cultured by Van der Eb, and transformed with sheared fragments of Ad5 comprising the early region 1 (E1) by Graham et al. (2). The original HEK 293 cell clone contained several nucleotides of

Stephan Sudowe and Angelika B. Reske-Kunz (eds.), *Biolistic DNA Delivery: Methods and Protocols*, Methods in Molecular Biology, vol. 940, DOI 10.1007/978-1-62703-110-3_10, © Springer Science+Business Media, LLC 2013

Ad5, including early region 1 (E1) transforming sequences, integrated into chromosome 19 of the host genome, a product of Frank Graham's 293rd experiment (3).

Researchers have been interested in HEK 293 cells since their original creation, because their maintenance costs are low, they are easy to work with, culture and transfect. They have previously been used to demonstrate the efficacy of many transfection reagents and techniques, including the lipid cell fusion reagent Lipofectamine (Invitrogen), the cationic polymer particle endocytosis reagent FuGene (Promega) and physical penetration by electroporation (4). The widespread use of this cell line has confirmed its high transfectability by various techniques (5–8).

HEK 293 cells have an interesting development history, which makes them a special tool in cellular biology. As an experimentally transformed cell line, HEK 293 cells should not be considered as a suitable model for fundamental research concerning the functions of kidney cells, cancer cells, or other "normal" cell lines. However, they can be used as a tool to evaluate methods for experiments in which the behavior of the cell itself is not of interest. Graham and coworkers (9) provided evidence that HEK 293 cells have the properties of immature neurons, suggesting that adenovirus was taken up by, and transformed a neuronal lineage cell from the original kidney derived cultures. A DNA microarray analysis of 293 cells showed that they are strongly and specifically stained with antibodies to neurofilament proteins NF-L, NF-M, and NF-H, which are generally thought to be excellent markers for neuronal lineage cells. These observations suggest that in many respects HEK 293 cells are closely related to neuronal cells and early differentiating neurons, and are therefore considered suitable for studies of biological aspects of neurobiology, rather than as an in vitro model for kidney cells (10, 11).

Gene gun bombardment does not show preferential selection of cell type, is therefore useful for hard-to-transfect cells and has been used to express exogenous genes in a variety of cultured mammalian cell types and for a variety of purposes (12). Biolistic transfection has been successfully developed for transfer of DNA and RNA for vaccination and gene therapy applications in live animals, via the skin, muscle and internal organs, including liver, pancreas, spleen, kidney, and brain (13–16).

Based on fluid dynamics, the spreading area of gold particles powered by air fluid at a high velocity is related to the distance the particles travel. The longer the path the particles travel, the larger the area they spread over. An inappropriate distance (longer than the standard spacer length) results in a risk of particles going out of the restricted area of the spacer end and into the sidewall of the well causing a low transfection efficiency. The standard spacer of the Helios gene gun cannot be placed into "wells" which have a diameter smaller than that of the spacer (22 mm) affecting the path length and the efficiency of transfection of cells cultured in multi-well plates (12 or more-wells/plate). A longer travel path will decrease the force of the particles penetrating the cells. Modification of the gene gun spacer is therefore necessary to control the travel path of the particles for various experimental conditions, as suggested by O'Brien (10, 17). The length of the standard Bio-Rad spacer was reduced to achieve optimal transfection efficiency in multi-well plate applications.

2. Materials

Water used in the experiments was purified deionised water at a sensitivity of 18 MΩ cm at room temperature and sterilized by autoclaving.

2.1. Gene Gun and Bullets

2.1.1. Basic Reagents

1. Gold microcarriers (0.6, 1.0, 1.6 μm) (Bio-Rad, Hercules, CA, USA).
2. Fresh absolute ethanol (see Note 1).
3. Spermidine: 0.05 M solution in water.
4. Calcium chloride ($CaCl_2$): 1 M solution in water.

2.1.2. Auxiliary Apparatus

1. Ultrasonic cleaner (Branson 1210).
2. Vortex mixer.
3. Analytical balance.
4. Microfuge.
5. 1.5 mL Microfuge tubes.
6. 15 mL Disposable polypropylene centrifuge tubes.
7. 5 and 10 mL Pipettes.
8. 1/8″ Open end or 10″ or 12″ (ca. 25 cm) adjustable wrench.
9. Helium gas cylinder (BOC, grade 4.5 or higher).
10. Nitrogen gas cylinder (BOC, grade 4.8 or higher).
11. Nitrogen regulator (Bio-Rad).

2.1.3. Gene Gun Assembly (all the Products Are from Bio-Rad Company)

1. Helios Gene Gun.
2. Cartridge holder and cartridge extractor tool.
3. Helium hose assembly.
4. Helium regulator.

5. Tubing Cutter and razor blades.
6. Tubing Prep Station (base, tubing support cylinder, and power cord).
7. Nitrogen hose (~4 m, Nalgene tubing 8000–0030, 3/16″ ID, 5/16″ OD).
8. 10 mL Syringes, Gold-Coat Tubing (~26 m).
9. A standard Bio-Rad spacer with length of 2.8 cm.
10. A modified spacer with length of 1.3 cm.

2.2. Reagents for Transfection, DNA and Protein Analysis

1. Plasmid DNA gWIZ-*luc* encoding luciferase (5 mg/mL in water) (Aldevron, Fargo, USA).
2. Plasmid DNA gWIZ-GFP encoding green fluorescence protein (5 mg/mL in water) (Aldevron, Fargo, USA).
3. Lipofectamine™ 2000 (Invitrogen Ltd. Paisley, UK).
4. TE buffer: 10 mM Tris–HCl, 1 mM EDTA, pH 8.0.
5. Tris-acetate-EDTA (TAE) buffer: 40 mM Tris-acetate, 1 mM EDTA.
6. Luciferase Assay System kit (Promega UK Ltd, Southampton, UK).
7. Micro bicinchoninic acid (BCA) Protein Assay Kit (including protein standard bovine serum albumin (BSA)) (Perbio Science UK Ltd., Northumberland, UK).
8. Molecular biology agarose (Bio-Rad, Hemel Hempstead, UK).
9. Ethidium bromide (10 mg/mL in water).
10. BSA.

2.3. Cell Culture

1. HEK 293 cell line: HEK 293 cells, an epithelial cell line (ATCC® Number: CRL-1573™ ATCC, Manassas, USA).
2. Culture medium: Dulbecco's Modified Eagle's Medium (DMEM) with 4.5 g/L D-glucose, 4 mM l-glutamine, 1.0 mM sodium pyruvate and 3.7 g/L sodium bicarbonate (Sigma, St. Louis, MO, USA) supplemented with 100 units/mL Penicillin, 100 μg/mL Streptomycin and 10% Fetal bovine serum (FBS).
3. Phosphate buffered saline (PBS): sterilized by autoclaving, pH 7.4.
4. Trypsin solution: 0.25% (w/v) trypsin in Hanks' balanced salt solution, supplemented with 1 mM ethylenediamine tetraacetate (EDTA) 4Na (Gibco®, Bethesda, MD, USA).
5. Glass-bottom tissue culture dish: FluoroDish™, dish Ø35 mm, glass Ø23 mm, glass thickness: 0.17 mm (World Precision Instruments, Inc., Sarasota, FL, USA).
6. Cell culture plate: 12-well plate (Nunc™).

2.4. Analysis and Assay Equipments

1. Confocal microscope: A Carl Zeiss LSM 510 Meta confocal microscope equipped with a 30 mW (lines at 458, 477, 488 (blue), and 514 nm) Argon ion laser, a 1 mW 543 nm (green) HeNe laser and a 5 mW 633 nm (red) HeNe laser using Zeiss LSM Image Browser version 3.2.0.70 and data processing LSM 510 software (CARL Zeiss, Germany).
2. Agarose gel electrophoresis equipment: Sub-Cell® GT and POWER PAC 1000 (Bio-Rad, Dorset, UK).
3. DNA quantitation: a DNA/RNA Calculator Genespec I (GeneQuant, Pharmacia, USA) and IBM-compatible computer assembly running the GeneSpec I software (Program No: 7A00541-02, Naka Instruments Co. Ltd., USA).
4. Microplate reader: Wallac Victor 2 1420 Multilabel counter (Wallac Oy, Turku, Finland).
5. 96-Well black plate: Sero-Wel® 96-well black plate (Bibby Sterulin Ltd., Staffs, UK).

3. Methods

3.1. Coating of DNA on Gold Microcarriers

1. Add a predetermined amount of gold particles (6.3, 12.5, 25, or 50 mg) of different sizes (diameter: 0.6, 1.0, or 1.6 μm) (see Note 2) and 100 μL of 0.05 M spermidine to a 1.5 mL microfuge tube and vortexe for a few seconds, then sonicate for 5 s to break up gold clumps.
2. Add the required amount (25, 50, 100, or 200 μg) of gWIZ/*luc* or gWIZ/GFP (5 mg/mL in stock) and vortex for 5 s. Add 100 μL of 1 M $CaCl_2$ dropwise by vortexing the mixture at a moderate rate (see Note 3).
3. Allow the mixture to settle at room temperature for 10 min (see Note 4).
4. After centrifugation at 5,000 rpm for 30 s, discard the supernatant and wash the pellet 3–4 times with 1 mL of fresh absolute ethanol.
5. Resuspend the pellet in 200 μL ethanol and transfer it to a 15 mL disposable polypropylene tube. Repeat this transfer with 200 μL ethanol until all the gold particles have been transferred, then make up to the necessary volume (3 mL) with ethanol (see Note 5).
6. In the case of non-coated gold particles, wash the required quantity (25 mg) 3–4 times with 1 mL of fresh absolute ethanol and then resuspend in 3 mL of fresh absolute ethanol.

3.2. Preparation of Cartridges

1. Vortex the DNA- coated or uncoated gold particle suspension to ensure an even suspension of gold particles.
2. Draw up the suspension into a 74–76 cm (29–30 in.) piece of Gold-Coat tubing (see Note 6); this has previously been dried in the Tubing Prep Station with a flow of nitrogen (0.3–0.4 L/min) for 20 min. To draw the gold particle suspension, use a 10 mL syringe attached to the end of the tubing.
3. Slid the loaded tubing, with syringe attached, into the Tubing Prep Station and allow to settle for 3 min.
4. Remove ethanol slowly from the tubing using the syringe. Then detach the syringe from the tubing.
5. Turn the tubing immediately 180° and allow to settle for a few seconds.
6. Rotate the tubing for 20–30 s to allow the gold particles to spread onto its inner surface.
7. Dry the gold/DNA particles in the tubing, using a flow of nitrogen (0.3–0.4 L/min) for 15–20 min, while it continues to rotate.
8. Cut off any unevenly coated sections, then cut the remaining tubing into 1.3 cm (0.5 in.) pieces using the Tubing Cutter, and store them desiccated at 4°C until required.

3.3. Unloading and Analysis of DNA and Gold Particles from the Cartridge

The extraction of DNA and gold particles from the cartridges is performed by incubation or sonication. In order to investigate the integrity of the coated plasmid DNA, the DNA in the eluting solution is assayed by agarose gel electrophoresis.

1. Put the cartridges into a microfuge tube and add 0.5–1.0 mL of TE buffer.
2. After vortexing, incubate at 37°C for 20 min while shaking at 150 rpm, or sonicate for 5 min 3 times at intervals of 5 min in order to avoid degrading DNA by high temperature.
3. Collect the supernatant to analyze the DNA in the solution after centrifugation at 5,000 rpm for 30 s.
4. Prepare agarose gels with a freshly prepared 0.7% (m/v) agarose solution in TAE buffer by heating and add ethidium bromide at a concentration of 1.0 μg/mL. Carry out electrophoresis at 100 V for 2.5 h using the electrophoresis equipments Sub-Cell® GT and POWER PAC 1000. The sample loaded in the well is 20 μL of eluting solution containing 0.1 μg of the DNA (Fig. 1).
5. Determine the amount of DNA in the eluting solution by UV spectroscopy using the DNA/RNA Calculator Genespec I and IBM-compatible computer assembly running the GeneSpec I software.
6. Determine the amount of gold particles in the cartridge by weighing the tubing before and after dislodging.

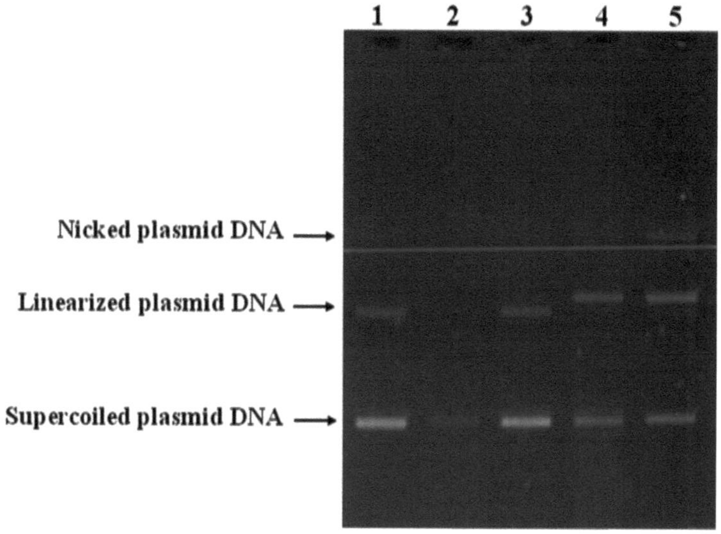

Fig. 1. 0.7% Agarose gel electrophoresis of gWIZ-*luc* eluted from gold particles with 1.0 μm in diameter in the cartridges. Gold particle loading: 0.5 mg/cartridge; DNA loading: 2 μg/mg gold particles. Lane 1: free gWIZ-*luc*; Lane 2: free gWIZ-*luc* after sonication; Lane 3: free gWIZ-*luc* after incubation; Lane 4: gWIZ-*luc* after elution from the cartridges by sonication; Lane 5: gWIZ-*luc* after elution from the cartridges by incubation. Sonication: for 5 min three times at intervals of 5 min; incubation: at 37°C for 20 min while shaking at 150 rpm. Reprinted from(6), Copyright 2009, with permission from Elsevier.

3.4. Cell Culture

Culture HEK 293 cells in culture medium at 37°C and 5% CO_2. Passage the cells at 2–3 day intervals in a 1:10 dilution in fresh media after removal of depleted medium, wash with PBS and digest with 0.25% (w/v) trypsin solution. They are used in transfection studies at passage numbers 68–74.

3.5. In Vitro Cell Transfection

1. The transfection of HEK 293 cells by gold particle bombardment can be carried out with cells cultured either in (a) single dishes, or (b) in multi-well plates, depending on the experimental purpose (see Note 7).

   One day before the transfection, seed HEK 293 cells in

   (a) A glass-bottom tissue culture dish for confocal or fluorescent microscopy examination with a concentration of 1×10^5 cells/dish and incubate at 37°C and 5% CO_2 overnight (see Note 8),

   or

   (b) A 12-well plate at a density of 4×10^5 cells/well (see Note 9).

2. Set the cartridges coated with gWIZ-*luc* (or gWIZ-GFP)-loaded gold particles on a cartridge holder which had twelve holes. Then insert the cartridge holder into the Helios Gene Gun.

3. Prior to bombardment attache the gene gun to the helium cylinder through the helium regulator.

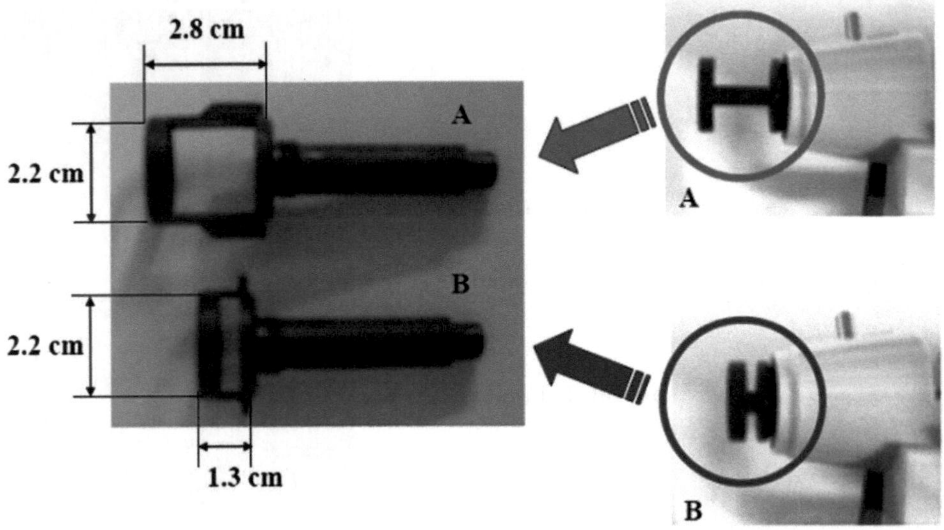

Fig. 2. Modification of the spacer of Helios Gene Gun. (**a**) Original standard spacer with a length of 2.8 cm, (**b**) Modified spacer with a length of 1.3 cm.

4. Carry out the bombardment of the gold particles into either (a) a single 35 mm dish using the standard spacer with a length of 2.8 cm; or (b) wells of 12-multi-well plate using a modified spacer with a shortened length of 1.3 cm (Fig. 2).

   In the case of (a), immediately after removing the media from a single circular 35 mm dish (gWIZ-GFP transfection), place the end of the plastic spacer as close as possible (about 3–5 mm) to the target HEK 293 cells (see Note 10) which are transfected by a single bombardment with gWIZ-GFP-coated gold particles using the gene gun under different conditions (see Note 11).

   In the case of (b), using 12-well plate (gWIZ-luc transfection), bring the end of the modified short plastic spacer to touch to the frame of the well when initiating the bombardment (Fig. 3) (see Note 12).
5. Initiate bombardment with the gold particles in the cartridge by pressing the trigger button which opens the main helium gas valve. The gold particles are accelerated to high velocity by a flow of helium gas.
6. After particle bombardment, add 1–2 mL of media and incubate the cell cultures for 48 h at 37°C and 5% CO_2.
7. Carry out transfection using Lipofectamine™ 2000, as a positive control, according to the manufacturer's instructions. One to two micrograms of DNA and 2.5–5 μL of Lipofectamine™ 2000 are applied to each well.

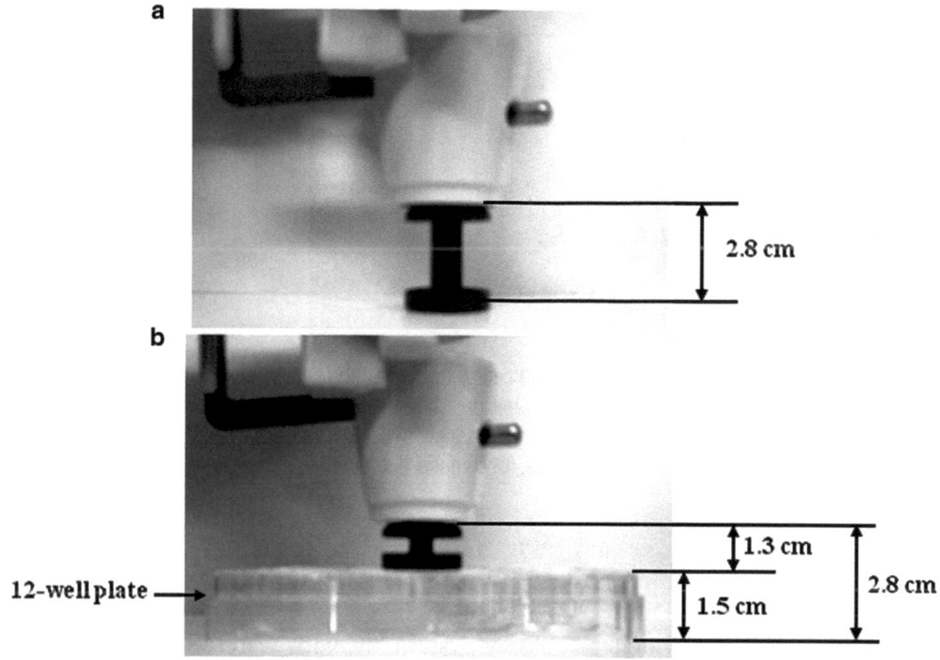

Fig. 3. The modified spacer is designed for use in combination with a 12-well plate. The position of the spacer touches the frame of the well. (**a**) Standard method, (**b**) Modified method.

3.6. Confocal Laser Scanning Microscopy

A visible evaluation of the transgene expression and cellular location of gold particles after bombardment can be performed with reporter plasmid gWIZ-GFP by confocal microscopy confocal laser scanning microscopy (CLSM), Zeiss LSM 510 Meta, equipped with a temperature and CO_2 level controllable chamber. The temperature is set at 37°C and the CO_2 level is set at 5% one h before the observation of the cells.

1. Remove the medium of the culture in the glass-bottom dish immediately before the confocal microscopy viewing, 2 days after the bombardment.
2. Wash the 293 cells in the dish transfected by bombardment of gWIZ-GFP coated gold particles with 500 μL of PBS.
3. Add 1 mL of fresh DMEM culture medium to the dish.
4. Observe the cells in the dish using the confocal microscope (Fig. 4) (see Note 13).

3.7. Preparation of Cell Lysate for Luciferase Assay

1. Forty eight hours post-transfection of HEK 293 cells, remove media from the wells.
2. Wash each well of cultured cells with 500 μL of PBS which is then removed from the well.

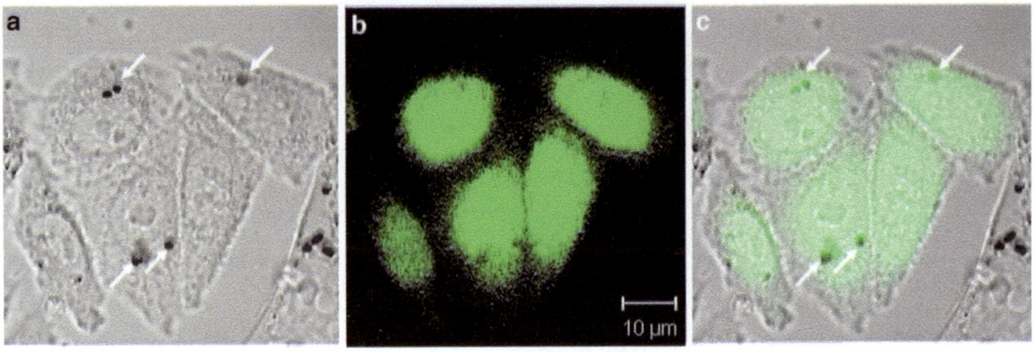

Fig. 4. Confocal laser scanning micrographs of HEK 293 cells 2 days after bombardment with gWIZ-GFP-coated gold particles with 1.0 µm in diameter at 100 psi of helium pressure showing the cells, gold particles (*white* arrows) and the expressed GFP (*green* color). Gold particle loading: 0.1 mg/cartridge; gWIZ-GFP loading: 2 µg/mg gold particle; The number of cells is 1×10^5 cells/dish. (**a**) transmitted light image, (**b**) confocal fluorescence image, (**c**) merged image.

3. Dilute Luciferase Cell Culture Lysis Reagent, 5×, from Luciferase Assay System kit, with water at a ratio of 1:10 to a 0.5× solution.
4. Add 200 µL of the working lysis solution (0.5×) to each well.
5. Leave the wells at room temperature for 30 min.
6. Then check the cells with a microscope to make sure that they have been lysed.
7. Transfer the cell lysate to a microfuge tube and then vortex.
8. After centrifugation at 12,000 × *g* for 2 min at 4°C, transfer the supernatant to a new tube.
9. Store the samples at −70°C prior to determination of luciferase expression and measurement of protein content.

3.8. Assay for Luciferase Reporter Gene Expression

1. After thawing and vortexing, place 20 µL of the cell lysate into a well of a 96-well black plate.
2. Add 50 µL of Luciferase Assay Reagent immediately before assay.
3. Assay the plate containing the cell lysates for luminescence intensity generated by the enzyme catalytic reaction of the luciferase produced in transfected cells and the luciferin from the reagent, using a microplate reader.
4. Luminescence is expressed as relative light units (RLU) accumulated over 10 s, for evaluation of the transfection efficiency. The RLU is normalized to protein amounts in the cell extracts (as measured by a BCA protein assay) and recorded as RLU/mg protein. Transfection efficiency is evaluated as the level of RLU/mg protein (Fig. 5) (see Note 14).

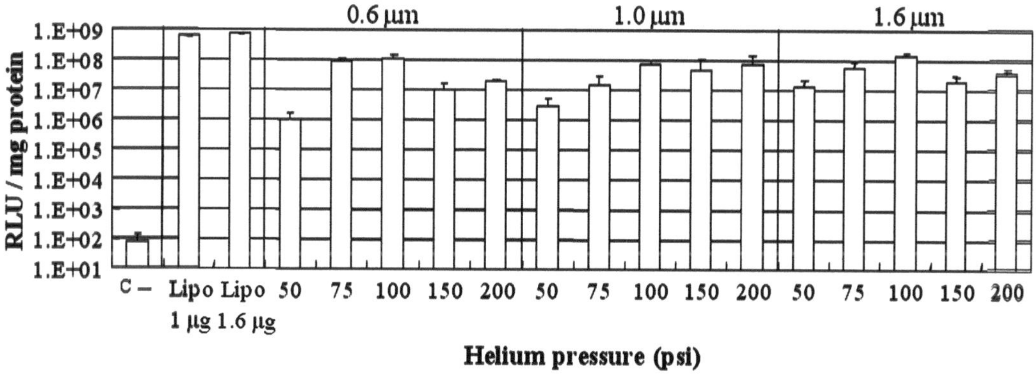

Fig. 5. The transfection efficiency of biolistics bombardment of plasmid gWIZ-*luc* coated gold particles at different particle size and helium pressure using Human Embryonic Kidney (HEK) 293 cells (4×10^5 cells/well, 12-well plate) at 48 h after bombardment transfection. Gold particles : 0.6, 1.0, and 1.6 μm, Pressure: 50, 75, 100, 150, 200 psi, Plasmid DNA : gWIZ-*luc*, Dose : 0.5 mg gold particles/1 μg DNA/shot (cartridge). Transfection efficiency compared with negative control (untransfected cells) and positive control (Lipofectamine™ 2000 as transfection reagent).

3.9. The Assay of Protein Content in Cell Lysates

1. After thawing and vortexing, dilute cell lysates one hundred times with water.
2. Place 150 μL of the diluted lysate in a well of a 96-well plate.
3. Add 150 μL of Micro BCA™ Working Reagent to each well.
4. Shake the plate for mixing and then incubate at 37°C for 2 h.
5. After the plate has cooled to room temperature, measure the absorbance at 570 nm in each well using the microplate reader.
6. Generate a standard curve using known concentrations of BSA.
7. Estimate the amount of protein in the cell lysate by comparison to the BSA standard.

4. Notes

1. It is extremely important that this be free of water; an unopened bottle should be used daily.
2. Gold particles come in a variety of sizes, and particle size affects the entry of DNA into cells. The choice of the particle sizes should be dependent on the instrument, sample origin and the transfection conditions. There should not be a fixed particle size.
3. DNA concentration must be appropriate for efficient transfection and must not be too low or too high; too low could result in an insufficient transfection and too high could cause agglomeration of the gold particles as well as material waste.

4. This procedure leads to the precipitation of plasmid DNA onto gold particles.
5. If necessary, polyvinylpyrrolidone (PVP, Mw 360,000, Bio-Rad) in fresh absolute ethanol is used. It serves as an adhesive during the cartridge preparation process. At higher discharge pressure, preparing cartridges with PVP can increase the total number of particles delivered. The optimum amount of PVP to be used must be determined empirically. Typically PVP concentrations range from 0.01 to 0.1 mg/mL.
6. The length of suspension in the Gold-Coat tubing is dependent on the amount of the gold particles, as well as the need of the transfection experiment. Normally the amount of gold particle coating in the tube should be 1 mg/in. to make a 0.5 mg of gold particles per shot (0.5 in. tube is used as a bullet set in the cartridge holder).
7. The choice of plate or dish for the cell culture should be dependent on the experimental aim. Normally, multi-well plate should be used, if the amount of transfection samples in the experiment is large and a high efficiency of the bombardment is required (see Note 12). Otherwise, separate dishes should be used for the gene gun transfection for the optimal results.
8. In the case of confocal evaluation, HEK 293 cells should be seeded in a tissue culture dish with cover glass bottom. This glass-bottom culture dish facilitates observation by live cell microscopy and recording of the state of the cells.
9. Luciferase gene transfection was used to evaluate the transfection efficiency, which is normally statistically quantified with multiple repeat test data at various conditions. The number of the cell samples for the quantification of transfection efficiency was often large, so that multi-well plates were needed. Moreover, the smaller size of the wells of the multi-well plate also ensures transfection of all cells in the well by gene gun bombardment.
10. When the bombardment was carried out using the 35 mm dish, the spacer should be as close as possible to the target cells. However, the end of the spacer should not touch the cells because this could result in damage or contamination of the cells. The distance between the end of the spacer and the target cells should be 3–5 mm.
11. The gas pressure used in the gene gun needs to be optimized for each particular instrument and biological system under investigation.
12. When transfection is carried out using a multiple well plate (over 6-well plate), the length of the spacer on the gene gun should be shorter than the standard Bio-Rad spacer. This is because of the size of the spacer and the sizes of the wells of the

plate (Fig. 2). The wells of the 12-well plate have a diameter of 2.2 cm, equal to that of the spacer. Therefore the spacer cannot be put into the well of the 12-well plate. The standard spacer has a longer distance than the modified spacer, for particle travel, and would therefore result in a larger spreading area and reduced power for the particles to get into the cells, potentially reducing transfection efficiency.

13. In our study, the confocal setting was as follows: Scan mode: Plane; Objective: Plan-apochromat 63×/1.4 oil dic; Excitation wavelength: 488 nm, Pinhole: 97 μm, Output: 5%; Detector gain: 1193, Carl Zeiss Laser Scanning Systems LSM510 using Zeiss LSM Image Browser Version 3.2.0.70. Data was processed with a desktop computer linked to an optical disk drive using LSM 510 software to merge transmitted light and confocal fluorescence images. The transmitted light and confocal fluorescence images were used to analyze the cellular location of gold particles and green fluorescence expression.
14. The efficiency of transfection is calculated for the distribution area of the bombarded gold particles. The area of the 35 mm dish is larger than the area of the particle spreading (normally in the spacer limited area) by the bombardment. The diameter of end of the spacer (about 2 cm) is almost the same as the diameter of a well in a 12-well plate but smaller than that of the 35 mm dish. This means that not all cells could be shot by the particles if the particle bombardment is for the cells in a 35 mm dish.

Acknowledgments

This work was supported by the BBSRC grant BB/D014204/1.

References

1. HEK cell-Wikipedia, http://en.wikipedia.org/wiki/HEK_cell
2. Graham FL et al (1977) Characteristics of a human cell line transformed by DNA from human adenovirus type 5. J Gen Virol 36:59–74
3. Louis N, Evelegh C, Graham FL (1997) Cloning and sequencing of the cellular viral junctions from the human adenovirus type 5 transformed 293 cell line. Virology 233:423–429
4. Neumann E et al (1982) Gene transfer into mouse lyoma cells by electroporation in high electric fields. EMBO J 1:841–845
5. Li XW et al (2003) Sustained expression in mammalian cells with DNA complexed with chitosan nanoparticles. Biochim Biophys Acta 1630:7–18
6. Uchida M et al (2009) Transfection by particle bombardment: delivery of plasmid DNA into mammalian cells using gene gun. Biochim Biophys Acta 1790:754–764
7. Kim H, Park Y (2002) Effect of lipid compositions on gene transfer into 293 cells using Sendai F/HN-virosomes. J Biochem Mol Biol 35:459–464
8. Corsi K et al (2003) Mesenchymal stem cells, MG63 and HEK293 transfection using chito-

san-DNA nanoparticles. Biomaterials 24:1255–1264

9. Shaw G et al (2002) Preferential transformation of human neuronal cells by human adenoviruses and the origin of HEK 293 cells. FASEB J 16:869–71
10. O'Brien JA, Lummis SCR (2006) Biolistic transfection of neuronal cultures using a hand-held gene gun. Nat Protoc 1:977–981
11. Thomas P, Smart TG (2005) HEK293 cell line: a vehicle for the expression of recombinant proteins. J Pharmacol Toxicol Methods 51:187–200
12. Johnston SA, Tang DC (1994) Gene gun transfection of animal cells and genetic immunization. Methods Cell Biol 43:353–365
13. Yang NS et al (1990) In vivo and in vitro gene transfer to mammalian somatic cells by particle bombardment. Proc Natl Acad Sci USA 87:9568–9572
14. Williams RS et al (1991) Introduction of foreign genes into tissues of living mice by DNA-coated microprojectiles. Proc Natl Acad Sci USA 88:2726–2730
15. Cheng L, Ziegelhoffer PR, Yang NS (1993) In vivo promoter activity and transgene expression in mammalian somatic tissues evaluated by using particle bombardment. Proc Natl Acad Sci USA 90:4455–4459
16. Sato H et al (2000) In vivo gene gun-mediated DNA delivery into rodent brain tissue. Biochem Biophys Res Commun 270:163–170
17. O'Brien JA et al (2001) Modifications to the hand-held gene gun: improvements for in vitro biolistic transfection of organotypic neuronal tissue. J Neurosci Methods 112:57–64

Chapter 11

Biolistic Transfection of Tumor Tissue Samples

Kandan Aravindaram, Shu-Yi Yin, and Ning-Sun Yang

Abstract

A nonviral method for gene transfer into mammalian cells has been developed using physical force which accelerates plasmid DNA-coated gold particles to high -speed and penetrate the mammalian cells. This technology of gene transfer via a biolistic transfection method has been shown to have multiple applications to mammalian gene transfer systems. This method has also been adapted for delivery of other macromolecules like RNA, microRNA, and proteins. A broad range of somatic cell types, including primary cell cultures and established cell lines, have been successfully transfected ex vivo or in vitro by using the gene gun technology, either as suspension or adherent cells in cultures. This chapter describes the general procedures for in vitro DNA transfection by particle-mediated delivery to nonadherent and adherent cells. These procedures can be readily employed by using the Helios gene gun system (Bio-Rad, Hercules, CA) based on the Accell design.

Key words: Cytokines, DNA vaccine, Gene gun, gp100, Immunity, Melanoma

1. Introduction

Genetic immunization using genes that code for tumor-associated antigens (TAAs) is evolving as a useful strategy for cancer therapy approaches. This vaccination strategy is formulated on the observations that a sustained local expression of transgenic TAAs may confer a high-level immunogenic presentation of tumor antigens, and this will in turn enhance in vivo sensitization and activation of T cells that are capable of recognizing the TAA-associated peptides on the tumor cell surface. Series of studies have demonstrated the T-cell-dependent efficacy of this genetic TAA immunization strategy in murine models (1–3). The approach of genetic vaccines in these preclinical murine studies has led to several clinical trials using genetic vaccination with TAA genes, including trials of MART-1 and gp100 DNA for melanoma patients (4).

Stephan Sudowe and Angelika B. Reske-Kunz (eds.), *Biolistic DNA Delivery: Methods and Protocols*,
Methods in Molecular Biology, vol. 940, DOI 10.1007/978-1-62703-110-3_11, © Springer Science+Business Media, LLC 2013

The biolistic transfection technology (3, 5–7) makes use of a burst of helium to accelerate DNA-coated microscopic gold particles into target cells. This results in transgene expression levels that are often comparable with and sometimes superior to those achieved by other DNA delivery systems (5). Potential advantages of the biolistic transfection approach (8, 9) include its ability to: (a) physically confer gene delivery into cells by nonviral means; (b) perform a transfection in several seconds; (c) transfect resting, nondividing cells, irrespective of cell lineage; (d) alter or modify levels of transgene expression over time via a multiple gene delivery regimen; and (e) simultaneously deliver multiple candidate therapeutic genes into targeted individual cells. The latter feature might be beneficial for in vivo skin DNA vaccination strategies when a TAA gene is used in combination with other immunostimulatory reagents, such as cytokine or chemokine genes. Vaccine strategies using tumor cells transfected with TAA and cytokine genes, including hgp100, IL-2, IL-4, IL-12, IFN-g, and GM-CSF, have been effective in mediating either T-cell-dependent or inflammatory responses that can lead to tumor regression (6, 7).

The Helios gene gun (Fig. 1a, b) (Bio-Rad, Hercules, CA) is commercially available to public users and today is still the only standardized and routinely applicable gene gun device that is available on the market. In principle, a helium gas shockwave is used to propel the high-density DNA-coated gold particles to a very high velocity, efficiently enough to penetrate cell membranes of targeted mammalian cells/tissues, resulting in a direct physical delivery of plasmid DNA, mRNA, microRNA, or other polymer (e.g., protein) molecules at high copy numbers in a dry form. In this chapter, we describe the protocols and techniques for the particle-mediated gene transfer method for gene transfection into suspension and

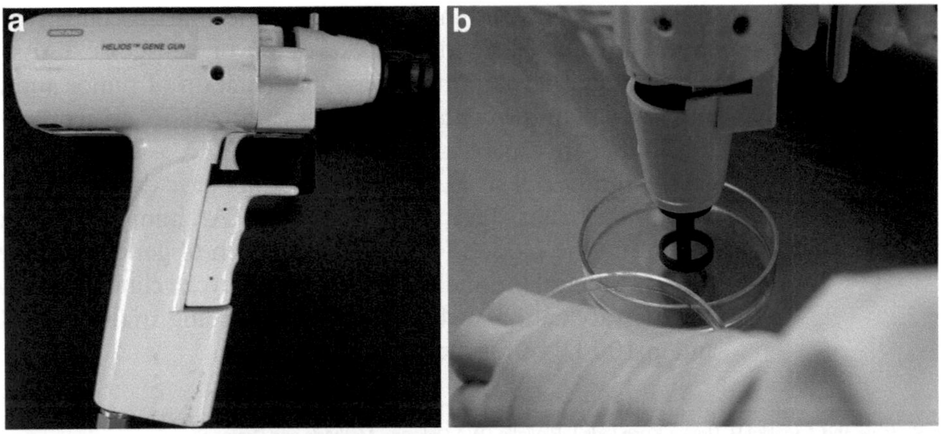

Fig. 1. (**a**) The Helios gene gun device is commercially available from Bio-Rad. (**b**) In vitro gene delivery into attached monolayer cells using the Helios gene gun.

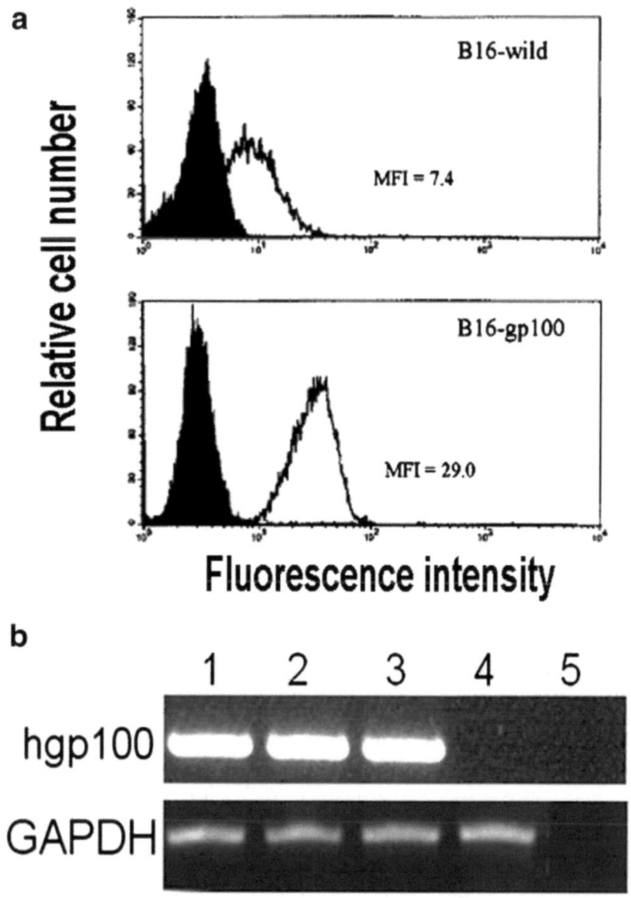

Fig. 2. Transgene expression of hgp100 cDNA vectors in vitro and in vivo. (**a**) Expression of hgp100 in B16-gp100 melanoma cells. Mouse B16 melanoma cells (*B16-wild*) or the cell clone derived from B16 cells after particle-mediated transfection with hgp100 DNA were used for flow cytometry. The cells were first permeabilized, then incubated with HMB-45 mAb reactive with human gp100, and stained with FITC-conjugated goat anti-mouse IgG (*open curve*). The control staining was performed with mouse IgG isotype control, followed by FITC-conjugated goat anti-mouse IgG (*dark curve*). Mean fluorescent intensity (*MFI*) values are shown. (**b**) The abdominal skin tissue of C57BL/6 mice was bombarded with gold coated with 3 μg of hgp100 plasmid DNA/mg gold; skin samples were removed at specific time intervals, and total RNA was extracted for RT-PCR. Lane 1, RNA from B16-hgp100 cells stably transfected with hgp100 cDNA; lanes 2 and 3, RNA from skin transfected with hgp100 cDNA at 24 and 48 h post bombardment, respectively; lane 4, RNA from skin transfected with empty vector; lane 5, negative control sample (without RNA). The housekeeping gene GAPDH was used as an internal control.

adherent tumor cells in vitro. On the other hand, using a similar protocol, an in vivo biolistic transfection technique can also be readily applied to experimental approaches for cancer gene therapy (6, 7). For instance, the gene-based, prime-vaccination against B16 melanoma in C57BL/6JNarl mice, followed by boosting with a viral vector expressing hgp100, together this procedure is called the "heterologous prime and boost system," has shown strong suppressive effect on growth of B16/hgp100 primary tumors and the metastasis into lung (Fig. 2a, b). The in vivo application of this technology therefore is also introduced and discussed here.

2. Materials

2.1. Instrumentation

1. The commercially available Helios gene gun (Bio-Rad, Hercules, CA). Instructions for the mechanical operation and/or manipulation of the gene gun and associated devices (e.g., cartridge holder, tefzel tubing, tubing cutter) are given in the manufacturer's manual.
2. Compressed helium of grade 4.5 or higher.
3. Hearing protection device.

2.2. Elemental Gold Particles

Microscopic gold particles (Degussa, South Plainfield, NJ, or Bio-Rad (see Notes 1–4)).

2.3. DNA Vectors

Dissolve a clean plasmid DNA, for example encoding human gp100 (a tumor-associated antigen (TAA) cDNA gene) and genes for a number of pro-inflammatory cytokines (TNF-α, NF-κB, GM-CSF (10, 11)), or luciferase (Luc) in TE buffer (10 mM Tris–HCl, pH 7.0, 1 mM EDTA), or distilled water. Cocktails of different DNA vector systems or preparations can be mixed in desired molar ratios in aqueous solution (see Note 5).

2.4. Coating of DNA onto Gold Particles for Gene Transfer

1. 0.05 M spermidine or polyethyleneglycol (PEG) in H_2O. Use fresh spermidine made weekly from a free-base solution (Sigma, St. Louis, MO).
2. 1 mg/mL polyvinylpyrrolidone (PVP; Sigma) in H_2O.
3. 2.5 M $CaCl_2$ in H_2O.
4. 100% ethanol kept at −20°C.
5. 1.5 mL microcentrifuge tube.
6. 15 mL culture tube.
7. Vortex mixer.
8. Sonicator.
9. Microcentrifuge.

2.5. Biolistic Transfection of Cells and Skin Tissue

1. Biosafety cabinet.
2. Ring stand clump.
3. 70% ethanol or 10% NaOCl (Sodium hypochloride).
4. Distilled water.
5. Primary tumor cell line of interest (cryopreserved) (e.g., mouse B16 melanoma cells) or nonadherent cells (such as peripheral blood mononuclear cells (PBMC)).
6. Water bath.
7. 15 mL conical centrifuge tube.

8. Centrifuge for sedimenting cells.
9. Appropriate culture medium for the cells used, 37°C.
10. Hemacytometer.
11. 1 M HEPES in normal saline.
12. Wide-bore pipet tips (Bio-Rad).
13. Pipettor.
14. 35-mm dish.
15. CO_2 incubator.
16. 5% Trypsin (Invitrogen, Grand Island, NY).
17. Mice of appropriate strain.

2.6. Analysis of Transgene Expression

2.6.1. Flow Cytometry

1. ORTHO PermaFix solution (Ortho Diagnostics Systems, Raritan, NJ).
2. Anti-gp100 monoclonal antibody (mAb), HMB-45 (Coulter-Immunotech, Westbrook, ME).
3. Mouse IgG.
4. FITC-conjugated goat anti-mouse IgG.
5. Cytofluorometer (e.g., FACScan, Becton Dickinson, San Jose, CA).

2.6.2. Reverse Transcriptase-Polymerase Chain Reaction

1. TRIzol reagent (Invitrogen, Carlsbad, CA).
2. Diethyl pyrocarbonate-treated water.
3. Access Quick RT-PCR system (Promega, Madison, WI).
4. Thermal cycler.
5. 1.5% agarose gel (Ultrapure, Invitrogen).
6. Molecular weight marker.
7. UV illumination.
8. Type 55 positive/negative film (Polaroid Corp., Cambridge, MA).

3. Methods

3.1. Coating DNA onto Gold Particles

1. Prepare DNA solution at a concentration of approx. 1 μg/μL and store at 4°C.
2. One cartridge of the Helios gene gun device (for one transfection) contains approx. 0.5 mg of gold particles.
3. Assuming that 40 cartridges will be needed, weigh 21 mg of gold particles into a 1.5 mL microcentrifuge tube.
4. Add 250 μL of 0.05 M spermidine or PEG to the tube with the gold particles (e.g., use 200–300 μL for 20–50 mg and 400–500 μL for 120 mg).

5. Vortex and sonicate for 3–5 s to break up gold clumps.
6. Add DNA at 2.5 μg DNA/mg gold particle to 21 mg of gold.
7. Add 250 μL of 2.5 M $CaCl_2$ (or the same volume as spermidine) dropwise while vortexing the tube at a low speed.
8. Incubate at room temperature for 10 min. While waiting, prepare a culture tube with 3 mL of 100% ethanol.
9. Microcentrifuge for 3–5 s. Remove and discard the supernatant. Break up the pellet by flicking the tube. Add 0.5 mL of cold 100% ethanol dropwise while gently vortexing the tube and then 0.5 mL more of 100% ethanol. Mix by inversion. Wash the pellet with cold ethanol three times.
10. Transfer particles into the culture tube containing 3 mL of 100% ethanol (see step 8 above) (7 mg gold particle/mL ethanol). Sonicate to disperse particles. Particle suspensions can be used immediately or stored at 4 or −20°C under stringent desiccation for 2–3 h in a conventional glass desiccator.

3.2. Preparation of DNA Cartridges

1. Sonicate DNA-coated gold particle suspension for 2–3 s, vortex, and add PVP at 0.01 mg/mL of H_2O, sonicate again, and immediately load the suspension into the Tefzel tubing following the manufacturer's instruction (Bio-Rad).
2. Discard unevenly coated ends or portion of the tubing and cut the tubing into 0.5-in. pieces using the tubing cutter. Cartridges may be stored desiccated in a tightly sealed container at 4°C for several weeks.

3.3. Biolistic Transfection of Gene(s) into Nonadherent Cells

1. Attach the gene gun device to a ring stand clump inside a biosafety cabinet, with the barrel of the device held vertical, facing down. Disinfect the exit nozzle spacer with 70% ethanol, or with 10% bleach followed by distilled water.
2. Rapidly thaw an appropriate quantity of cryopreserved cells by swirling vial in a 37°C water bath. Disinfect outer surface of vial with 70% ethanol.
3. Add cells to 10 mL of 37°C culture medium in a 15 mL conical centrifuge tube and mix gently. Centrifuge cells 5 min at 200 × *g*, room temperature.
4. Remove supernatant, tap tube to resuspend pellet, and add 10 mL of culture medium.
5. Determine the number of viable cells present using a hemacytometer. Calculate the resuspension volume necessary to give a cell concentration of 50×10^6 cells/mL (1×10^6 cells per 20 μL of target cell suspension). Centrifuge cells 5 min at 200 × *g*, room temperature (see Note 6).

6. Remove supernatant and resuspend cells to the volume determined in step 5, using 37°C culture medium supplemented with 25 mM HEPES (see Note 7).
7. Tap tube to mix cell suspension. Using a wide-bore pipet tip, smear 20 μL of cell suspension into a 1.5-cm circle in a 35-mm dish.
8. Center the exit nozzle over the cell circle and press tightly against the plastic dish. Immediately transfect the target cells by triggering the gun to release the gold particles into targeted test cells. Then add exactly 2 mL culture medium to the target cells. Promptly return culture to the CO_2 incubator. (Smear and transfect only one target at a time to prevent the cells from drying out).
9. Repeat steps 7 and 8 as required.
10. Culture the cells for the appropriate time interval, and collect supernatant and/or cells for determination of transgene activity.

3.4. Particle Bombardment of Gene(s) into Adherent Monolayer Cells

1. Attach the device to a ring stand clamp inside a biosafety hood, with the barrel of the device held vertical, facing down. Disinfect the exit nozzle spacer with 70% ethanol, or with 10% bleach followed by distilled water.
2. Trypsinize cells from a log-phase cell culture, or thaw cryopreserved cells and resuspend them in culture medium. Centrifuge cells 5 min at 200 × *g*, room temperature.
3. Remove supernatant and tap the tube to resuspend the cell pellet. Resuspend cells in culture medium and determine the number of viable cells present using a hemacytometer. Centrifuge cells 5 min at 200 × *g*, room temperature.
4. Resuspend cell pellet at a density of 2×10^6 cells/ml in culture medium buffered with 25 mM HEPES.
5. Use 100 μL of the cell suspension to create a 1.5-cm-diameter circle in the center of a 35-mm dish, being careful not to break the meniscus. It is useful to draw a 1.5-cm-diameter circle on an index card for use as a template under the culture dish.
6. Do not disturb target plates! Let the targeted cells sit in the tissue culture hood until attached (30–60 min). (see Note 8).
7. Gently add 1 mL of culture medium to each culture dish and place it in a CO_2 incubator. Incubate at least 4 h. (Cells should be allowed to regain their usual morphology before transfection).
8. To transfect test cells, remove great majority (95%) of the medium from dish and center the cell target directly under the exit nozzle spacer. With the opening of the device touching the culture dish, press the discharge button. Immediately

add 2 mL culture medium, and promptly return cultures to the CO_2 incubator. Repeat operation with further cells as required (Transfect one dish at a time to prevent the cells from drying out).

9. Culture the cells for appropriate time intervals, and collect culture supernatant and/or cell samples for determination of transgene activity.

3.5. Particle Bombardment of Gene(s) into Skin Tissue

Transgene expression under in vivo conditions was tested by bombardment of plasmid cDNA on mouse abdominal skin, using a Helios helium-pulse gene gun with slight modifications as described previously (5).

1. Bombard animal skin samples in different experimental groups in vivo with DNA-coated gold particles carrying a specific gene or Luc cDNA expression vectors. Each gene delivery treatment may employ 3–4 gene transfections (5 μg plasmid DNA/treatment) with a 350–400 psi helium gas pulse.
2. For tumor therapy, transfect tissues overlaying and surrounding the target tumor tissues in vivo and in situ with specific gene expression vectors starting from day 7 after intradermal injection of test tumors. Among the 3–4 transfections, perform first transfection directly over the tumor site, and evenly space others treatments around the circumference of the tumor in a triangle pattern (5).

3.6. Analysis of Transgene Expression

3.6.1. Flow Cytometry

1. After TAA or cytokine gene tranfection, permeabilize tumor cells in ORTHO PermaFix solution for 1 h (here we used TAA human gp100 in B16 melanoma cells).
2. Incubate cells with anti-gp100 mAb, HMB-45, for 40 min at 4°C. As a negative control, use mouse IgG.
3. After washing, stain the cells with FITC-conjugated goat anti-mouse IgG for 40 min at 4°C. Analyze stained cells using a cytofluorometer. Collect data for 10,000 events/samples.

3.6.2. Reverse Transcriptase-Polymerase Chain Reaction

1. Forty-eight hours after gene transfection, collect tumor cells and extract the total RNA using TRIzol reagent and resuspend in 25 μL of diethyl pyrocarbonate-treated water.
2. Add 1 μg of total RNA from each sample to the reaction mixture containing 1× AccessQuick master mixture (*Tfl* DNA polymerase, avian myeloblastosis virus/*Tfl* reaction buffer, 25 mM $MgSO_4$, and 10 mM dNTP mixture), a 10 μM concentration of each of specific sense and antisense primers, five units of avian myeloblastosis virus reverse transcriptase, and nuclease-free water to obtain a final volume of 50 μL. Incubate reactions at 48°C for 60 min, and carry out PCR amplification after denaturing at 95°C for 2 min.

3. Separate the PCR products on 1.5% agarose gels at 55 V for 75 min along with a molecular weight marker, and visualize by UV illumination after staining with 0.5 μg/mL ethidium bromide solution.
4. Photograph gels with type 55 positive/negative film.

4. Notes

1. The gene gun method as a nonviral system for gene transfer can be applied to a broad range of tissues and cell types; a key feature of this technology is its applicability to in vivo gene transfer to various tissues and organs, especially the epidermal skin cell layers (8, 12–14). The latter case thus permits a powerful nucleic acid or other gene-based vaccine strategy.
2. Two key technical advantages have been observed for the particle-mediated gene delivery method: (a) a very wide range for DNA load (1 ng–10 μg/dose/transfection site) can be delivered into a 3–5 cm^2 surface area of targeted tissues, resulting in different efficacies of transgene expression, depending on target cells, tissues, or organ types as well as the in vitro, ex vivo, or in vivo experimental conditions; and (b) there is little or no restriction on the molecular size or form of testing DNA, at least at a molecular size ≤40 kb as double-stranded DNA (15, 16).
3. Gold particles of 0.7–10.0-μm-diameter are in general used for in vitro transfection of various leukocytes, lymphocytes, and small tumor cells. 2-μm particles were found to be the best for in vivo gene transfection into skin (15), and 1–2-μm gold particles for adhesive monolayer cells or cell aggregates and tissue clumps in culture.
4. It is important that these gold particles be obtained as elemental gold, not as gold salt, or colloidal gold. If necessary, the gold particles can be washed and cleaned by rinsing them in distilled water, 70% ethanol, and 100% ethanol in sequence prior to use, and they can also be sterilized in phenol or $CHCl_3$ if necessary. It is important to examine each lot of newly purchased gold particle preparation microscopically, making sure that the lot, particle size, and form are correct and appropriate as desired by commercial suppliers and for test systems.
5. Besides DNA (7, 8), RNA (15), and peptide nucleic acid (PNA) (17) may also be effectively employed as a vector for gene-based vaccination using gene gun delivery. For exploratory gene transfer experiments, convenient and sensitive reporter gene systems that have low or no endogenous activity

background (e.g., green fluorescence protein and luciferase) are recommended for verification of transient gene expression systems.

6. To calculate the resuspension volume, divide the total number of cells in the tube by 10^6, and multiply this result by 20. This will give the correct resuspension volume in microliters. For many applications, 1×10^6 cells per transfection yields a sufficiently high transgene expression. When larger numbers of transfected cells are required for a single test sample, 2×10^7 cells may be transfected in a dense cell suspension as a single target. The optimal cell density and volume of the target cell smear must be determined for each cell line and application.
7. To obtain an accurate final volume, place 2 mL medium in a 35-mm dish and add 50 μL of 1 M HEPES in normal saline. Tap the tube containing the target cells to resuspend pellet, then add medium from the 35-mm dish to give one-half of the total desired resuspension volume. Set pipettor to full resuspension volume. Draw the cells into the pipet tip until all the cell suspension is in the tip, but no additional air. Place the tip into the medium in the 35-mm dish, and fill the tip to the full resuspension volume. Return cell suspension to the 15 mL tube.
8. For consistency, it is important that the cell monolayer be uniform. Minimize movement of the target plates until the cells have attached to the plastic substratum. Most cells will adhere within 60 min. If cells require a longtime to reattach, carefully move the cells to the CO_2 incubator, or consider transfecting them as a suspension.

References

1. Chen PW et al (1996) Therapeutic antitumor response after immunization with a recombinant adenovirus encoding a model tumor-associated antigen. J Immunol 156:224–231
2. Warnier G et al (1996) Induction of a cytolytic T-cell response in mice with a recombinant adenovirus coding for tumor antigen P815A. Int J Cancer 67:303–310
3. Aravindaram K et al (2009) Transgenic expression of human gp100 and RANTES at specific time points for suppression of melanoma. Gene Ther 16:1329–1339
4. Mahvi DM et al (2002) Immunization by particle-mediated transfer of the granulocyte-macrophage colony-stimulating factor gene into autologous tumor cells in melanoma or sarcoma patients: report of a phase I/IB study. Hum Gene Ther 13:1711–1721
5. Rakhmilevich AL, Yang NS (1997) Particle-mediated gene delivery system for cancer research. In: Strauss M, Barranger JA (eds) Concepts in gene therapy. Walter de Gryuter, New York, pp 109–120
6. Yang NS, Sun WH (1995) Gene gun and other non-viral approaches for cancer gene therapy. Nat Med 1:481–483
7. Rakhmilevich AL et al (2001) Effective particle-mediated vaccination against mouse melanoma by coadministration of plasmid DNA encoding gp100 and granulocyte-macrophage colony-stimulating factor. Clin Cancer Res 7:952–961
8. Cheng L, Ziegelhoffer PR, Yang NS (1993) *In vivo* promoter activity and transgene expression in mammalian somatic tissues evaluated by using particle bombardment. Proc Natl Acad Sci USA 90:4455–4459

9. Sun WH et al (1995) *In vivo* cytokine gene transfer by gene gun reduces tumor growth in mice. Proc Natl Acad Sci USA 92:2889–2893
10. Staniforth V, Chiu LT, Yang NS (2006) Caffeic acid suppresses UVB radiation-induced expression of interleukin-10 and activation of mitogen-activated protein kinases in mouse. Carcinogenesis 27:1803–1811
11. Chiu SC, Yang NS (2007) Inhibition of tumor necrosis factor-alpha through selective blockade of Pre-mRNA splicing by shikonin. Mol Pharmacol 71:1640–1645
12. Yang NS et al (1990) *In vivo* and *in vitro* gene transfer to mammalian somatic cells by particle bombardment. Proc Natl Acad Sci USA 87:9568–9572
13. Jiao S et al (1993) Particle bombardment mediated gene transfer and expression in rat brain tissues. Biotechnology (N Y) 11:497–502
14. Burkholder JK, Decker J, Yang NS (1993) Transgene expression in lymphocyte and macrophage primary cultures after particle bombardment. J Immunol Methods 165: 149–156
15. Qiu P et al (1996) Gene gun delivery of mRNA in situ result in efficient transgene expression and immunization. Gene Ther 3:262–268
16. Yang NS, Ziegelhoffer P (1994) The particle bombardment system for mammalian gene transfer. In: Yang NS, Christou P (eds) Particle bombardment technology for gene transfer. Oxford University Press, New York, pp 117–141
17. Liang KW, Hoffman EP, Huang L (2000) Targeted delivery of plasmid DNA to myogenic cells via transferrin-conjugated peptide nucleic acid. Mol Ther 1:236–242

Chapter 12

Biolistic Transfection of Freshly Isolated Adult Ventricular Myocytes

David F. Steele, Ying Dou, and David Fedida

Abstract

Transfection of mammalian cells has long been an extremely powerful approach for the study of the effects of specific gene expression on cell function. Until recently, however, this approach has been unavailable for the study of gene function in adult cardiac myocytes. Here, an adaptation of the biolistic method to the transfection of adult cardiac myocytes is described. DNA is precipitated onto gold particles in the absence of PVP and the particles are biolistically delivered to freshly isolated adult rat cardiomyocytes via a Bio-Rad Helios System gene gun. The myocytes are cultured in the absence of bovine serum albumin and expression of the introduced genes, in phenotypically intact myocytes, is robust within 12–24 h.

Key words: Transfection, Cardiomyocytes, Heart, Gene gun, Biolistic

1. Introduction

While transfection of rat neonatal cardiomyocytes has long been feasible via standard liposome-mediated protocols (1–3), the endogenous currents expressed in these cells are quite different from those of adult cardiomyocytes (4, 5). Thus, the validity of these cells as models of their adult counterparts is questionable. These neonatal systems have been popular, nevertheless, largely because transfection of adult cardiomyocytes has been an intractable problem. Until recently, the introduction of exogenous genetic material into adult cardiac myocytes had proven to be practical only with retroviruses (e.g., see ref. (6)). While these viral transduction methods are effective in adult cardiomyocytes, they require sophisticated containment facilities and prolonged culture of the myocytes, during which time substantial de-differentiation can

Stephan Sudowe and Angelika B. Reske-Kunz (eds.), *Biolistic DNA Delivery: Methods and Protocols*, Methods in Molecular Biology, vol. 940, DOI 10.1007/978-1-62703-110-3_12, © Springer Science+Business Media, LLC 2013

occur (7, 8). We have recently developed an adaptation of biolistic transfection methods that reliably yields transfectants at good efficiencies in acutely isolated adult rat ventricular myocytes (9). The freshly isolated myocytes are adhered to laminin-coated coverslips and transfected with the use of a Bio-Rad Helios System gene gun approximately 1 h after isolation. To prevent de-differentiation, they are cultured in media lacking bovine serum albumin. Expression of the introduced genes is robust within 24 h of transfection, a time frame during which, in control, untransfected- and GFP-transfected myocytes, myocyte morphology, and currents remain essentially unchanged. Thus, these transfected cells likely recapitulate closely the regulatory mechanisms extant in intact cardiomyocytes. This method is outlined here in detail.

2. Materials

Prepare all solutions using ultrapure water (~18 MΩ resistance).

2.1. Myocyte Isolation Solutions and Equipment

A Langendorff apparatus hooked up to a carbogen (95% O_2, 5% CO_2) tank (with a secondary line attached for carbogen through buffers) and a 37°C circulating bath (e.g., Haake model DC 30) are required for cardiomyocyte isolation as described herein. These are available from several manufacturers.

1. Base Ca^{2+}-free Solution: 121 mM NaCl, 5 mM KCl, 24 mM $NaHCO_3$, 5.5 mM glucose, 2.8 mM sodium acetate, 1 mM $MgCl_2$, 1 mM Na_2HPO_4.
2. Perfusion Solution: Base Ca^{2+}-free Solution plus 1.5 mM $CaCl_2$. Important: Prior to addition of Ca^{2+}, Base Ca^{2+}-free Solution must be bubbled with 95% O_2, 5% CO_2 until pH7.4 is achieved. Failure to do so will result in precipitation of the added Ca^{2+}.
3. Low-Ca^{2+} Perfusion Solution: base Ca^{2+}-free Solution plus 5 μM $CaCl_2$.
4. Pre-digestion Solution (prepare fresh on day of use): base Ca^{2+}-free Solution plus 20 mM taurine, 12.5 U/mL type II collagenase (Worthington, Lakewood, NJ), 0.1 U/mL type XIV protease (Sigma-Aldrich), 40 μM $CaCl_2$.
5. Digestion Solution (prepare fresh on day of use): base Ca^{2+}-free Solution plus 20 mM taurine, 240 U/mL type II collagenase, 2.0 U/mL type XIV protease, 10 mg/mL essentially fatty acid-free Bovine Serum Albumin (BSA, Sigma-Aldrich), 100 μM $CaCl_2$.

6. Storage Solution (prepare fresh on day of use): base Ca^{2+}-free Solution plus 20 mM taurine, 10 mg/mL essentially fatty acid-free BSA, 100 μM $CaCl_2$.
7. Supplemented M199: M199 media, pH7.4 (Gibco), supplemented with 2 mM EGTA, 0.6 μg/mL insulin, 5 mM creatine, 2 mM DL-carnitine, 2 mM glutamine, 5 mM taurine, 50 U/mL penicillin, 50 μg/mL streptomycin.
8. Low-speed swinging bucket centrifuge that accepts 15 mL conical base tubes.
9. 35 mm Sterile Petri dishes.
10. 22 mm Glass coverslips.
11. Laminin (e.g., Sigma).
12. 3-0 Surgical thread.
13. Heparin (Hepalean, 1,000 U.S.P. units/mL).
14. Pentobarbital (65 mg/mL).
15. Forceps, curved surgical scissors (to cut skin), straight scissors to cut bone, and fine scissors to cut aorta.
16. 25 mL Erlenmeyer flask.
17. 15 mL Polypropylene centrifuge tubes with screw caps and conical base.

2.2. Biolistic System Solutions and Equipment

1. Bio-Rad Helios Gene Gun System with low-pressure regulator (see Note 1) including BioRad Helios Gene Gun System Prep Station assembled as per manufacturer's instructions.
2. Helium tank (Helium grade 4.5 or higher).
3. Nitrogen tank (Nitrogen grade 4.8 or higher).
4. Bio-Rad Econo Pump peristaltic pump or equivalent.
5. Branson 1210 sonicating bath or equivalent.
6. Analytical balance.
7. 37°C, 5% CO_2 cell culture incubator.
8. Microcentrifuge (e.g., IEC Micromax).
9. Variable speed vortexer.
10. 0.6 μm Gold beads (BioRad).
11. Silica gel beads (8 Mesh; EMD Chemicals, Gibbstown, NJ, or equivalent).
12. Fresh 100% ethanol, analytical grade.
13. 1.5 mL Microcentrifuge tubes.
14. 50 mM Spermidine.
15. Plasmid DNA encoding the gene(s) of interest in a mammalian expression vector (see Note 2).

16. 1 M $CaCl_2$.
17. 15 mL Polypropylene centrifuge tubes with screw caps and conical base.
18. BioRad Gold-Coat Tubing.
19. M199 Media.
20. Supplemented M199: see step 7 in Subheading 2.1.
21. 25 μM Blebistatin (Tocris Bioscience; optional).

3. Methods

3.1. Preparation of DNA-Coated Beads

This is a variation on the method described in the Bio-Rad Helios System Instruction Manual (10), modified to maximize transfection of adult cardiomyocytes.

1. At least 2 days prior to the preparation of DNA-coated beads, half-fill a 100–250 mL bottle with 8 Mesh Silica Gel beads (EMD Chemicals, Gibbstown, NJ, Catalog number DX0013-1 or equivalent) and fill the bottle to the top with 100% ethanol. Cap tightly, shake well to mix and allow the silica gel to settle. On the following day, half-fill a second, smaller bottle and transfer the ethanol from the first bottle to the second, mix as before and allow to settle for another day. This procedure will ensure that the ethanol used in the bead preparation is water-free (see Note 3).
2. Weigh out 25 mg of 0.6 μm gold beads (see Note 4) into a 1.5 mL microfuge tube and add 50 μL of 50 mM spermidine.
3. Vortex the gold/spermidine mixture at a high setting for 5 s then sonicate for 5 s in a Branson 1210 sonicating bath (or equivalent).
4. Add 1–2 μg (in a volume of 10 μL or less) of the plasmid DNA to be transfected to the gold/spermidine mix and vortex for 5 s.
5. While vortexing at a middle setting on a variable speed vortexer, add 50 μL of 1 M $CaCl_2$ dropwise to the mixture.
6. Leave the mixture for 10 min at room temperature without shaking.
7. Pellet the gold-DNA complex by spinning 15 s in a microcentrifuge. Discard the supernatant.
8. Vortex briefly to resuspend the gold/DNA in the residual supernatant and wash three times with 1 mL of fresh 100%

ethanol (previously dried over silica—see above). Spin ca. 5 s after each wash, discard the supernatants.

9. Resuspend the final pellet in 200 μL of silica-dried 100% ethanol and transfer to a 15 mL polypropylene centrifuge tube with a screw cap. Rinse the microcentrifuge tube four times with 200 μL of silica-dried 100% ethanol, transferring the rinse each time to the polypropylene centrifuge tube (see Note 5).
10. Add an additional 2 mL silica-dried 100% ethanol to the polypropylene centrifuge tube, bringing the final volume of the ethanol-gold-DNA mix to 3 mL (again, see Note 5).
11. Load the gold-DNA complex into Gold-Coat Tubing as described in the BioRad Helios Gene Gun System Instruction Manual (10).
12. Remove the tubing from the Prep Station and cut to 0.5″ lengths (using the Bio-Rad tubing cutter).
13. Use immediately (see below) or transfer to 35 mm Petri dishes, wrap in parafilm and store under dessication at 4°C (see Note 6).

3.2. Myocyte Preparation

1. Prior to performing the myocyte isolation procedure, coat 22 mm glass coverslips with laminin by diluting laminin to 25 μg/mL in M199 media and covering the individual coverslips, in individual 35 mm petri dishes, with this solution (see Note 7). Leave these coverslips thus covered until it is time to plate the myocytes on them.
2. Prepare Base Ca^{2+}-free Solution, Perfusion Solution, Low Ca^{2+} Perfusion Solution, Pre-digestion Solution, Digestion Solution, and Storage Solution. Base Ca^{2+}-free solution should be gassed with 95% O_2, 5% CO_2 until pH = 7.4. Perfusion Solution, Low Ca^{2+} Solution, Pre-digestion Solution, and Digestion solution should be warmed to 37°C prior to use.
3. Prepare 30 mL cold (4°C) Perfusion Solution in a beaker for heart collection.
4. Prepare the heart Langendorff apparatus. Set the water bath so that the outflow from the tip of the cannula is 37°C. Rinse columns three times with 2 L ultrapure water. Add 500 mL of each of Perfusion, and Low Ca^{2+} Perfusion and 100 mL of Pre-digestion Solution in different columns and bubble each with carbogen. Ensure that the cannula and tubes connecting the aortic cannula with the columns are filled and free from air bubbles. Make sure the perfusion is at a constant hydrostatic pressure of 82 cm H_2O.
5. Place 3-0 surgical thread around the cannula and tie loosely.

6. Inject a ~250 g rat with 500 units (0.5 mL) heparin intraperitoneally (i.p.). Ten minutes later, anesthetize the animal by i.p. injection with 0.5 mL of 65 mg/mL (~ 125 mg/kg) pentobarbital. Test for complete anesthesia testing for absence of the toe pinch withdrawal reflex. If complete loss of this reflex does not occur within 6–10 min, administer additional pentobarbital (see Note 8).
7. Once the rat is fully anesthetized, cut open the thorax and expose the heart. Using fine scissors, cut the aorta at about 3 mm from its entry into the heart, and rapidly excise heart (see Note 9), removing extraneous tissues, if necessary. Rinse the heart briefly in cold Perfusion Solution before mounting on the cannula.
8. Cannulate the aorta and attach a small clip at the end of the cannula to prevent the heart from falling. Start the perfusion with Perfusion Solution immediately and tie the aorta to the cannula with 3-0 thread. The heart should be beating vigorously at this point.
9. Once the perfusate is clear, switch to Low Ca^{2+} Perfusion Solution and perfuse the heart for 8–10 min. The heart should stop beating soon after switching to this solution.
10. Switch to Pre-digestion Solution and perfuse the heart for 4–8 min.
11. Cut the heart from the cannula before it turns whitish–grey. Cut the ventricles into small pieces (approximately 2 mm^2) and place them into a 25 mL flask containing approximately 10 mL Digestion Solution. Swirl gently at 37°C and periodically check cells that come off into the Digestion Solution under a microscope at low magnification (e.g., ×32).
12. Discard the solution that contains mainly rounded myocytes obtained at the beginning of the digestion process. Once intact myocytes start to appear in the Digestion Solution, periodically decant ca. 1 mL of the Digestion Solution into a 15 mL conical tube and fill with Storage Solution (see Note 10). Typically, 8–10 conical tubes containing viable myocytes are obtained.
13. When myocyte collection is complete (see Note 11), centrifuge the collected myocytes at ca. 40 g for 5 min in a swinging bucket rotor, discard the supernatant and resuspend the pellet in each tube in 10 mL Storage Solution.
14. Re-centrifuge as above and remove all but approximately 0.5–1 mL of the supernatant.
15. Resuspend the pelleted myocytes in the residual supernatant and plate them at high density (Fig. 1) on the laminin-coated glass coverslips in 35 mm petri dishes, removing the M199-laminin immediately prior to plating the myocytes. Usually

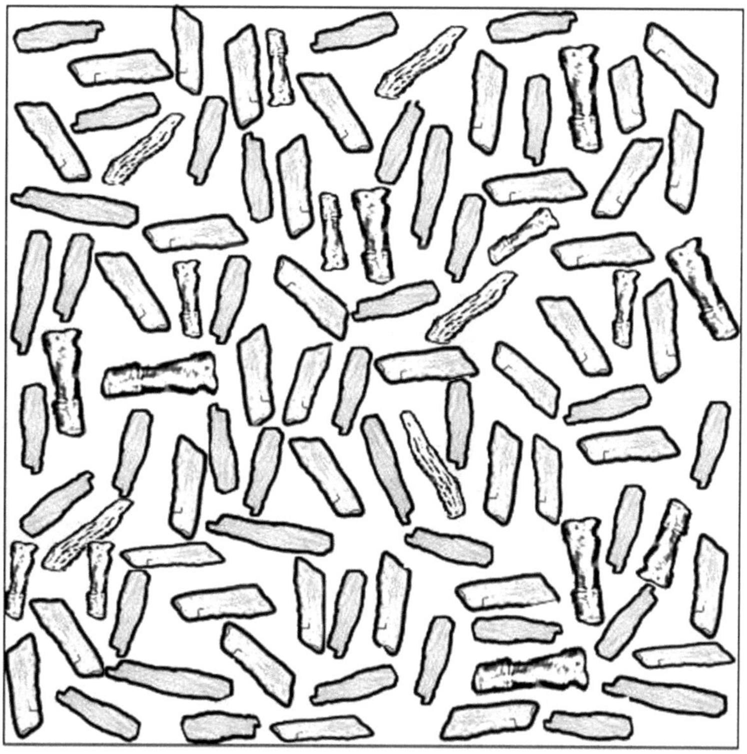

Fig. 1. Graphic representation of appropriate cardiomyocyte plating density for effective transfection. Myocytes should be plated to this density or higher to ensure that sufficient numbers are present to yield surviving transfectants.

2–3 coverslips can be plated per tube of isolated myocytes (see Note 12). Add sufficient unsupplemented M199 media to ensure that the myocytes will remain covered by liquid and incubate at room temperature for 30 min.

16. Remove the unsupplemented M199 and replace with 1–2 mL Supplemented M199. Transfer to a 37°C, 5% CO_2 incubator and incubate for at least 1 h prior to biolistic transfection.

3.3. Transfection of Ventricular Myocytes Using the Prepared Cartridges and a Bio-Rad Helios Gene Gun

1. Activate Helios Gene Gun using an empty cartridge as described in the Bio-Rad Helios Gene Gun System Instruction Manual (Subheading 5.3) (10).
2. Load the Helios Gene Gun with the appropriate DNA/gold-coated cartridge(s) (see Note 2).
3. Set the helium pressure to the predetermined appropriate pressure, generally a value between 90 and 110 psi, using the

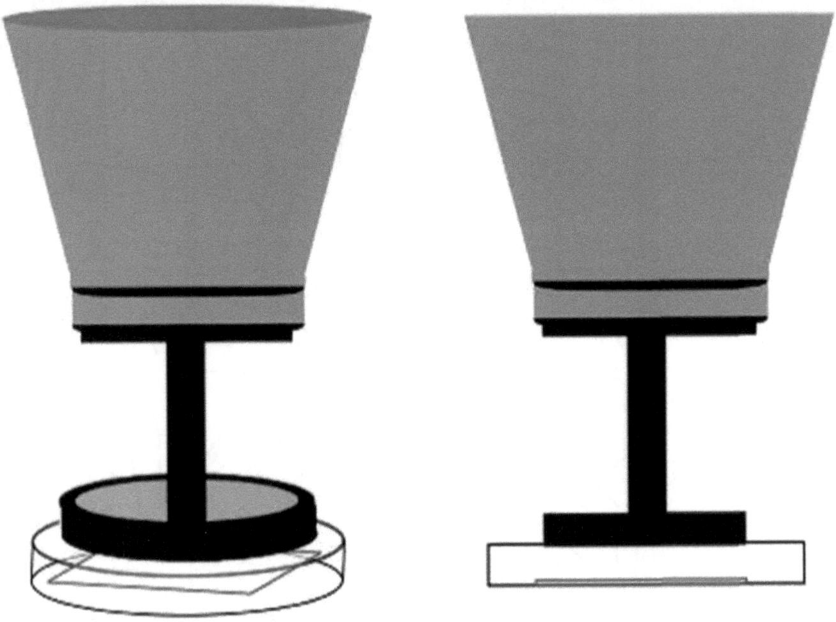

Fig. 2. Proper positioning of the gene gun nozzle for effective and reproducible cardiomyocyte transfection. The nozzle should be held parallel to and even with the top of Petri dish walls as illustrated from two perspectives here.

regulator adjustment knob. It is important to determine this optimal pressure for efficient transfection empirically (see Note 13). We find that the appropriate pressure reading varies among different pressure regulators.

4. Complete the following separately for each dish of myocytes to be transfected:
   (a) Remove the lid and the supplemented M199 media from the dish of plated myocytes.
   (b) Immediately place the nozzle of the gene gun even with the top of the dish walls, parallel to bottom of the dish and fire the gene gun (Fig. 2).
   (c) Immediately add 1 mL Supplemented M199 and transfer the transfected myocytes to a 37°C, 5% CO_2 incubator.
5. Incubate overnight. We find that expression of transfected potassium channels is generally robust by 16–24 h post-transfection (Fig. 3).
6. Where longer expression times are desired or required, inclusion of 25 μM blebistatin in the supplemented M199 media will maintain viability of the myocytes with intact morphology for several days.

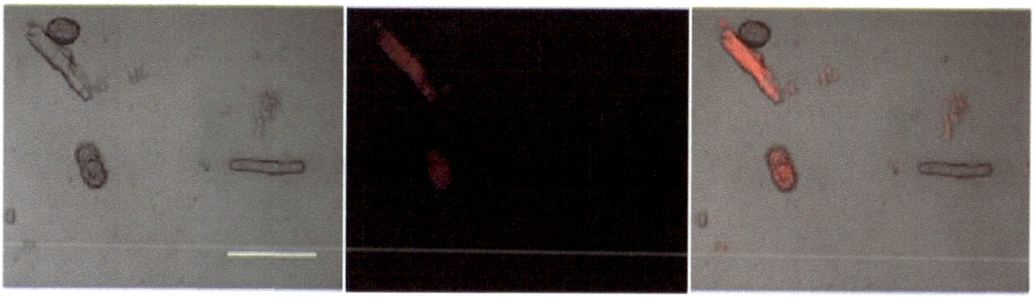

Fig. 3. A myocyte transfected with an expression vector encoding mCherry. Freshly isolated rat ventricular myocytes were transfected with pcDNA3-mCherry coated 0.6 μm gold beads at a pressure of 90 psi. After bombardment, the myocytes were cultured for 24 h in Supplemented M199 media then visualized using an Olympus Fluoview 1000 laser scanning confocal microscope. The transfected myocyte, visible at upper left in each frame, glows brightly in the *red* (*grey* in the greyscale image). Also visible is a nearby untransfected myocyte and normal detritus resulting from the biolistic transfection. *Left*, bright field image. *Middle*, fluorescence image. *Right*, combined bright field and fluorescence image.

4. Notes

1. While a standard Bio-Rad Helios System helium tank regulator may be used, the low pressure regulator allows more precise adjustment of the pressures employed, allowing for better day to day reproducibility of the transfection procedure.
2. The choice of the marker for transfected myocytes is important. Isolated myocytes autofluoresce strongly in the green. Thus, unless expression is expected to be high (e.g., where myocytes are cotransfected with pEGFP plus the gene of interest), GFP is not always a good marker of transfection. Where its expression is low, it can be difficult to distinguish transfected from untransfected myocytes. Cardiomyocytes autofluoresce only weakly in the red. Thus, the use of mCherry as a marker of transfection, either as a cotransfectant or as a gene fusion, yields readily identifiable transfectants. It is also advisable, of course, to avoid green fluorescent antibody markers for imaging studies of proteins expressed at low levels in the transfected myocytes.
3. It is very important that the ethanol be water free. If water is present, the beads will neither distribute evenly on the Gold-Coat tubing nor eject efficiently from the Gene Gun.
4. The use of larger gold particles is not recommended. In our hands, particles larger than 0.6 μm resulted in near-complete killing of the plated myocytes.

5. In order to quantitatively transfer the DNA-gold mixture it is necessary to rinse repeatedly. Depending on the vigor with which one does this, more than 4 rinses may be required to fully accomplish this transfer.
6. In our hands, DNA-coated beads stored sealed and dessicated remain useful for transfection for approximately 6 months. As the preparations age, the efficiency of their ejection by the gene gun falls.
7. Ideally, the laminin-M199 mixture should form a meniscus to the edges of the coverslip. This will facilitate the concentrated plating of the myocytes, once prepared. When the laminin-M199 overflows the coverslip onto the bottom of the Petri dish, the myocyte suspension subsequently added will overflow also, substantially reducing the density of the cells obtainable on the coverslip.
8. In our hands, the initial dose of pentobarbital has been uniformly sufficient to achieve loss of the toe pinch reflex within less than 10 min of administration.
9. It is extremely important to excise the heart as rapidly as possible. Failure to do so results in substantial necrosis and low levels of myocyte viability.
10. Myocytes should be collected in this way once the test aliquots show rod-shaped, striated myocytes. Where large numbers of rounded myocytes are present, plating efficiencies will be low.
11. Myocytes should be collected as long as the ratio of rod-shaped to rounded myocytes is increasing. Once the number of rounded cells begins to increase, collection should be stopped as the viability of the rod-shaped cells from these isolates will be low upon plating.
12. Achieving a high plating density is very important. In successful transfection experiments, greater than 90% of the myocytes will be killed and a variable proportion of the surviving cells, generally 10–20%, will express the introduced transgene. In our hands, where less than 90% of myocytes are killed by the biolistic procedure, transfection rates are extremely low.
13. When first employing a new regulator, test transfections with an easily assayed transgene (e.g., mCherry) at pressures from 80 to 120 psi in 5 psi increments are advisable. We have noted very substantial differences in optimal biolistic pressures from one regulator to another. As the regulator wears, we find that pressure settings become inaccurate and readings as high as 135 psi have been necessary to achieve transfection. We recommend replacement of the regulator at this stage since the pressures actually delivered appear to vary substantially from day to day.

Acknowledgments

This work was supported by funding to D.F. from the Canadian Institutes for Health Research and the Heart and Stroke Foundation of British Columbia and the Yukon.

References

1. Shubeita HE et al (1992) Transcriptional activation of the cardiac myosin light chain-2 and atrial natriuretic factor genes by protein kinase C in neonatal rat ventricular myocytes. Proc Natl Acad Sci USA 89:1305–1309
2. Komuro I et al (1990) Stretching cardiac myocytes stimulates protooncogene expression. J Biol Chem 265:3595–3598
3. Parker TG et al (1990) Differential regulation of skeletal alpha-actin transcription in cardiac muscle by two fibroblast growth factors. Proc Natl Acad Sci USA 87:7066–7070
4. Quignard JF et al (1996) Absence of calcium channels in neonatal rat aortic myocytes. Pflugers Arch 431(5):791–793
5. Walsh KB, Parks GE (2002) Changes in cardiac myocyte morphology alter the properties of voltage-gated ion channels. Cardiovasc Res 55:64–75
6. Kirshenbaum LA et al (1993) Highly efficient gene transfer into adult ventricular myocytes by recombinant adenovirus. J Clin Invest 92: 381–387
7. Aikawa R, Huggins GS, Snyder RO (2002) Cardiomyocyte-specific gene expression following recombinant adeno-associated viral vector transduction. J Biol Chem 277: 18979–18985
8. Zhou YY et al (2000) Culture and adenoviral infection of adult mouse cardiac myocytes: methods for cellular genetic physiology. Am J Physiol Heart Circ Physiol 279: H429–H436
9. Dou Y et al (2010) Normal targeting of a tagged Kv1.5 channel acutely transfected into fresh adult cardiac myocytes by a biolistic method. Am J Physiol Cell Physiol 298:C1343–C1352
10. Bio-Rad Laboratories (1999) Helios gene gun system instruction manual, Rev C. Bio-Rad Laboratories, Hercules, CA

Chapter 13

Biolistic Transfection of Neurons in Organotypic Brain Slices

John A. O'Brien and Sarah C.R. Lummis

Abstract

Transfection of postmitotic neurons is one of the most challenging goals in the field of gene delivery. Currently most procedures use dissociated cell cultures but organotypic slice preparations have significant advantages as an experimental system; they preserve the three-dimensional architecture and local environment of neurons, yet still allow access for experimental manipulations and observations. However exploring the effects of novel genes in these preparations requires a technique that can efficiently transfect cells deep into tissues. Here we show that biolistic transfection is an effective and straightforward technique with which to transfect such cells.

Key words: Organotypic brain slices, Biolistic transfection, Neurons, DNA microprojectiles, Microparticles, Nanometer gold particles, Micrometer gold particles

1. Introduction

Many different methods have been developed to incorporate DNA into cells; these include microinjection, electroporation, calcium-phosphate transfection, lipofection, and viral transfection (1–3). Gene transfer using these methods has become a routine tool for studying a range of scientific problems such as gene regulation and function (4–7). A number of cell types, however, particularly postmitotic neurons, are not easily transfected by these routes. One solution is to use biolistic transfection, which was originally designed to circumvent difficulties in transfecting plant cells whose cell walls present a physical barrier to conventional transfection techniques (8–10); in the last 10 years it has also proved to be an effective procedure in transfecting a wide variety of animal tissues (11, 12).

Stephan Sudowe and Angelika B. Reske-Kunz (eds.), *Biolistic DNA Delivery: Methods and Protocols*,
Methods in Molecular Biology, vol. 940, DOI 10.1007/978-1-62703-110-3_13, © Springer Science+Business Media, LLC 2013

Biolistic transfection is a physical method of transfection in which gold or other particles are coated with plasmid DNA and then accelerated to high velocity so that they penetrate the target cells or tissue, which can then transcribe the DNA as in other transfection procedures (13, 14). For the biolistic device discussed in this protocol (the Helios Gene Gun system; Bio-Rad), high-pressure helium provides acceleration for the gold particles. Because biolistic transfection relies on physical penetration for successful transfection, it is possible to use this technique to transfect cells that are resistant to transfection by other means, such as mammalian neurons in primary culture and organotypic slice cultures.

To explore the effectiveness of biolistic transfection, we investigated the 3D-architecture and morphology of neurons in living brain slices using fluorescently labeled cells in organotypic hippocampal brain slice cultures. The method is efficient, reliable, and does not require sophisticated facilities for its application (15–17). Here we describe the procedure.

2. Materials

1. Helios gene gun system (Bio-Rad, Hercules, CA, USA).
2. Tefzel tubing (Bio-Rad, Hercules, CA, USA).
3. Gold particles (Bio-Rad, Hercules, CA, USA).
4. Polyvinylpyrrolidone (PVP) (Bio-Rad, Hercules, CA, USA).
5. Cell Culture Inserts (Millicell, Billericay, MA, USA).
6. Tissue culture plates (Costar, Corning, NY, USA).
7. DNA midiprep kits (QIAGEN GmbH, Germany).
8. Cryogenic tubes (Corning, NY USA).
9. Vectashield mounting medium (Vector laboratories, CA, USA).
10. Filter units (Gibco, Invitrogen, Paisley, UK).
11. Cell culture medium: Dulbecco's Modified Eagle Medium (DMEM), 10% fetal calf serum, 1,100 U/mL Penicillin-Streptomycin.
12. Standard laboratory reagents (Sigma Aldrich; St. Louis, MO, USA).
13. HEPES-buffered saline: 140 mM NaCl, 5 mM KCl, 1 mM $MgCl_2$, 24 mM Glucose, 1 mM $CaCl_2$, 10 mM HEPES, in millipure H_2O, pH 7.2 with NaOH. Filter with a 0.2 μm sterile filter unit. Store at 4°C.
14. Phosphate-buffered saline (PBS): 137 mM NaCl, 2.7 mM KCl, 8.1 mM Na_2HPO_4, 1.76 mM, KH_2PO_4, in millipure H_2O, pH to 7.4 with HCl. Sterilize by autoclaving. Store at room temperature.

15. Neuronal medium: Minimum Eagles Medium (MEM) supplemented with 25 mM HEPES, 10%, Horse serum 30 mM Glucose, 1:100 diluted N2 supplement, 1,100 U/mL Penicillin-Streptomycin; pH to 7.2 with NaOH. Filter with a 0.2 μm sterile filter unit. Store at 4°C.
16. Spermidine stock solution: 0.05 M Spermidine in millipure H_2O; pH to 7.2 with NaOH. Store at room temperature.
17. PVP stock solution: 20 mg PVP (Bio-Rad) in 1 mL 100% EtOH. Aliquot and freeze at −20°C.
18. PVP working solution: 0.05 mg/mL PVP in EtOH (Add 10 μL PVP stock solution to 3.5 mL 100% EtOH.).
19. $CaCl_2$ stock solution: 1 M $CaCl_2$ in millipure H_2O.
20. Plasmid-DNA encoding the gene of interest.
21. Dissecting microscope.
22. Spatula and fine forceps.
23. Tissue chopper (e.g., McIlwain).
24. 1.5 mL-Microfuge tubes.

3. Methods

3.1. Brain Organotypic Slices

1. Remove the brain using lateral skull cuts starting at the foramen magnum and ending at the olfactory lobes (see Note 1). Gently lift the skull from the rear exposing the brain. Make cuts between the olfactory lobes and the frontal cortex, and rostral to the cerebellum. Gently lift the rostral part of the brain and cut the optic nerves. Finally remove the brain and place in a tissue culture dish filled with ice-cold HEPES-buffered saline.
2. Using a dissecting microscope, peel the pia using fine forceps, carefully remove blood vessels and meninges around the hypothalamus, and gently straighten out the residual optic nerves. Cut away all cortex on both sides lateral to the hypothalamus (from a ventral view) and make a clean horizontal cut 2–3 mm above the third ventricle (coronal view) to form a block of tissue.
3. Place the tissue block onto the tissue chopper disc membrane so that the hypothalamus is ventral side down and the optic nerves are abutting the chopper blade (see Note 2).
4. Cut slices 350–400 μm thick for rats and 300 μm for mice, using a blade speed of 40 strokes per min. Avoid excess fluid on chopper disc membrane, as this will cause the tissue to be picked up by the blade.
5. When finished, carefully remove the membrane disc, with the slices, from the tissue chopper.

6. Remove the slices from the disc using ice-cold HEPES-buffered saline to slide them into a new tissue culture dish.
7. Gently separate the slices with a small spatula and forceps (see Note 3).
8. Place filter inserts into an appropriate single or multi-well tissue culture plate (e.g., a 6 multi-well dish).
9. Place tissue slices to be cultured onto filters inserts using two small spatulas touching each other, and allowing the capillary forces between them to transfer the slices. Ensure the tissue rests flat on the filters.
10. Using a Pasteur pipette, remove as much HEPES-buffered saline as possible from the membrane without touching the slices (see Note 4).
11. Place 1 mL of culture media under each insert; the medium should wet the insert without causing the slices to float. Remove any bubbles from beneath the insert.
12. Maintain slices in a 5% CO_2/37°C incubator, changing the culture media every 2–4 days. The slices can survive for 1–2 months, but do become thinner over such long time periods. For studying transfected cells deep in the slices we transfect the organotypic slices 4–6 days after being prepared (Fig. 1).

3.2. Preparation of DNA-Coated Microprojectiles

For investigating cortical and/or hippocampal neurons, we recommend 1.0 μm gold particles. We observed transfection efficiency was significantly lower using 1.6 μm particles (which caused significant tissue damage) or 0.6 μm particles (which resulted in agglomeration). We have also had reasonable success with 40 nm particles.

1. Add 10 mg gold particles and 50 μL spermidine stock solution to a 1.5 mL-microfuge tube.
2. Vortex for 5 s.
3. Add 10 μg of plasmid DNA (1 mg/mL) to the microfuge tube and vortex for 5 s (see Note 5).
4. Add 50 μL of $CaCl_2$ stock solution dropwise.
5. Incubate the gold and plasmid DNA at room temperature for 10 min, flicking the tube periodically to mix the contents.
6. Spin the tube for 5 s at 1,000 rpm to pellet the gold.
7. Remove the supernatant and store a sample (see Note 6).
8. Wash the gold particles three times with 1 mL of 100% EtOH.
9. Resuspend the gold pellet in 200 μL PVP working solution from a 3.5 mL aliquot, and then transfer the suspension to a 5 mL tube. Repeat this transfer with 200 μL aliquots of PVP until all the gold particles have been transferred, and then add the remaining PVP solution.

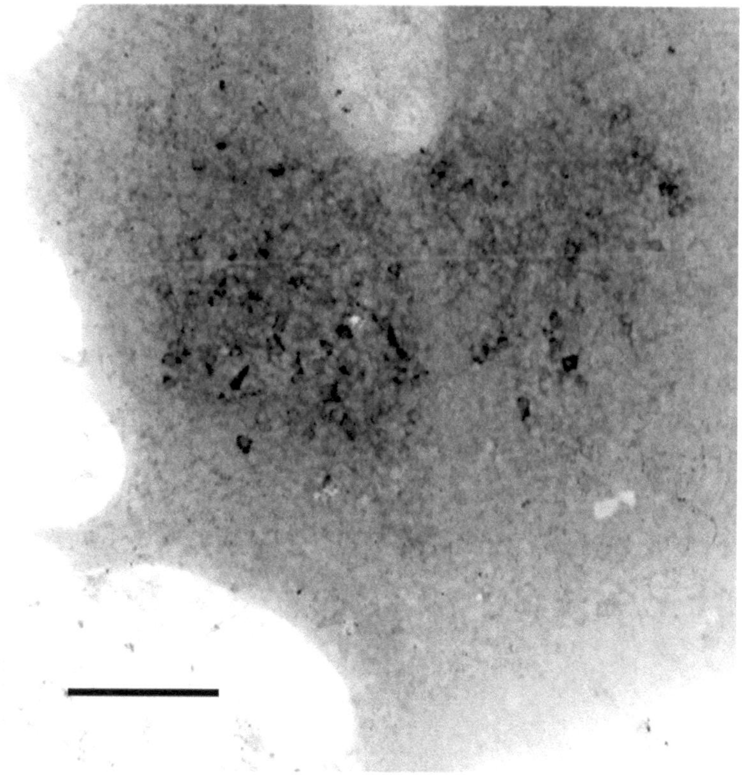

Fig. 1. A typical example of a 300 μm thick organotypic mouse brain slice cut using the McIlwain tissue chopper and viewed using light microscopy with a ×40 objective lens. Scale bar = 100 μm.

10. Insert 75 cm Tefzel tubing into the tubing prep station and purge with nitrogen at 0.35 L per minute (LPM) for 30 min.
11. Attach a plastic 10 mL syringe to one end of the Tefzel tubing.
12. Pulse vortex the gold slurry prepared above.
13. Using the syringe, insert 3.5 mL of the DNA-PVP gold slurry into the open end of the Tefzel tubing. Fill the tubing steadily until the gold slurry is approximately 1 cm away from the syringe.
14. Keeping the syringe attached and the tubing horizontal, insert the loaded tubing into the tubing prep station until the end is through the O-ring.
15. Allow the gold to settle for 1–2 min, keeping the syringe attached.
16. Slowly and steadily suck out the solution using the 10 mL syringe, leaving the settled gold undisturbed. Detach the syringe.

17. Immediately turn the tubing 180° using the switch on the prep station, allowing the gold to coat the tubing, then switch off and leave it for 5 s.
18. Rotate the tubing for a further 30 s to allow an even distribution of gold around the Tefzel tube.
19. Flow nitrogen (0.35 LPM) over the gold for 5 min.
20. Stop rotation. Remove the tubing.
21. Cut the tubing into 1 cm cartridges using a tubing cutter. Remove any pieces of tubing that are not uniformly coated with gold.
22. Store the gold-DNA cartridges ("bullets") in a dry environment. These cartridges (Fig. 2) can be stored for up to 6 months at 4°C.

3.3. Biolistic Transfection of Slices with DNA-Coated Gold microprojectiles

The following steps should take place within a sterile laminar flow hood.

1. Insert a 9 V battery and an empty cartridge holder into the Helios Gene Gun.
2. Attach the Gene Gun to the helium tank with the helium hose. Fire 2–3 shots at 80 psi to pressurize the helium hose and the reservoirs in the gun.

Fig. 2. A bullet. A piece of Tefzel tubing showing the DNA-coated gold particles. These are evenly distributed through the tubing. Should large clusters or bare areas be observed, the bullet should be discarded. Insert is of a higher magnification of the same image showing the individual gold particles. Scale bar = 1 μm.

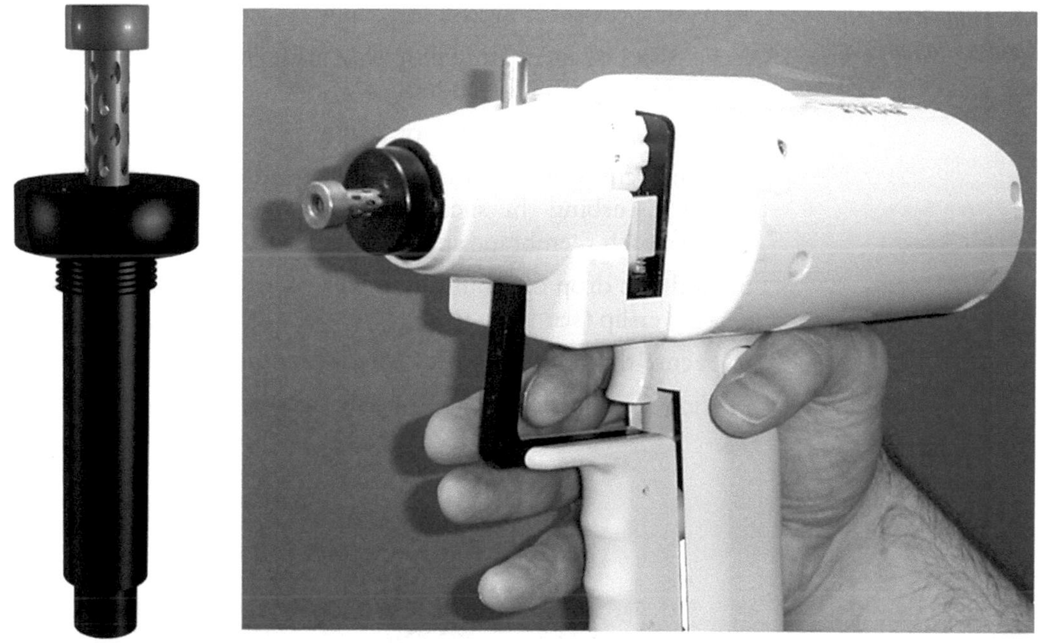

Fig. 3. The Bio-Rad gene gun and the modified gene gun barrel that can be easily attached to the gene gun.

3. Load the cartridges containing the DNA-coated gold into the cartridge holders and load the holder into the Gene Gun.
4. Attach the modified gene gun barrel to the Gene Gun (Fig. 3; see Note 7).
5. Set the helium gas pressure regulator to 50 psi (see Note 8).
6. Using sterile forceps, place a filter insert containing slice cultures onto a sterile plastic dish.
7. Immediately place the gun above the slice perpendicular to the cells with the end of the barrel level with the top of the filter insert (see Note 8).
8. Discharge the gun.
9. Replace the insert in its original media.
10. Repeat steps 6–9 until all slices are transfected. Then return slices to the incubator.
11. Close the helium tank, release the pressure from the Gene Gun, and detach the Gene Gun from the helium tank.
12. Remove the cartridge holder and discard the cartridges.
13. Check for successful tissue penetration by visualizing the slices using an inverted microscope. The gold particles should be evenly dispersed; there should be no dense areas of gold particles seen on the slice.
14. Incubate slices for desired time; we suggest 2–5 days.

3.4. Fixing and Visualization of Slices

1. Wash slices twice in PBS for 2 min.
2. Fix slices by incubating in freshly made, ice-cold 4% paraformaldehyde in PBS for 20 min.
3. Wash slices twice in PBS for 2 min.
4. Gently cut round the membrane supports using a scalpel, without disturbing the slices mounted on them. Use forceps to place each membrane support/slice on a microscope slide.
5. Add one drop of mounting media on top of the slice and add a coverslip (see Note 9).
6. Secure the coverslip with a thin layer of nail varnish.
7. View the slices on an appropriate microscope (Fig. 4).

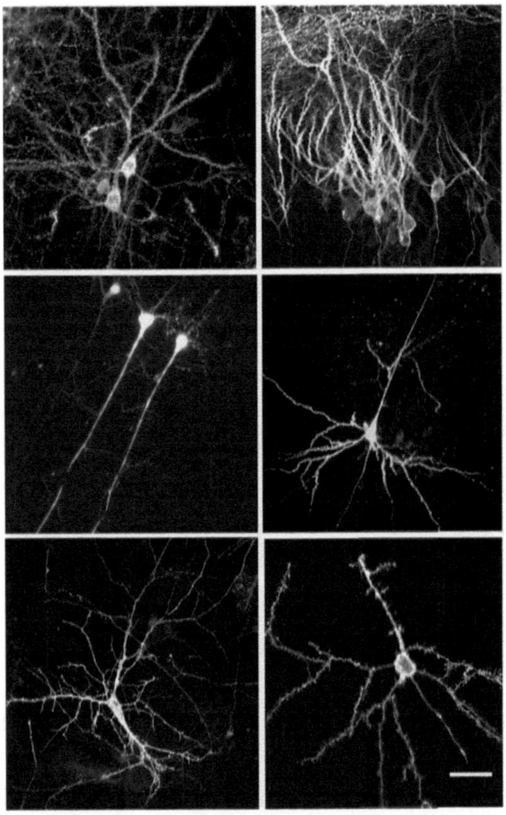

Fig. 4. Images of hippocampal organotypic slices biolistically transfected using the modified gene gun. Left panels: Confocal images of slices transfected with plasmid-DNA encoding yellow fluorescent protein using 1 μm gold particles, and fixed after incubation for 5 days. Right panels: Confocal images of slices transfected with DNA encoding yellow fluorescent protein using 40 nm gold particles, and fixed after incubation for 5 days. Scale bar = 10 μm.

4. Notes

1. Wear gloves, wipe hands with 70% EtOH often, and wear a facemask. Keep all solutions ice-cold as the tissue is more solid and easier to handle, and the slices are healthier.
2. For brain slices we use a McIlwain tissue chopper, which is especially advantageous for slicing small or irregular shaped specimens (e.g., as one might obtain from biopsies). The brain is placed on the circular white disc on the stainless steel table and is transversed automatically from left to right at an adjustable speed. At the same time, the blade is raised and dropped at speeds varying from zero to over 20 strokes per min. Slices are collected and are placed into membrane inserts.
3. It is critical for the survival of the tissue to be extremely gentle in transferring and manipulating the slices.
4. Too much fluid on the filter will prevent tissue from adhering to the filter. If this happens, remove the excess fluid or replate.
5. Prepare plasmid DNA using a Qiagen Plasmid Midi Kit (or equivalent) according to the manufacturer's instructions.
6. To determine the efficiency of DNA deposition, a sample of the supernatant (from step 8) can be run on an agarose gel and its concentration estimated. We usually observe close to 100% efficiency.
7. For a description of the modified gene gun barrel see www.genegunbarrels.com. This barrel allows deeper transfection of tissue slices than the original barrel (see ref. (6)). The latter has a spacer attached, which is designed to spread the gold/DNA particles superficially over a wide area. The modified barrel no longer has a spacer and the cone-shaped barrel is replaced with an external barrel with a reduced diameter.
8. A number of factors affect the success of biolistic transfection in a given tissue. These parameters must be optimized for each cell and tissue type, as transfection rates can vary according to the amount of gold used in each blast, the efficiency of the expression construct, and the exact pressure used. The amount of DNA required for biolistics is also important as both too high or too low concentrations can decrease efficiency. We suggest 1 μg DNA per shot is a good starting concentration.
9. We use Vectashield mounting medium containing 4′,6-diamidino-2-phenylindole (DAPI) for easy visualization of cells.

Acknowledgments

The work was supported by the Medical Research Council and the Wellcome Trust. S.C.R.L. holds a Wellcome Trust Senior Research Fellowship in Basic Biomedical Science.

References

1. Felgner PL et al (1987) Lipofection: a highly efficient, lipid-mediated DNA-transfection procedure. Proc Natl Acad Sci USA 84:7413–7417
2. Monkkonen J, Urtti A (1998) Lipid fusion in oligonucleotide and gene delivery with cationic lipids. Adv Drug Deliv Rev 34:37–49
3. Ross PC, Hui SW (1999) Lipoplex size is a major determinant of in vitro lipofection efficiency. Gene Ther 6:651–659
4. Witkowski PT et al (2009) Gene gun-supported DNA immunisation of chicken for straightforward production of poxvirus-specific IgY antibodies. J Immunol Methods 341: 146–153
5. O'Brien JA, Lummis SCR (2006) Biolistic transfection of neuronal cultures using a hand-held gene gun. Nat Protoc 1:977–981
6. O'Brien JA et al (2001) Modifications to the hand-held Gene Gun: improvements for in vitro biolistic transfection of organotypic neuronal tissue. J Neurosci Methods 112:57–64
7. Sato H et al (2000) In vivo gene gun-mediated DNA delivery into rodent brain tissue. Biochem Biophys Res Commun 270:163–170
8. Klein TM et al (1987) High-velocity microprojectiles for delivering nucleic acids into living cells. Nature 327:70–73
9. Johnston SA (1990) Biolistic transformation: microbes to mice. Nature 346:776–770
10. Sanford JC, Smith FD, Russell JA (1993) Optimizing the biolistic process for different biological applications. Methods Enzymol 217:483–509
11. Cui Z, Mumper RJ (2003) The effect of co-administration of adjuvants with a nanoparticle-based genetic vaccine delivery system on the resulting immune responses. Eur J Pharm Biopharm 55:11–18
12. Cheng L, Ziegelhoffer PR, Yang NS (1993) In vivo promoter activity and transgene expression in mammalian somatic tissues evaluated by using particle bombardment. Proc Natl Acad Sci USA 90:4455–4459
13. O'Brien JA, Lummis SCR (2011) Nano-biolistics: a method of biolistic transfection of cells and tissues using a gene gun with novel nanometer-sized projectiles. BMC Biotechnol 11:66–76
14. Lo DC (2001) Neuronal transfection using particle-mediated gene transfer. Curr Protoc Neurosci. doi:10.1002/0471142301 UNIT3.15
15. Smith FD, Harpending PR, Sanford JC (1992) Biolistic transformation of prokaryotes: factors that affect biolistic transformation of very small cells. J Gen Microbiol 138:239–248
16. Barry MA, Singh RAK, Andersson HA (2003) Gene Gun technologies: applications for gene therapy and genetic immunization. In: Smyth-Templeton N (ed) Gene therapy: therapeutic mechanisms and strategies. Marcel Dekker Inc., New York, pp 263–285
17. Donnelly JJ, Wahren B, Liu MA (2005) DNA vaccines: progress and challenges. J Immunol 175:633–639

Part IV

Biolistic DNA Delivery in Mammalian Cells In Vivo

Chapter 14

Biolistic DNA Delivery to Mice with the Low Pressure Gene Gun

Meng-Chi Yen and Ming-Derg Lai

Abstract

Biolistic DNA delivery is an approach to deliver plasmid to culture cells, plants, or animals. Plasmid DNA is usually transferred through bombardment of DNA-coated particles by highly pressurized gas in various kinds of delivery vehicles. The low pressure gene gun can deliver plasmid at lower pressure. Here, we describe methods of biolistic DNA delivery to mice using the low pressure gene gun.

Key words: Low pressure gene gun, DNA vaccine, Biolistic DNA delivery, Mouse, Naked DNA

1. Introduction

Biolistic DNA delivery is a physical method of gene transfer into culture cells, plants, and animals. Plasmid DNA is coated on gold particles and then is transferred to target cells or tissues by gas discharge (1). The technique can be used in vivo and in vitro. Therefore, it is widely applied in various fields, including genetic vaccination (2) and agricultural technology (3). In most of biolistic vehicles, plasmid DNA is transferred by a highly pressurized helium pulse (usually at 400 psi). In contrast, the low pressure gene gun is applicable to deliver plasmid DNA to target cells at relative low pressure (50 psi). When the helium flow travels from the inside to the outside of the spray nozzle (Fig. 1a), the gas flow accelerates to supersonic speed to deliver plasmid DNA. This vehicle has been demonstrated to transfer plasmids to cells in culture (4), to animals (5, 6), and to plant cells (7).

The other characteristic of this gene gun is to deliver gold particle-coated plasmids as well as non-particle-coated plasmids

Stephan Sudowe and Angelika B. Reske-Kunz (eds.), *Biolistic DNA Delivery: Methods and Protocols*,
Methods in Molecular Biology, vol. 940, DOI 10.1007/978-1-62703-110-3_14, © Springer Science+Business Media, LLC 2013

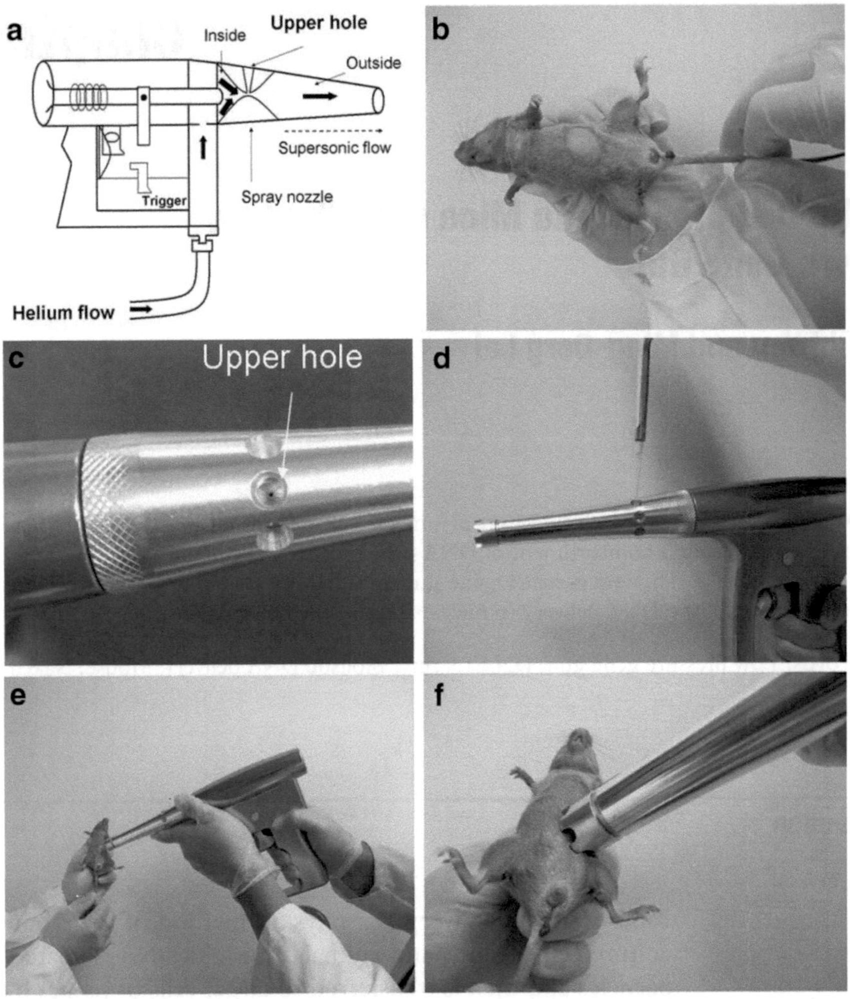

Fig. 1. Low pressure gene gun. (**a**) Schematic drawing depicting the features of the gene gun. Plasmid is loaded into the upper hole. When pressing the trigger, the helium flow travels from the inside of the spray nozzle to the outside and the pressure difference results in supersonic flow to carry particles-coated plasmids or the plasmid solution to target cells. (**b**) Shaved abdominal skin of mice. (**c**) Photograph showing the upper hole of the low pressure gene gun. (**d**) Loading of plasmid DNA into the upper hole of the low pressure gene gun. (**e**) Biolistic DNA delivery onto a mouse. One person holds the mouse and another person holds the gene gun which was loaded with plasmid DNA. (**f**) Gene gun bombardment on the shaved abdomen.

through the supersonic flow. For delivery of gold particle-coated plasmid DNA, the supersonic flow can carry the particles to penetrate through the membranes of cells. In addition, plasmids without coating on metal particles can be also delivered to target cells. The plasmid solution is sprayed out through the sprayer by pressurized gas flow. The efficacy of DNA delivery in vivo is illustrated in Fig. 2. The naked plasmid DNA delivery has been demonstrated to induce antitumor immunity in mice (8–10). Furthermore, a previous study has demonstrated that different delivery routes

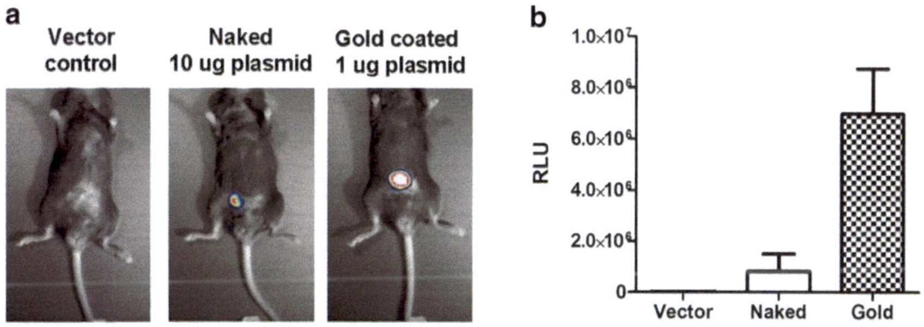

Fig. 2. Evaluation of efficiency of plasmid DNA delivery with the low pressure gene gun. (**a**) Empty plasmid vector (pGL3-Basic), 10 μg luciferase plasmid without particles and 1 μg luciferase plasmid coated on gold particles, respectively, was bombarded to C57BL/6 mice. After 48 h, luciferase activity was detected by in vivo image system. (**b**) Quantification of luciferase activity. 48 h after biolistic transfection, the skin from the plasmid-bombarded mice was homogenized. The luciferase activity of the skin lysate was determined.

result in different immune polarization (8). Th1 responses are induced by particle-free plasmid DNA delivery, whereas Th2 responses are induced by biolistic immunization using gold particle-coated plasmid DNA. It implies that the immunological polarization can be manipulated by appropriate plasmid delivery routes. Here, we describe the methods of biolistic DNA delivery with the low pressure gene gun.

2. Materials

2.1. Chemicals and Other Components

All solutions and plasmid DNA were prepared by using ultrapure water.

1. 1.0 μm gold particles (Bio-Rad, Hercules, CA).
2. 0.05 M spermidine (Sigma-Aldrich): dissolve 0.0726 g spermidine in 10 mL sterile water and then sterilize with a 0.22 μm filter. Store at 4°C.
3. 2.5 M calcium chloride ($CaCl_2$) (Sigma-Aldrich): dissolve 2.77 g $CaCl_2$ in 10 mL sterile water. Store at 4°C.
4. Absolute ethanol (Merck).
5. Vortex mixer.
6. 1.5 mL microcentrifuge tube.
7. Ultrasonic cleaner.
8. Low pressure gene gun (BioWare Technologies Co. Ltd, Taiwan).
9. Helium gas (99.995% pure).

2.2. Preparation of Plasmid DNA

Prepare plasmids by using a QIAGEN endofree mega kit (Qiagen, Chatsworth, CA, USA). Adjust the stock concentration of plasmid DNA to 1 mg/mL.

3. Methods

3.1. Delivery of Plasmid DNA Precipitated onto Gold Particles

3.1.1. Coating Gold Particles with Plasmid DNA

1. Estimate the amount of plasmid in each experimental group (see Note 1).
2. Weigh appropriate gold particles and add into a 1.5 mL microcentrifuge tube.
3. Add appropriate amount of plasmid DNA into the 1.5 mL microcentrifuge tube.
4. Add sterile water to 50 μL in the 1.5 mL microcentrifuge tube (see Note 2).
5. Mix the gold particles and plasmid solution by pipeting.
6. Move the tube to an ultrasonic cleaner for 3–5 s.
7. Add 75 μL of 0.05 M spermidine into the solution with continuous vortexing.
8. Add 75 μL of 2.5 M $CaCl_2$ into the solution with continuous vortexing.
9. Move the tube to the ultrasonic cleaner for 3–5 s (see Note 3).
10. Place the tube on ice for 10 min.
11. Collect the gold particle after centrifugation (10,000 × g) for 1 min.
12. Remove supernatant.
13. Wash gold particles with 500 μL of absolute ethanol by vortexing 3–5 s.
14. Remove ethanol after centrifugation (10,000 × g) for 1 min.
15. Repeat steps 13 and 14, wash gold particles twice (see Note 4).
16. Resuspend the gold particles in appropriate volume of absolute ethanol in the tube (see Note 5).
17. Pipet the ethanol and gold particles several times and aliquot 100 μL of the absolute ethanol/gold particles mixture to new tubes immediately. Each tube is placed on ice for bombardment.

3.1.2. Delivery of Gold-Coated Plasmid with the Low Pressure Gene Gun

1. Shave abdominal region of mouse (see Fig. 1b).
2. Set the helium pressure at 50 psi.
3. Hold the gene gun and then load 20 μL of the mixture of absolute ethanol/gold particles to the upper pore of gene gun after pipetting (see Note 6).

4. Prepare mouse for bombardment (see Note 7).
5. Hold gene gun directly against the shaved abdomen of mouse and then press the trigger of gene gun (Fig. 1f).
6. Repeat steps 4 and 5 for five times (see Note 8).

3.2. Delivery of Naked Plasmid DNA Without Particles

1. Shave abdominal region of mice.
2. Set the helium pressure at 50 psi
3. Dilute stock plasmid DNA with sterile water to the appropriate concentration (see Note 9).
4. Hold the gene gun and load 20 μL of diluted plasmid solution to the upper pore of the gene gun (see Note 6).
5. Prepare mouse for bombardment (see Note 7).
6. Hold the gene gun directly against the shaved abdomen of the mouse and then press the trigger of the gene gun (see Note 10).

4. Notes

1. Each mouse is bombarded with 1 μg plasmid. In addition, the plasmid DNA and gold is at the ratio of 1 μg plasmid per mg gold particle.
2. For example, five mice are bombarded with plasmid DNA. 5 mg of gold particles is added into the tube and then 5 μL of the plasmid DNA solution (1 mg/mL) is added into the same tube. 45 μL of sterile water is added to a total volume of 50 μL.
3. The continuous vortexing is necessary when spermidine and $CaCl_2$ are added to the tube drop by drop. The sonication can avoid aggregation of gold particles.
4. To check the coating efficiency, collect the absolute ethanol-washed gold particles by centrifugation and remove absolute ethanol. Add 20–30 μL of sterile water to the gold particles to dissolve DNA and load the solution to agarose gel for electrophoresis. The visible band can be observed when a successful coating was done.
5. 1 mg of gold particles is resuspended with 100 μL of absolute ethanol. For example, add 500 μL of absolute ethanol to 5 mg of gold particles in the tube.
6. Two persons are required for gene gun bombardment. One person holds gene gun and the other one holds the mouse. In addition, the mixture of absolute ethanol/gold particles must be

loaded on the upper hole of spray nozzle and then delivered by supersonic flow. Thus, hold gene gun horizontally (Fig. 1c, d).

7. The second person holds the mouse. The abdominal region of mouse is turned towards the gene gun (Fig. 1e).
8. 100 μL mixture of absolute ethanol/gold particles is bombarded to five different regions of shaved abdominal skin. Besides, gold particles should be delivered equally in five shots. The gold particles must be mixed well to prevent the gold particles from precipitation in the bottom of the 1.5 mL tube.
9. For example, for delivery of 5 μg of plasmid DNA to a mouse the working plasmid DNA solution is diluted to 0.25 μg/μL with sterile water and then 20 μL of working plasmid solution is loaded into the gene gun and bombarded to a mouse.
10. For delivery of plasmid DNA without particles, the mouse is bombarded only once on the shaved abdominal skin.

Acknowledgments

This work is supported by Grant NSC99-2323-B006-004 from National Science Council, Taiwan, Republic of China.

References

1. Williams RS et al (1991) Introduction of foreign genes into tissues of living mice by DNA-coated microprojectiles. Proc Natl Acad Sci USA 88:2726–2730
2. Larregina AT et al (2001) Direct transfection and activation of human cutaneous dendritic cells. Gene Ther 8:608–617
3. Liu CW et al (2007) Stable chloroplast transformation in cabbage (Brassica oleracea L. var. capitata L.) by particle bombardment. Plant Cell Rep 26:1733–1744
4. Wang BW et al (2007) Mechanical stretch enhances the expression of resistin gene in cultured cardiomyocytes via tumor necrosis factor-alpha. Am J Physiol Heart Circ Physiol 293:2305–2312
5. Tu CF et al (2007) Autologous neu DNA vaccine can be as effective as xenogenic neu DNA vaccine by altering administration route. Vaccine 25:719–728
6. Hsieh CY et al (2007) IL-6-encoding tumor antigen generates potent cancer immunotherapy through antigen processing and anti-apoptotic pathways. Mol Ther 15:1890–1897
7. Kao CY, Huang SH, Lin CM (2008) A low-pressure gene gun for genetic transformation of maize. Plant Biotechnol Rep 2:267–270
8. Lin CC et al (2008) Delivery of noncarrier naked DNA vaccine into the skin by supersonic flow induces a polarized T helper type 1 immune response to cancer. J Gene Med 10:679–689
9. Yen MC et al (2009) A novel cancer therapy by skin delivery of indoleamine 2,3-dioxygenase siRNA. Clin Cancer Res 15:641–649
10. Huang CH et al (2010) Promoting effect of *Antrodia camphorata* as an immunomodulating adjuvant on the antitumor efficacy of HER-2/neu DNA vaccine. Cancer Immunol Immunother 59:1259–1272

Chapter 15

Chemokine Overexpression in the Skin by Biolistic DNA Delivery

Ahmad Jalili

Abstract

Chemokines are a family of small, secreted proteins that function in leukocyte and tumor cell trafficking and recruiting. CC chemokine ligand 21 (CCL21)/secondary lymphoid chemokine (SLC) belongs to the inflammatory subgroup of chemokines and is expressed by stromal cells in the T-cell-rich zones of peripheral lymph nodes, afferent lymphatic endothelial cells and high endothelial venules. CCR7 (both in human and mouse) and CXCR3 (in mouse) are expressed by the most potent antigen-presenting cells (dendritic cells), naïve/central memory, and effector T cells, respectively. Inflammation in the skin can induce expression of CCL21 which is subsequently drained into loco-regional lymph nodes responsible for co-localization of antigen-presenting cells and T lymphocytes, a prerequisite for induction of adaptive immune responses. Here, skin functions as a remote control for induction of targeted cell migration in vivo. This chapter describes Gene Gun administration of plasmid DNA expressing functionally active CCL21 (as an example of a chemokine) into the skin in mice and subsequent functional evaluation of the transgene expression in vivo.

Key words: Skin, Chemokine, CCL21, Plasmid DNA, Gene Gun, Immunohistochemistry, ELISA

1. Introduction

Chemokines are a family of small proteins secreted by cells (1). They have the ability to induce directed chemotaxis in nearby responsive cells (1) and share structural characteristics such as small size (they are all approximately 8–10 kDa in size), and the presence of four cysteine residues in conserved locations (2). They can be classified as homeostatic or inflammatory. Homeostatic chemokines are constitutively expressed in a certain tissue or organ, suggesting a specific function involving cell migration. The inflammatory chemokines in contrast are strongly upregulated by inflammatory

Stephan Sudowe and Angelika B. Reske-Kunz (eds.), *Biolistic DNA Delivery: Methods and Protocols*,
Methods in Molecular Biology, vol. 940, DOI 10.1007/978-1-62703-110-3_15, © Springer Science+Business Media, LLC 2013

or immune stimuli in various cell types such as macrophages, fibroblasts, T cells, etc. The latter are likely to participate in the development of immune or inflammatory reactions and are involved in controlling the migration of cells during normal processes of tissue maintenance or development (1).

CCL21 [secondary lymphoid organ chemokine (SLC), 6Ckine] is a ligand for CCR7 and CXCR3 (in mouse but not in human) and guides the interactions between CCR7+ T cells (naïve and central memory T cells) and antigen-presenting cells (APCs, especially dendritic cells) and CXCR3+ effector T cells needed for T cell education and priming. This is necessary for both triggering adaptive immunity and maintaining peripheral tolerance (3). CCL21 belongs to the family of pro-inflammatory chemokines (3) and is expressed by stromal cells in the T-cell-rich zones of peripheral lymph nodes, afferent lymphatic endothelial cells and high endothelial venules. It has been demonstrated that the expression of pro-inflammatory chemokine CCL21 can be induced by inflammation in the skin. Secreted CCL21 can be subsequently drained through the afferent lymphatic vessels located in the dermis into loco-regional lymph nodes (4). This proves that the skin can function as a remote control for the induction of targeted cell migration into secondary lymphoid organs (SLOs). After injection of recombinant CCL21 into the skin in mice, CCL21 is drained into loco-regional lymph nodes (4). Here, CCL21 is expressed on the surface of high endothelial venules (HEVs) where trans-endothelial migration of cells expressing respective chemokine receptors occurs. Increased expression of CCL21 in the skin and its accumulation in the draining lymphoid organs is capable of bringing distinct cellular populations needed for induction of adaptive immune responses in a particular immunological compartment such as lymph node. Injection of recombinant chemokines into the skin is not feasible as there is a rapid protein turnover in the skin and proteins are very costly.

Plasmid DNAs (pDNAs) expressing transgenic chemokines are suitable alternatives. They are easy to manufacture and cost-effective. However, in vivo injection of naked pDNAs in the skin in mice results in low expression of the transgene. Here, Gene Gun can be an attractive strategy. Successful delivery of pDNAs expressing antigens such as β-galactosidase into the skin by using Gene Gun (particle bombardment) has been previously reported (5, 6).

Among the advantages of particle bombardment for in vivo gene transfer is that it is an easy to use, rapid versatile gene delivery system, useful for both transient and stable expression, which requires only small amounts of DNA and may obtain high levels of the transgene expression.

We have recently shown that biolistic-mediated delivery of pDNA expressing an inflammatory chemokine CCL21 into the skin of mice results in transient expression of the transgene in the epidermis, and subsequent drainage of transgenic CCL21 into draining lymph nodes where it attracts target cells into this compartment (7). Such a strategy could have a potential application for design and modification of immunotherapies. In this chapter this methodology is explained in detail including dermal administration of pDNA into mouse skin using the helium-driven biolistic Gene Gun system and evaluation of the transgene expression.

2. Materials

Prepare all solutions using ultrapure water (ddH2O) and analytical grade reagents. Prepare and store all reagents at room temperature unless indicated otherwise. The Helios Gene gun is available from Bio-Rad Laboratories (Bio-Rad, Hercules, CA) including helium regulator, tubing prep station, and tubing cutter.

2.1. Preparation of Bullets and Gene Gun Application of the Bullets onto Murine Skin

2.1.1. Plasmids

In our experiments we have used pVR1012-mCCL21-IRES-EGFP (allowing separate translation of murine CCL21 and EGFP from a single bicistronic mRNA), pVR1012-IRES-EGFP and pVR1012 (parental vector provided by B. Zaugg, Vical, San Diego, CA) pDNA vectors.

2.1.2. Helios Gene Gun Kit

1. Helios Gene Gun.
2. 1× Battery, 9 V.
3. 5× Cartridge holders.
4. 5× Barrel O-rings.
5. 5× Barrel liners (four plus one installed in Gene Gun).
6. Cartridge extractor tool.
7. Helium hose assembly.
8. Helium regulator.
9. Tubing Prep Station, including Tubing Prep Unit (base, tubing support cylinder, and power cord), Nitrogen hose [12 ft, (~4 m), Nalgene tubing 8000-0030, 3/16″ ID, 5/16 ″ OD], 3/16″ barb-to-male Luer-Lok fitting, 10 cc syringe sleeve, 5× O-rings, 2 1/8″ barb-to-male Luer fittings, 5/64″ Allen wrench, Syringe Kit (5× 10 cc syringes, 5× 1/8″ barb-to-female Luer fittings, 1× syringe adaptor tubing [silicone, 5 ft, (~2.6 m), 0.104″ ID, 0.192″ OD]).
10. Tubing cutter and ten razor blades.
11. The optimization kit for the Helios Gene Gun system which includes one vial each of the three sizes of gold microcarriers (0.6, 1, and 1.6 μm, each 0.25 g) along with a cartridge kit

that includes 50 ft Tefzel tubing (ca. 15 m), 5 cartridge collection/storage vials, five desiccant pellets (store tightly sealed), and 0.5 g polyvinylpyrrolidone (PVP). These are used in the sample tubing coating procedure and are sufficient for preparing nearly 1,000 Gene Gun samples.

2.1.3. Other Equipments

1. Ultrasonic cleaner (e.g., Fisher FS3, Branson 1210).
2. Vortex mixer.
3. Analytical balance microfuge.
4. Peristaltic pump capable of pumping 5–8 mL/min (e.g., Bio-Rad Econo Pump).
5. 1.5 mL Microfuge tubes.
6. 20, 200, and 1,000 μL Micropipettors and tips.
7. 5 and 10 mL Pipettes and pipettors.
8. 14 and 50 mL Falcon™ tubes with screw caps.
9. Lab timer.
10. Ear protection.
11. 1 1/8″ Open end or 10″ or 12″ (ca. 25 cm) adjustable wrench.
12. Helium tank (grade 4.5 or higher).
13. Nitrogen tank (grade 4.8 or higher).
14. Nitrogen regulator.
15. Scissors.
16. Marking pen.
17. Sonication device.
18. Oster clippers with a # 40 surgical blade.

2.1.4. Reagents

1. 100% Ethanol (see Note 1).
2. PVP, MW = 360,000: prepare a stock solution of 20 mg/mL PVP in 100% ethanol. This stock solution can be diluted with 100% ethanol to prepare PVP solutions at the desired concentration. The stock solution should be sealed with parafilm. PVP can be stored at −20°C.
3. Spermidine: prepare a 0.05 M solution in ddH2O. Spermidine can be stored at −20°C.
4. Calcium chloride (CaCl2), prepare a 1 M solution in ddH2O and store at room temperature as stock.
5. KetalarTM (ketamine)/RompunTM (xylazine) anesthesia [1 mL KetalarTM; Parke-Davis, and 1 mL 2% RompunTM TS; Bayer, in 10 mL physiologic saline; 7 μL/gr mouse (e.g., 210 μL for 30 gr weight)].

2.2. CCL21 Immunohistochemistry and ELISA

1. Scalpel and tweezer.
2. Staining jar with cover.
3. PAP pen.
4. Aqua-Mount™.
5. Optimal Cutting Temperature compound (OCT) (e.g., : Tissue-Tek® O.C.T. Compound).
6. Cryo-cassette and CryostatTM.
7. Liquid nitrogen.
8. Glass slides.
9. Nair hair removal.
10. 70% Ethanol.
11. 4% Paraformaldehyde (PAF): Dissolve 4 g of PAF in 90 mL of 1×PBS. Heat gently to 58–60°C under a hood. Do not heat over 60°C (PAF dissociates > 60°C). Add 10N NaOH to clear the solution (pH 10 dissolves the PAF), usually 5–10 drops. Remove from heat and measure pH. Carefully bring pH to pH 7.2–7.4. Bring to final volume (100 mL) with 1×PBS (for a final concentration of 4 g in 100 mL of 1×PBS). Filter sterilize through a 0.22 μm filter. Keep in 4°C refrigerator (no more than 1 week).
12. Phosphate buffered saline (PBS) without Ca/Mg.
13. 0.5% Triton X-100 in PBS.
14. Levamisole (e.g., Sigma-Aldrich).
15. Blocking solution: PBS/10% goat serum/0.1% natrium azide.
16. Polyclonal goat anti-mouse CCL21 antibody (we use AF457 clone from RnD Systems, the isotype control is goat IgG).
17. Alkaline phosphatase-conjugated rabbit anti-goat secondary antibody (we use R21458 clone from Molecular Probes).
18. Fuchsin Substrate-Chromogen (Dako).
19. Mayer's Hematoxylin: Dissolve 50 g aluminum potassium sulfate (alum) in 1,000 mL ddH20. When alum is completely dissolved, add 1 g hematoxylin. When hematoxylin is completely dissolved, add 0.2 g sodium iodate and 20 mL acetic acid. Bring solution to boil and cool, and filter.
20. RPMI-1640 medium with 25 mM HEPES.
21. RPMI-1640 medium with 25 mM HEPES containing 10% FCS, 100 U/mL penicillin, 100 μg/mL streptomycin, 50 μM 2-mercaptoethanol and 2 mM L-glutamine.
22. Collagenase/dispase (Roche Diagnostics).
23. BD Falcon™—Cell Strainer, 70 μm.
24. 12-Well cell culture plate.
25. Flat bottom 96-well plates.

26. ELISA blocking buffer: 1% BSA, 5% Sucrose in PBS with 0.05% NaN3.
27. Reagent diluent: 1% BSA in PBS.
28. Wash buffer: 0.05% Tween-20 in PBS.
29. HRP-substrate: 1:1 mixture of ready-to-use TMB and H2O2 solutions from R&D Systems, DY999.
30. 2N H2SO4.
31. ELISA reader with 450 and 570 nm wavelengths.
32. Mouse CCL21/6Ckine DuoSet kit (DY457, R&D Systems, other ELISA kits for respective chemokines can be also used on below principle):
    (a) Capture Antibody: 720 μg/mL of polyclonal goat anti-mouse CCL21 IgG antibody (AF457, R&D Systems) reconstituted with PBS. After reconstitution, store at 2–8°C for up to 60 days or aliquot and store at −20°C to −70°C for up to 6 months. Dilute to a working concentration of 4.0 μg/mL in PBS without carrier protein.
    (b) Detection Antibody: 9 μg/mL of biotinylated goat anti-mouse 6Ckine when reconstituted with 1.0 mL of reagent diluent. After reconstitution, store at 2–8°C for up to 60 days or aliquot and store at −20°C to −70°C in a manual defrost freezer for up to 6 months. Dilute to a working concentration of 50 ng/mL in reagent diluent.
    (c) Standard: 115 ng/mL of recombinant mouse 6Ckine when reconstituted with 0.5 mL of reagent diluent. Allow the standard to sit for a minimum of 15 min with gentle agitation prior to making dilutions. Aliquot and store reconstituted standard at −70°C for up to 2 months. A seven point standard curve using twofold serial dilutions in reagent diluent, and a high standard of 1,000 pg/mL is recommended.
    (d) Streptavidin-HRP: 1.0 mL of streptavidin conjugated to horseradish-peroxidase. Store at 2–8°C for up to 6 months after initial use. Do not freeze. Dilute to the working concentration specified on the vial label using reagent diluent.

3. Methods

Carry out all procedures at room temperature unless otherwise specified.

3.1. Preparation of Gene Gun Bullets

We routinely use 0.5″ = ca. 1.3 cm cartridges containing 1 μg of pDNA coated onto 0.5 mg gold microcarriers. Each preparation results in 50 cartridges (50 μg of pDNA and 25 mg of gold).

Table 1
Microcarriers and DNA Required for Various Microcarrier Loading Quantities (MLQ) and DNA Loading Ratios (DLR) 1. Adapted with permission from Bio-Rad Laboratories, Inc.

Calculated particle delivery conditions			Materials required for selected MLQ's and DLR's			
MLQ	DLR		Gold	DNA	Final volume	Tubing
(mg/shot)	(μg/mg gold)	(μg/shot)	(mg)	(μg)	(ml)[2]	(total in)[3]
0.5	2	1	50	100	6.0	50
0.125	8	1	12.5	100	6.0	50
0.25	4	1	25	100	6.0	50
0.75	1.33	1	75	100	6.0	50
1.0	1	1	100	100	6.0	50
0.5	0.002	0.001	50	0.1	6.0	50
0.5	0.02	0.01	50	1	6.0	50
0.5	0.2	0.1	50	10	6.0	50
0.5	10	5	20	200	2.4	20

[a] For most applications with mammalian cells, in initial experiments, use an MLQ of 0.5 and a DLR of 2

[b] Based on loading 1 mL of the DNA-coated microcarriers suspended in ethanol in 8.5 in. (22 cm) of Gold-Coat tubing

[c] Various lengths of tubing may be prepared. Adjust amounts of gold, DNA volume of ethanol in proportion to the change from the length of tubing listed above for each desired MLQ and DLR. Approximately 25 in. of tubing can be prepared in the Tubing Prep Station at one time; 50 in. of tubing will usually yield 80–90 cartridges

1. Weigh 1 μm gold microcarriers into a 1.5 mL microfuge tube (at a density of 2 mg of pDNA per μg of gold particles. We usually use 25 mg for 50 cartridges) (see Table 1 and Note 2).
2. Add 100 μL of 0.05 M spermidine (stock is dissolved by keeping it at room temperature).
3. Briefly vortex and sonicate the mixture to break up any gold clumps using ultrasonic cleaner (each 5 s).
4. Add the required volume of plasmid to achieve the desired DNA Loading Ratio (Table 1).
5. Mix DNA, spermidine, and gold by vortexing for 5 s.
6. While vortexing, slowly add 100 μL of 1 M calcium chloride dropwise to the mixture.
7. Continue vortexing for 5–10 s.
8. Allow the mixture to precipitate at room temperature for 10 min.

9. Microfuge at 19,500 × *g*, for 15 s to pellet the gold.
10. Remove and discard the supernatant.
11. Resuspend and wash the pellet three times with fresh 100% ethanol, spin briefly and discard the supernatant between each wash.
12. After the final ethanol wash, resuspend the pellet in 200 μL of the ethanol solution containing the appropriate concentration of PVP (0.01 mg/mL is a good starting point) (see Note 3).
13. Transfer the gold-ethanol slurry into a 15 mL tube with a screw cap.
14. Rinse the microfuge tube once with the same ethanol/PVP solution to collect any remaining sample.
15. Add the required volume of ethanol/PVP solution to the centrifuge tube to bring the DNA/microcarrier solution to the desired Microcarrier Loading Ratio (3 mL prepared for 70 cm tube, see Table 1 and Note 4). The pDNA-coated microcarrier suspension can be stored for several weeks at −20°C. Seal with parafilm before freezing.
16. The suspension is now ready for cartridge preparation:
17. Install the 70 cm tube in the Tubing Prep Station.
18. Connect nitrogen hose to the nitrogen regulator and allow the nitrogen flow 0.35 LPM (Liters Per Min.) for 15 min to dry the tube (see Note 5).
19. Afterward, put one end of the tube into the microcarrier solution (delicately vortex and invert before doing this), connect the other end to the syringe and aspirate the solution to the tube (try to avoid bubbles), remove tube from the solution and air suck for about 3–5 cm.
20. Invert the tube horizontally and insert into the Tubing Prep Station, wait 3–5 min, remove ethanol using the syringe with the speed of 0.5–1.0"/s (it should last 30–45 s, see Note 6).
21. Detach the syringe, rotate the tube 180°, and allow the gold to coat the tube for 3–5 s.
22. Turn on the Tubing Prep Station, let it run for 20–30 s, open the valve of the nitrogen tank and let it flow for 3–5 min at the pressure of 0.35–0.4 LPM while rotating.
23. Turn off the motor and nitrogen flow.
24. Prepare the cartridges using the tube cutter as below:
25. Check Gold-Coat tubing whether microcarriers are evenly distributed over the length of the tubing. Ideally, the gold should be spread uniformly over the entire inside surface of the tubing (as long as there are no clumps or bare sections, the tubing can be used for cartridges).

26. Using scissors cut off and discard sparsely and unevenly coated tubing from one of the ends.
27. Use the Tubing Cutter to cut the remaining tubing into 0.5″ (bullets, 0.5″= ca. 1.3 cm each) pieces [cartridges can be stored in a dry place (place them in 50 mL Falcon tubes sealed with parafilm) at 4°C for up to 8 months].

3.2. Gene Gun Application of DNA-Coated Gold Particles onto Murine Skin In Vivo

Coating of gold microcarriers with pDNA, loading into tubes, and preparing cartridges should be performed prior to day of in vivo Gene Gun application. We have used *plt-/plt-* mice on BALB/c background which are deficient in endogenous expression of the CCL21 in secondary lymphoid organs (8).

1. Check helium supply (50 psi in excess of desired delivery pressure). We use 350 psi helium pressure (300 + 50) for in vivo pDNA application. Clean and/or sterilize the Gene Gun, tube holders, and barrel liners and connect the Gene Gun to the helium source.
2. Activate the Gene Gun by turning on the flow of helium to the desired pressure and with an empty cartridge holder in place, make 2–3 "pre-shots" by engaging the safety interlock and firing the trigger. Each cartridge holder has slots for 12 cartridges. The numbers correspond to the firing order. The cartridge holder should be loaded with cartridges beginning with position 1, then clockwise through position 12.
3. Load up to 12 cartridges into the cartridge holder. Invert the cartridge holder and push the cartridges against a flat surface so that they are flush with the numbered side of the cylinder. Insert the loaded cartridge holder into the Helios Gene Gun. When the LED on the back of the Gene Gun indicates that the first cartridge is in firing position the device is ready to deliver DNA.
4. For safe handling animals can be anesthetized if necessary. For narcosis, we use intraperitoneal injections of KetalarTM (ketamine)/RompunTM (xylazine) solution. As animals are anesthetized, use eye-droplets to avoid drying and conjunctivitis and keep them under steady-state temperature (see Note 7).
5. We routinely use abdomen as target area. In this way, one can conveniently hold the mice and shoot.
6. Clip fur as closely as possible over the abdomen using Oster clippers with a # 40 surgical blade and brush or vacuum fur off. If the target site is wet or dirty, clean with 70% ethanol and let it dry.
7. Put on hearing protection.
8. Touch the target area with the spacer so that the spacer is flush and the Gene Gun is perpendicular to the target surface. Activate the safety interlock switch and press the trigger button

to deliver the DNA/microcarriers to the target. Apply three nonoverlapping shots into the abdominal area skin. In BLAB/c mice skin in the area of the applied microcarriers (beads) gets darker reflecting the penetration of gold particles. After each shooting, ratchet to the next cartridge by pulling in and releasing the cylinder advance lever (after approximately 5 s, the Gene Gun is ready to deliver the next cartridge).

9. After all tubes have been discharged, remove the cartridge holder.
10. After bombardment remove cartridge holder from Gene Gun and disconnect the helium hose and Gene Gun.

3.3. Evaluation of Transgene Expression in the Skin by CCL21 Immunohistochemistry

In order to see the exact skin localization of transgene expression in vivo one can use EGFP expressing vectors. Using above-mentioned settings of Gene Gun and helium pressure, the expression of EGFP was restricted to epidermis in our hands using pVR1012-mCCL21-IRES-EGFP vector (as shown in immunofluorescent microscopy in Fig. 1a). The highest expression of murine CCL21 after Gene Gun administration of pDNA-CCL21 is on day 1 with gradual decrease toward day 3 and 7. CCL21 immunohistochemistry is to be performed during the first 3 days.

1. Euthanize mice by cervical dislocation (although not required, the use of sedation or anesthesia prior to euthanasia is encouraged, see Note 8) or overdose of anesthetics.
2. Remove abdominal area skin using a scalpel and a tweezer, cover with OCT compound and snap freeze by immersing the closed cryo-cassette in liquid nitrogen.
3. Prepare cryosections measuring 10 μm in thickness using Cryostat™ machine.
4. Transfer the cryosections onto glass slides.
5. Fix in 4% freshly prepared PAF in PBS for 15 min (use staining jar with cover) and then wash three times with PBS.
6. Permeabilize in 0.5% Triton X-100 in PBS for 5 min (use staining jar with cover) and then wash three times with PBS.
7. Block unspecific binding sites with blocking solution containing 0.24 mg/mL of levamisole (to block endogenous alkaline phosphatase, see Note 9) for 1 h (use staining jar with cover).
8. Circle the sections on the glass slide with PAP pen (see Note 10).
9. Treat sections with either isotype control or goat anti-mouse polyclonal CCL21 antibody at the concentration of 2 μg/mL in blocking solution and incubate overnight at 4°C.
10. Wash three times in PBS and incubate with 1:50 concentration of secondary antibody in blocking solution for 30 min.
11. Wash three times in PBS and develop reactions using Fuchsin Substrate-Chromogen (Dako) according to the manufacturer's protocol.

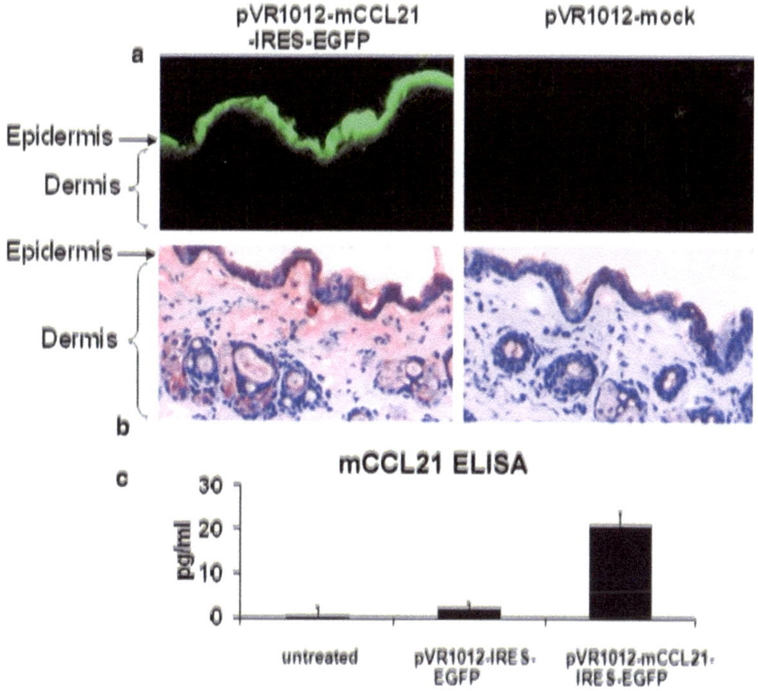

Fig. 1. Epidermal expression of transgenic CCL21 in the skin upon gene gun-mediated administration of pVR1012-mCCL21-IRES-EGFP in vivo and its subsequent drainage into PLNs. (**a**) Expression of EGFP is restricted to the epidermis, and no fluorescence is detected after application of the pVR1012-mock vector. Skin sections were obtained 24 h after pDNA administration. (**b**) Day-1 cryosections were immunostained using a polyclonal goat anti-mouse CCL21 Ab. Hematoxylin counterstaining. Note that CCL21 immunoreactivity in the epidermis is similar to the EGFP expression pattern. In addition, CCL21 immunoreactivity can be observed in the dermis, most probably representing CCL21 secreted from the epidermis into the dermis. No immunoreactivity after application of pVR1012-mock vector. (**c**) In vitro cultured whole single-cell suspension of skin 3 days after pVR1012-mCCL21-IRES-EGFP application. Secretion of mCCL21 as measured by mCCL21 ELISA in supernatants. Adapted with modification and permission from the Journal of Investigative Dermatology.

12. Counterstain sections with hematoxylin for 30 s (use staining jar with cover).
13. Wash slides three times in PBS and mount/coverslip using Aqua-Mount™.
14. View the slides under microscope and take appropriate pictures (Fig. 1b).

3.4. Evaluation of Transgene Expression in the Skin by CCL21 ELISA

Euthanize mice as mentioned above 24 h after pDNA administration.

1. Depilate the Gene Gun treated area using a commercial depilatory such as Nair to completely remove the animal's fur.
2. Disinfect the skin with 70% ethanol and subsequently cut the treated area using tweezer and scalpel.
3. Mince in RPMI-1640 containing 25 mM HEPES.

4. Digest with 2 mg/mL of collagenase/dispase (Roche Diagnostics) in RPMI-1640 containing 25 mM HEPES for 60 min at 37°C and with light agitation (see Note 11).
5. Filter the digested skin through cell strainer to generate a single-cell suspension. Wash cell strainer with RPMI-1640 medium and collect them in 50 mL Falcon tube.
6. Centrifuge cells at 300 × *g*, 10 min and 4°C.
7. Resuspend the cells in 1 mL of RPMI-1640 containing 10% FCS, 100 U/mL penicillin, 100 μg/mL streptomycin, 50 μM 2-mercaptoethanol and 2 mM L-glutamine.
8. Culture cells in 12-well plate for 72 h at 37°C and 5% CO2 in the incubator.
9. Collect supernatants for CCL21 ELISA using a mouse CCL21/6Ckine DuoSet kit:
10. Dilute the capture antibody to the working concentration in PBS without carrier protein (coating solution).
11. Immediately coat a 96-well microplate with 100 μL per well of the diluted capture antibody. Seal the plate and incubate overnight at room temperature.
12. Discard coating solution, add ELISA blocking buffer and incubate for 90 min.
13. Prepare CCL21 standards by diluting serially in reagent diluent in duplicates.
14. Discard blocking buffer and wash plates three times with wash buffer.
15. Add 100 μL of standard or sample in Reagent Diluent in duplicates and incubate for 2 h.
16. Wash three times as above.
17. Add 100 μL of detection antibody and incubate for 2 h.
18. Wash three times as above.
19. Add 100 μL of streptavidin-HRP diluted 1:200 in reagent diluent and incubate for 30 min.
20. Wash four times as above.
21. Add 100 μL of substrate solution and incubate in the dark.
22. Develop for 15 min (watch the plate from time to time to stop the reaction if the reaction goes faster).
23. Stop the reaction with 2N H2SO4 (50 μL) and read the plates with ELISA reader at 450 nm with reference at 570 nm.
24. For calculation of results average the duplicate readings for each standard, control, and sample and subtract the average zero standard optical density (O.D.). Standard curve can be created by using Microsoft ExcelTM by plotting the mean

absorbance for each standard on the y-axis against the concentration on the x-axis to draw a best fit curve through the points on the graph. The data may be linearized by plotting the log of the 6Ckine concentrations versus the log of the O.D. and the best fit line can be determined by regression analysis. If samples have been diluted, the concentration read from the standard curve must be multiplied by the dilution factor (Fig. 1c).

4. Notes

1. It is important to use an unopened bottle of 100% ethanol each day this step is performed. Opened bottles of ethanol absorb water and the presence of water in the tubing while drying will lead to streaking, clumping, and uneven coating of the microcarriers over the inner surface of the Gold-Coat tubing, resulting in poor or unusable cartridges. All ethanol solutions should be opened only briefly when in use and kept tightly capped when not in use.
2. Any pDNA vector expressing a desired chemokine can be used. However, we have used pcDNA™3.1 (Invitrogen), VR1012 (provided by B Zaugg, Vical, San Diego, CA) or pCMV-EGFP (Addgene Cambridge, MA) pDNA vectors with success (7). A clean preparation of pDNA resuspended in ddH2O at the concentration of 1 μg/mL should be used. We routinely use Qiagen columns without having problems. Gold particles are available in three different sizes of 0.6, 1, and 1.6 μm. With the same helium pressure the smaller the size is, the deeper the penetration of the particles in skin in vivo. In our hands 1 μm gold particles are the ideal size for penetration into the epidermis.
3. PVP serves as an adhesive during the cartridge preparation process. Preparing cartridges with PVP can increase the total number of particles delivered especially at higher discharge pressures. The optimum amount of PVP to be used must be determined empirically (usually range from 0.01 to 0.1 mg/mL).
4. It is important to calculate the amount of DNA and gold required for each transformation. DNA loading ratio (DLR) is the amount of DNA loaded per mg of microcarriers. DLRs are usually in the range of 1 and 5 μg DNA/mg gold. More DNA tends to cause agglomeration (as a result of DNA binding to more than one particle). Microcarrier Loading Quantity (MLQ) is the amount of microcarriers delivered per target. MLQs usually range between 0.25 and 0.5 mg/cartridge for in vivo delivery to epidermal cells. Delivering 0.5 mg of gold and 1 μg of pDNA per target is a suitable starting point. At a

MLQ of 0.5 mg/cartridge, a DLR of 2 μg DNA/mg gold results in loading 1 μg of DNA/cartridge and in delivery of 1 μg of DNA per target. The concentration of DNA should be approximately 1 μg/μL and the volume of DNA should not exceed the volume of spermidine. If the DNA is too dilute it can be concentrated by ethanol precipitation.

5. It is very important to dry the tubing first. The main reason behind non-even loading of the tubing with the gold particles is the residual water in the ethanol. Seal PVP and microcarriers with 100% ethanol with parafilm. Wait until they acquire the room temperature before opening. Use fresh 100% ethanol whenever possible.
6. It is very important to follow the speed velocity mentioned. Faster or slower speeds result in unequal distribution of the gold/pDNA particles on the inner side of the tube.
7. This step is not mandatory as some mice strains (e.g., BALB/c) are calm in behavior and do not need narcosis.
8. The animal is held by its tail and placed on a surface that it can grip, and then it will stretch itself out so that a pencil or similar object can be placed firmly across the back of the neck. A sharp pull on the base of the tail will then dislocate the neck.
9. DAKO Alkaline Phosphatase Inhibitor solution containing levamisole can be an alternative.
10. By using PAP pen one can avoid using staining jar and significantly save the amount/volume of antibody solution needed.
11. We use laboratory incubator shakers with temperature set on 37°C.

References

1. Rossi D, Zlotnik A (2000) The biology of chemokines and their receptors. Annu Rev Immunol 18:217–242
2. Colditz IG et al (2007) Chemokines at large: in-vivo mechanisms of their transport, presentation and clearance. Thromb Haemost 97: 688–693
3. Ebert LM, Schaerli P, Moser B (2005) Chemokine-mediated control of T cell traffic in lymphoid and peripheral tissues. Mol Immunol 42:799–809
4. Stein JV et al (2000) The CC chemokine thymus-derived chemotactic agent 4 (TCA-4, secondary lymphoid tissue chemokine, 6Ckine, exodus-2) triggers lymphocyte function-associated antigen 1-mediated arrest of rolling T lymphocytes in peripheral lymph node high endothelial venules. J Exp Med 191:61–76
5. Stoecklinger A et al (2011) Langerin+ dermal dendritic cells are critical for CD8+ T cell activation and IgH gamma-1 class switching in response to gene gun vaccines. J Immunol 186:1377–1383
6. Nagao K et al (2009) Murine epidermal Langerhans cells and langerin-expressing dermal dendritic cells are unrelated and exhibit distinct functions. Proc Natl Acad Sci USA 106:3312–3317
7. Jalili A et al (2010) Induction of targeted cell migration by cutaneous administration of a DNA vector encoding a biologically active chemokine CCL21. J Invest Dermatol 130:1611–1623
8. Gunn MD et al (1999) Mice lacking expression of secondary lymphoid organ chemokine have defects in lymphocyte homing and dendritic cell localization. J Exp Med 189:451–460

Chapter 16

Enhancement of Gene Gun-Induced Vaccine-Specific Cytotoxic T-Cell Response by Administration of Chemotherapeutic Drugs

Steve Pascolo

Abstract

Because they specifically kill dividing cells, untargeted chemotherapeutic drugs such as platin derivatives, antimetabolites or topoisomerase inhibitors for example impact the immune system resulting in more or less profound transient lympho- and/or myelo-ablations in treated patients. Although this side effect of chemotherapeutic regimens could be assumed as immunosuppressive, it surprisingly appeared to eventually potentiate vaccination. By demonstrating that regulatory T-cells that mediate inhibition of immune responses proliferate more than other CD4-positive T-cells, we identified a possible mechanism underlying the vaccine-enhancing feature of certain chemotherapeutic anticancer regimen. The combination of cytostatic drugs and Gene Gun vaccination is of great interest in particular for the enhancement of antitumor, including "anti-self," vaccination strategies to treat cancer. Here we describe the effect of Gemcitabine, a standard chemotherapeutic drug, on human and mouse regulatory T-cells in vivo and present the methods allowing to trigger and detect an enhanced cytotoxic T-cell immune response using Gene Gun vaccination after Gemcitabine administration.

Key words: Gene gun, PMED, Chemotherapy, Gemcitabine, Regulatory T-cells

1. Introduction

Antitumor T-cell responses can help in controlling cancer development (1, 2). Such immune responses can appear spontaneously due to local tumor inflammation, after radio-chemo-therapies that induce tumor cell death and be induced by specific vaccination. One very efficacious, safe, versatile and reliable vaccination method is Gene Gun or Particle Mediated Epidermal Delivery (PMED). It allows the transient expression in the dermis of

Stephan Sudowe and Angelika B. Reske-Kunz (eds.), *Biolistic DNA Delivery: Methods and Protocols*,
Methods in Molecular Biology, vol. 940, DOI 10.1007/978-1-62703-110-3_16, © Springer Science+Business Media, LLC 2013

proteins encoded by minimal nucleic acid vectors (MNAVs) such as messenger RNA (mRNA) and plasmid DNA (pDNA) which are easy to produce under GMP conditions on a large scale, and are considered safe (review by Weide et al. (3)). One important feature of this vaccination strategy is that the tumor antigen being encoded by the delivered nucleic acid is produced by the host's cells. That means that Major Histocompatibility (MHC)-associated peptides, in particular MHC class I (MHC I) peptides produced from the vaccine antigen are the same as those produced from the natural endogenous protein in cancer cells. This contrasts with protein vaccination where the antigen is taken up and processed through the "exogenous" MHC pathway, leading to the presentation of different MHC I peptides compared to the "endogenous" (intracytosolic) pathway (4). We have shown recently for the broadly expressed tumor specific antigen of the cancer testis family: NY-ESO-1 (review by Gnjatic et al. (5)), that the endogenous production of the protein thanks to genetic vaccines was needed to trigger a relevant anticancer immune response. Injection of NY-ESO-1 proteins in mice did not induce the relevant immunity (cytotoxic T-cells against an immunodominant MHC I epitope from NY-ESO-1) and failed in protecting against cancer progression (6). However, in humans (clinical phase I study), Gene Gun vaccination of cancer patients with plasmid coding NY-ESO-1 gave disappointing clinical results: The vaccine-induced anti-NY-ESO-1 cytotoxic T-cell response was transient, mild and detected in only a few (5 out of 16) patients (7). Regulatory T-cells were found in this study to diminish PMED-induced antigen-specific T-cell response. Thus, it can be hypothesized that abrogating regulatory T-cell activity prior to vaccination may lead to better antitumor response in cancer patients.

Untargeted chemotherapeutic drugs interfere with metabolic processes necessary for cell division and in particular with DNA synthesis pathways. Within this class of drugs are included antimetabolites (e.g., Gemcitabine), antitumor antibiotics (e.g., Bleomycin), alkylating agents/platinum coordination compounds (e.g., Cyclophosphamide and Cisplatin), topoisomerase inhibitors (e.g., Etoposide) or antimicrotubule agents (e.g., Vincristine or Docetaxel). They penetrate in all cells and by compromising duplication of DNA induce cell death in dividing cells. Homeostasis of the immune system requires intensive and constant cell division in particular in bone marrow. Thereby, the untargeted chemotherapies have transient immunodepleting side effects. Depending on regimen and patients, this affects mostly myeloid (e.g., monocytes) or mostly lymphoid (e.g., lymphocytes) or both compartments. For this reason, experimental immunotherapies were tested outside of the periods of chemotherapeutic treatments. However, it was found in animal models and proven recently in a clinical study (8) that some chemotherapeutic regimen actually enhances the triggering of a specific cytotoxic T-cell immune response following peptide-based vaccination. Using Gene Gun vaccination, we could also document this synergy in an animal model: Gemcitabine (2′,2′-difluorodeoxycytidine: inhibition of ribonucleotide reductase that results in a reduction of the dNTP pool and lethal incorporation into DNA in place of CTP) treated animals developed a stronger cytotoxic immune response against NY-ESO-1 after PMED vaccination than untreated animals (9). Concurrently, we observed in human and mice that within the CD4 compartment regulatory T-cells (Tregs) proliferate more than nonregulatory T-cells (Fig. 1). Dividing Tregs like all dividing cells are depleted by Gemcitabine treatment (Fig. 2). Thereby, within the CD4 compartment, Tregs are more depleted than non-Tregs in both human (Fig. 3) and mice (Fig. 4) and remain few days at lower level than during homeostasis. Thus, the induction of a "low-Treg" status can be a side effect induced by untargeted chemotherapeutic drugs. It offers an ideal vaccination window where (auto-) immune responses induced by vaccination, including PMED vaccination, can be favored (Fig. 5).

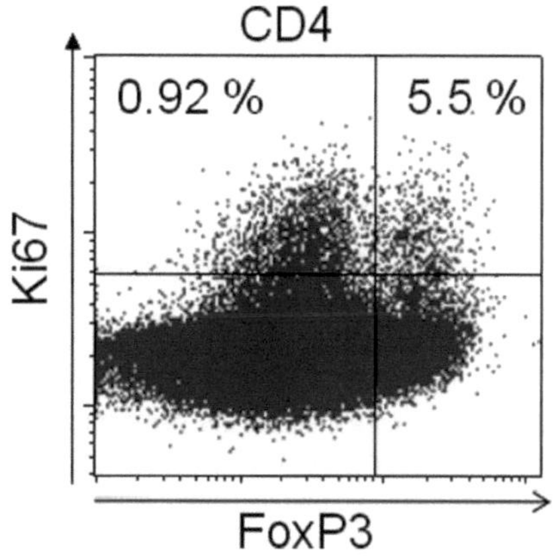

Fig. 1. Peripheral blood mononuclear cells (PBMCs) from a healthy donor were stained ex vivo using CD4, the marker for helper T-cells, CD19 (B-cells and exclusion marker), intracellular FoxP3, the marker for regulatory T-cells and intracellular ki67, the marker for proliferative cells. The numbers indicate the percentage of proliferating T-cells in the nonregulatory (FoxP3-negative) or regulatory (FoxP3-positive) T-cell compartments. The results show that within the CD4 compartment, regulatory T-cells proliferate more (ki67 high) than nonregulatory T-cells.

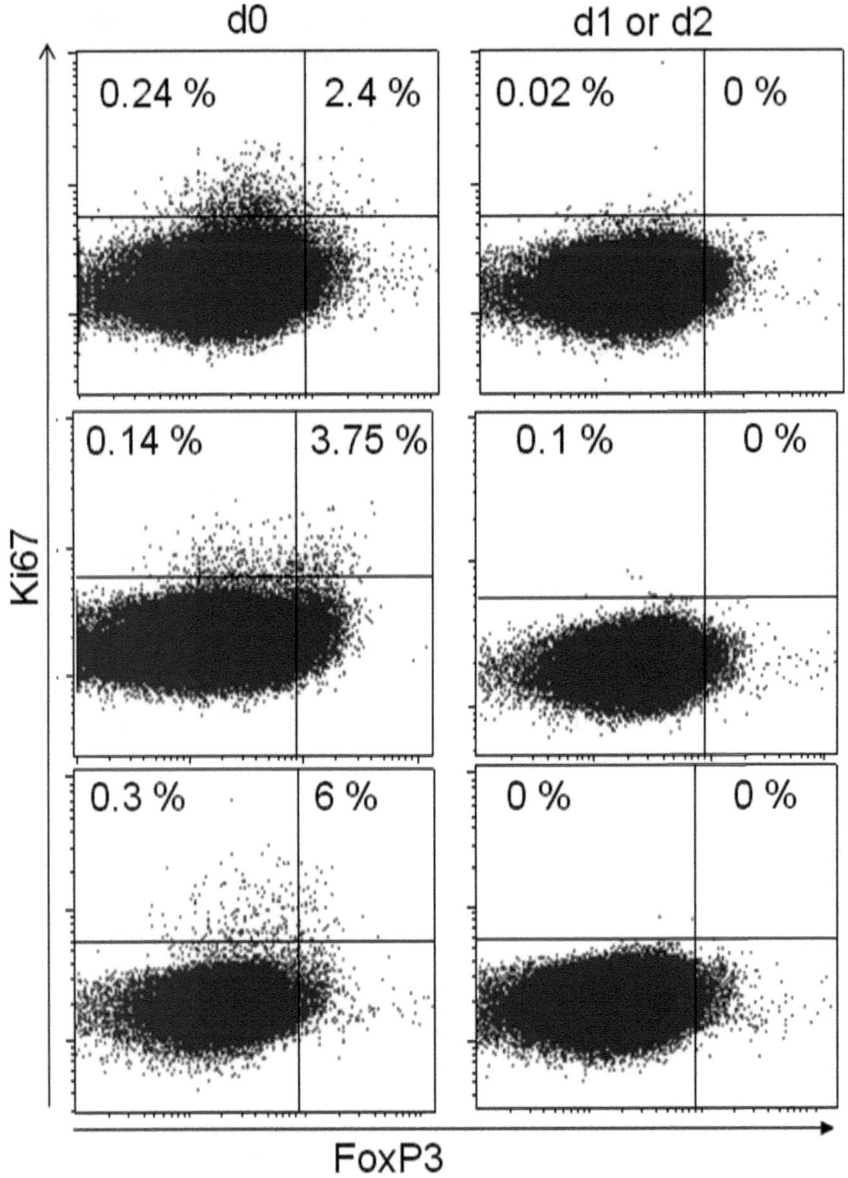

Fig. 2. PBMCs were prepared from the peripheral blood drawn from three cancer patients directly before Gemcitabine administration at day 0 and also either 1 or 2 days afterwards, and stored frozen. Thawed cells were stained as in Fig. 1 and analyzed by FACS. The numbers indicate the percentage of proliferating T-cells (ki67 high) in the nonregulatory (FoxP3-negative) or regulatory (FoxP3-positive) T-cell compartments. The results show that all proliferating cells are depleted by Gemcitabine treatment and that this effect lasts at least 2 days.

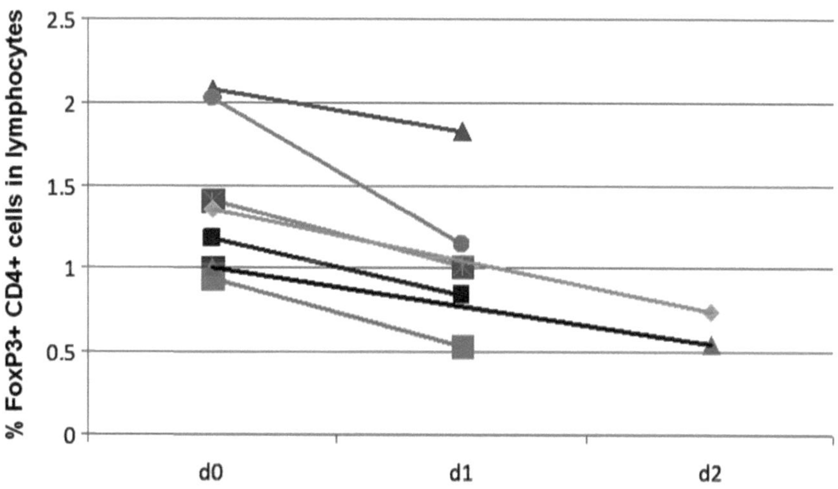

Fig. 3. PBMCs were prepared from peripheral blood drawn from seven cancer patients directly before Gemcitabine administration at day 0, and also either one (five patients) or two (two patients) days afterwards, and stored frozen. Thawed cells were stained as in Fig. 1 and analyzed by FACS. The graph indicates the percentage of regulatory T-cells (FoxP3- positive) at day 0, 1, or 2. The results indicate that the percentage of regulatory T-cells decreases in all patients treated by Gemcitabine.

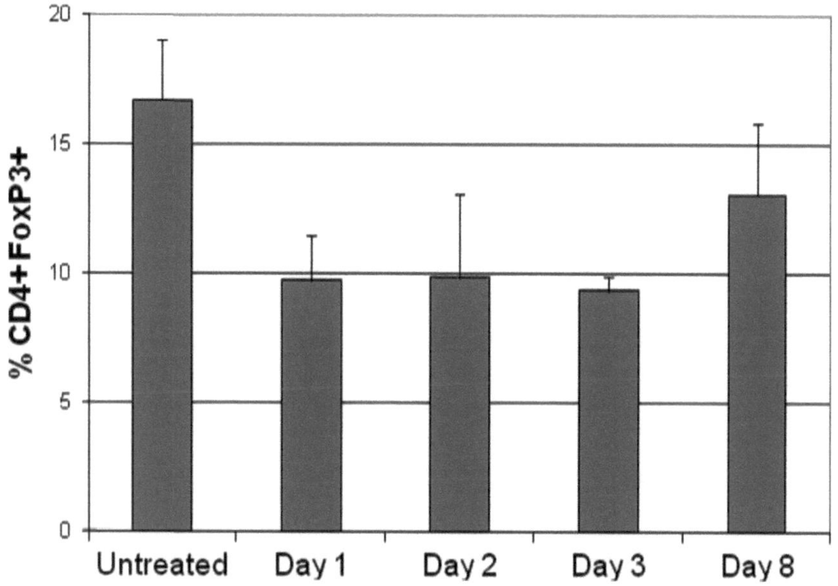

Fig. 4. Mice (three per group) were injected intraperitoneally with Gemcitabine and blood regulatory T-cells analyzed on day 1, 2, 3, or 8. Untreated mice were used as control. The data present the percentage of FoxP3-positive T-cells within the CD4 T-cell compartment. The results demonstrate that following Gemcitabine injection, the relative frequency of regulatory T-cells is reduced in mice during a few days.

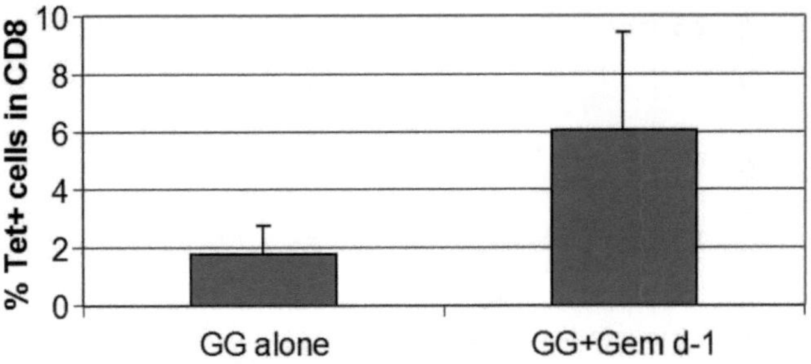

Fig. 5. Mice were vaccinated with PMED using pDNA encoding NY-ESO-1. The vaccine was given alone ("GG alone") or 1 day after injection of Gemcitabine ("GG + Gem d-1"). Two weeks after a boost immunization made under the same conditions as the prime (alone or combined with a Gemcitabine injection), splenocytes were analyzed for the presence of NY-ESO-1-specific cytotoxic T-cells using fluorescent MHC tetramers. The results show that preconditioning the mice with Gemcitabine allows the triggering of a higher cytotoxic T-cell response against the vaccine antigen.

2. Materials

1. BALB/c mice.
2. Gene Gun equipment: Helios Gene Gun System (BioRad). For handling and preparation of cartridges follow the manufacturer's instructions.
3. Plasmid: The cDNA corresponding to the NY-ESO-1 mRNA, cloned into the expression vector pBK-CMV (Stratagene). A CMV promoter directs the constitutive transcription of the cDNA and guaranties the strong intracellular production of the NY-ESO-1 protein in any cell type (10).
4. Phosphate-buffered saline (PBS).
5. Gemcitabine: Drug Gemzar® (Eli Lilly, Switzerland): resuspend by addition of 5 mL of sterile PBS to the 200 mg of lyophilized product (see Note 1). The final drug concentration is thus 40 mg/mL.
6. Basic cell-handling medium: add 50 mL of Fetal Calf Serum to 500 mL of RPMI 1640.
7. Cell Strainer, 70 μm (BD-Biosciences).
8. Ficoll density 1.077.
9. 5 mL FACS tubes, 15 mL tubes.
10. Antibodies: anti-mouse CD4-Fluoresceine (FITC) or –Phycoerythrin (PE), anti-mouse CD8-Allophycocyanin (APC), anti-mouse CD19-PercP, antihuman CD4-PE, antihuman

CD19-PercP, anti-mouse FoxP3-APC and antihuman FoxP3-APC, antihuman Ki67-Fluoresceine.

11. MHC tetramers, obtained from Dr. Immanuel Luescher (Ludwig Institute for Cancer research, Switzerland); they consist of H2-D^d monomers folded around the RGPESRLL synthetic peptide (dominant $H2^d$ epitope from NY-ESO-1 (11)) and tetramerized with streptavidin labeled with phycoerythrin (PE).
12. Permeabilization solution: eBioscience.
13. Wash solution: eBioscience.
14. Cyan FACS (Beckman Coulter) and Flowjo software (Treestar) for data analysis.
15. Centrifuge.

3. Methods

3.1. Pretreatment of Mice with Gemcitabine

One day before Gene Gun vaccination (see Note 2), inject mice intraperitoneally with 200 μL (see Note 3) of Gemcitabine solution at 40 mg/mL.

3.2. Vaccination with the Gene Gun

1. One day before vaccination (i.e., at the day of Gemcitabine injection), remove belly fur using a razor blade (see Notes 4 and 5).
2. For vaccination, immobilize each mouse with the belly up (with the left hand hold the mouse by the skin of the neck between thumb and index and by the tail between pinky and palm). Apply the Gene Gun on the exposed skin.
3. Deliver particles using 250 psi (see Note 6). Use three cartridges for each mouse (that is a total of 3 μg of plasmid per mouse: 1 μg per cartridge, 3 cartridges per mouse) (see Note 7).
4. Two weeks later, repeat this treatment (shaving, injection of Gemcitabine and 1 day later Gene Gun vaccination to deliver 3 μg of plasmid) (see Note 8).

3.3. Immunomonitoring of Vaccinated Mice

1. Ten to fifteen days after the boost, kill mice by CO_2.
2. Remove spleens and put them in a 15 mL tube containing 7 mL of RPMI/10%FCS. Smash them through a cell strainer.
3. Slowly pipet the filtered 7 mL solution on top of 5 mL of Ficoll in a 15 mL tube (see Note 9). Centrifuge the tubes during 20 min at 863 × g without break.
4. Discard the red supernatant (RPMI/10%FCS) by aspiration using a 5 mL pipette. Collect the interface using the same emptied 5 mL pipette (see Note 10) and dilute in 10 mL of PBS placed in a 15 mL tube.

5. Centrifuge tubes 8 min at 485 × *g*. Discard the supernatant, break the cell pellet by kicking the tube's extremity and add 3 mL of RPM/10% FCS.
6. Place 200 μL of this cell suspension (approximately two million cells) in a 5 mL FACS tube.
7. Stain first with the tetramer at the final concentration of 2 μg/mL. Incubate the tube 20 min at 37°C.
8. Thereafter, directly add mouse CD4-FITC and mouse CD8-APC antibodies at 2 μg/mL final concentration to the cell suspension (see Note 11). Incubate for 30 min at 4°C.
9. Wash cells twice by addition of 4 mL of PBS and centrifugation for 5 min at 485 × *g*. Resuspend the cell pellet in 300 μL of PBS.
10. Analyze the cells by FACS. As shown in Fig. 5, mice pretreated with Gemcitabine have more cytotoxic T-cells against the relevant NY-ESO-1 epitope than unconditioned mice.

3.4. Preparation of Human Peripheral Blood Mononuclear Cells (PBMCs)

1. Collect 7 mL of heparinized blood.
2. Slowly pipett the blood on top of 5 mL of Ficoll in a 15 mL tube. Centrifuge the tubes during 20 min at 863 × *g* without break.
3. Discard the yellow supernatant (plasma) by aspiration using a 5 mL pipette. Collect the interface using the same emptied 5 mL pipette (see Note 10) and dilute in 10 mL of PBS placed in a 15 mL tube.
4. Centrifuge tubes 8 min at 485 × *g*. Discard the supernatant, break the cell pellet by kicking the tube's extremity and add 1 mL of RPM/10% FCS.

3.5. Quantification of Tregs in Mouse Splenocytes and Human Peripheral Blood Mononuclear Cells (PBMCs) by Flow Cytometry

1. Place 200 μL (approximately two million cells) of the cell suspensions obtained in 3.3 (mouse) or 3.4 (human) in a 5 mL FACS tube.
2. Add CD4-PE and CD19-PercP antibodies at 2 μg/mL final concentration to the cell suspension. Incubate for 30 min at 4°C.
3. Wash cells twice by addition of 4 mL of PBS and centrifugation for 5 min at 485 × *g*. Discard the supernatant, break the cell pellet by vortexing the tubes and add 1 mL of permeabilization solution. Incubate 30 min at 4°C.
4. Wash the cells by adding 2 mL of wash solution and centrifugating for 5 min at 485 × *g*. Add 100 μL of wash solution containing 1 μL of the anti-FOXP3 antibody and 1 μL of the anti-ki67 antibody. Vortex the tubes. Incubate 30 min at 4°C.
5. Wash the cells by adding 2 mL of wash solution and centrifugating for 5 min at 485 × *g*. Repeat this step. Resuspend the cells in 300 μL of PBS.

6. Analyze the cells by FACS (see Note 12). As shown in Fig. 1, a larger frequency of FoxP3-positive cells compared to FoxP3-negative cells are dividing (ki67 high). As shown in Figs. 2, 3, and 4 proliferating cells are depleted following administration of Gemcitabine.

4. Notes

1. Although it is not recommended to store the drug after reconstitution in buffer, we found that Gemzar reconstituted with sterile PBS in strictly sterile conditions (under laminar flow) could be stored at least 2 weeks in the fridge (4°C) retaining its cytostatic capabilities in vitro (death of cultured cells).
2. Injection of Gemcitabine at the day but not 1 day after vaccination was found to enhance the efficacy of the vaccine. It can be expected according to the slow recovery (more than 3 days) of the homeostatic regulatory T-cell frequencies (Fig. 4) that Gemcitabine could be injected 2 or 3 days before vaccination and still significantly enhance the efficacy of the vaccine.
3. Lower doses of Gemcitabine (down to 75 μL of the 40 mg/mL solution) can also be used to enhance the efficacy of the subsequent vaccine.
4. For an efficacious, easy and safe shaving, mice are anesthetized by intraperitoneal injection of the standard Xylazine and Ketamine mixture. At least 1 h after they have recovered from the anesthesia, mice receive intraperitoneally the Gemcitabine solution.
5. Removing fur 1 day before vaccination guaranties that at the day of Gene Gun treatment, the skin will be healthy. Eventual irritations and superficial cuts made during shaving are healed within the 24 h period.
6. Because gold particles may create some local aggregates in some parts of the tubes/cartridge, we shoot twice through each cartridge. This guaranties that the whole gold beads content from each cartridge is delivered to the skin.
7. Since there are two shots for each cartridge, each mouse gets a total of six shots.
8. One vaccination cycle is not enough to detect ex vivo the cytotoxic T-cell response using tetramer staining. The vaccine boost is required.
9. Purifying cells on a ficoll gradient is preferred to simple red cell lysis of the splenocyte suspension since it allows to get rid of dead cells thereby strongly reducing backgrounds in FACS

analysis and consequently enhancing sensitivity of the MHC tetramer staining assay.

10. Circular movements of the pipette tip guaranty to take a maximum of cells in the interface. Two to three milliliter of ficoll phase can be taken up.
11. Using this tetramer, reliable stainings are best obtained with this protocol where first the tetramer alone is added on the cells and incubated at 37°C before the two antibodies are added for further 30 min incubation in the fridge. The analysis (CD8-tetramer dot plot) should be gated on CD4-negative (FITC-negative) cells. This allows to exclude dead cells that eventually unspecifically bind antibodies and tetramers.
12. Dead cells eventually bind unspecifically antibodies. Thus, using adequate gates, CD19-positive cells should be excluded. The whole analysis (CD4-FoxP3-ki67 dot plots) should be gated on CD19-negative cells.

References

1. Zitvogel L et al (2008) Immunological aspects of cancer chemotherapy. Nat Rev Immunol 8:59–73
2. Zitvogel L et al (2008) Immunological aspects of anticancer chemotherapy. Bull Acad Natl Med 192:1469–1487
3. Weide B et al (2008) Plasmid DNA- and messenger RNA-based anti-cancer vaccination. Immunol Lett 115:33–42
4. Robson NC et al (2010) Processing and cross-presentation of individual HLA-A, -B, or -C epitopes from NY-ESO-1 or an HLA-A epitope for Melan-A differ according to the mode of antigen delivery. Blood 116:218–225
5. Gnjatic S et al (2006) NY-ESO-1: review of an immunogenic tumor antigen. Adv Cancer Res 95:1–30
6. Parvanova I et al (2011) The form of NY-ESO-1 antigen has an impact on the clinical efficacy of anti-tumor vaccination. Vaccine 29:3832–3836
7. Gnjatic S et al (2009) NY-ESO-1 DNA vaccine induces T-cell responses that are suppressed by regulatory T cells. Clin Cancer Res 15: 2130–2139
8. Nistico P et al (2009) Chemotherapy enhances vaccine-induced antitumor immunity in melanoma patients. Int J Cancer 124:130–139
9. Rettig L et al (2011) Gemcitabine depletes regulatory T-cells in human and mice and enhances triggering of vaccine-specific cytotoxic T-cells. Int J Cancer 129:832–838
10. Chen YT et al (1997) A testicular antigen aberrantly expressed in human cancers detected by autologous antibody screening. Proc Natl Acad Sci USA 94:1914–1918
11. Muraoka D et al (2010) Peptide vaccine induces enhanced tumor growth associated with apoptosis induction in CD8+ T cells. J Immunol 185:3768–3776

Chapter 17

Dendritic Cell-Specific Biolistic Transfection Using the *Fascin* Gene Promoter

Yvonne Höhn, Stephan Sudowe, and Angelika B. Reske-Kunz

Abstract

The transcriptional targeting of gene expression to selected cells by cell type-specific promoters displays a fundamental tool in gene therapy. In immunotherapy, dendritic cells (DCs) are pivotal for the elicitation of antigen-specific immune responses following gene gun-mediated biolistic transfection. Here we report on transcriptional targeting of murine skin DCs using plasmids which include the promoter of the gene of the cytoskeletal protein fascin to control antigen production. Fascin, which is mandatory for the formation of dendrites, is synthesized among the hematopoietic cells exclusively by activated DCs. The activity of the promoter of the *fascin* gene reflects the endogenous production of the protein, being high in mature DCs but almost absent in immature DCs or other cutaneous cells. Here we describe the analysis of transgene-specific immune responses after DC-focused biolistic transfection. In conclusion, the murine *fascin* promoter can be readily used to target DCs in DNA immunization approaches and thus offers new opportunities for gene therapy.

Key words: Gene therapy, Biolistic DNA vaccination, Gene gun, Dendritic cells, Fascin, Transcriptional targeting, Type 1 responses, Mouse model

1. Introduction

Biolistic bombardment of skin with DNA-coated gold particles using the helium-powered gene gun has been used as a genetic vaccination approach for protective and therapeutic purposes in numerous pre-clinical animal models of viral, bacterial, and parasitic infections, but also as a novel form of immunotherapy against cancer and allergic diseases. Because of their function as professional antigen-presenting cells (APCs) with primary stimulatory capacity, dendritic cells (DCs) are of crucial importance for the initiation of transgene-specific immune responses by particle-mediated transfection of the skin (1–3),

Stephan Sudowe and Angelika B. Reske-Kunz (eds.), *Biolistic DNA Delivery: Methods and Protocols*, Methods in Molecular Biology, vol. 940, DOI 10.1007/978-1-62703-110-3_17, © Springer Science+Business Media, LLC 2013

although it has been controversially discussed whether epidermal Langerhans cells (LCs) or dermal DC populations are the principal APCs under these circumstances (4–7).

In line with the great importance of DCs for the elicitation of antigen-specific immune responses after gene gun immunization, the identification of regulatory gene elements, which allow the limitation of transgene expression specifically to DCs, has become a matter of great interest. We have shown that in contrast to biolistic transfection with conventional plasmid vectors which usually contain ubiquitously active promoters such as the cytomegalovirus immediate early promoter (pCMV) and which allow for antigen production by every successfully transfected somatic cell, transgene expression after gene gun-mediated delivery of plasmids which include the promoter of the *fascin* gene (pFascin) is focused primarily to mature skin-derived DCs (8). The actin-bundling protein fascin is not produced by immature DCs as represented by epidermal LCs. However, during the differentiation process to fully activated DCs, production of fascin, which is mandatory for the formation of dendrites and the migratory capacity of DCs, is strongly enhanced (9–11). Consequently, *fascin* gene expression is considerably upregulated during DC maturation (10–12). Apart from mature DCs, formation of fascin is restricted to a few nonhematopoietic cell types such as neuronal/glial cells and capillary endothelial cells and to some transformed cells (10, 13–15). Accordingly, we demonstrated that gene gun-mediated DNA immunization with antigen-encoding plasmids under the control of the fascin promoter resulted in antigen expression restricted to directly transfected DCs (8).

The transfection of only a few DCs was sufficient to induce potent transgene-specific immune responses, thus making the *fascin* gene promoter a suitable tool for DC-focused DNA immunization. In contrast to the immune response induced by biolistic transfection using conventional pCMV plasmids, which represents a mixed type 1/type 2 response, gene gun-mediated transfection with the DC-targeting vector pFascin propagates a potent and distinct type 1-biased cellular immune response, which was characterized by the prevalent production of antigen-specific IgG2a by B cells (Fig. 1) and the generation of substantial numbers of IFN-γ-producing $CD4^+$ T helper (Th) 1 cells as well as $CD8^+$ cytolytic T lymphocytes both in spleen and local draining lymph nodes (Figs. 2 and 3) (16–19).

Apart from the divergent differentiation pathways of $CD4^+$ T cells, the use of pFascin in biolistic DNA immunization approaches has several other advantages over the use of pCMV: (1) the risk of interference of viral elements is reduced by using a nonviral promoter; (2) the risk of unfavorable and/or unforeseen immune reactions is reduced by limiting transgene expression mainly to DCs; this excludes tolerizing effects that may result from transfec-

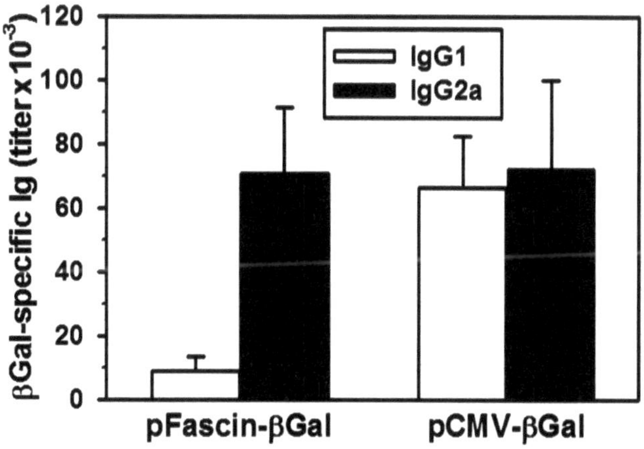

Fig. 1. Qualitative differences in the IgG profile following biolistic transfection with pFascin-βGal or pCMV-βGal. BALB/c mice ($n=4$) were immunized by three gene gun-mediated immunizations with pFascin-βGal and pCMV-βGal, respectively, in weekly intervals. Thirty-five days after the first immunization sera were collected and levels of βGal-specific IgG1 (*open bars*) and IgG2a (*closed bars*) were determined.

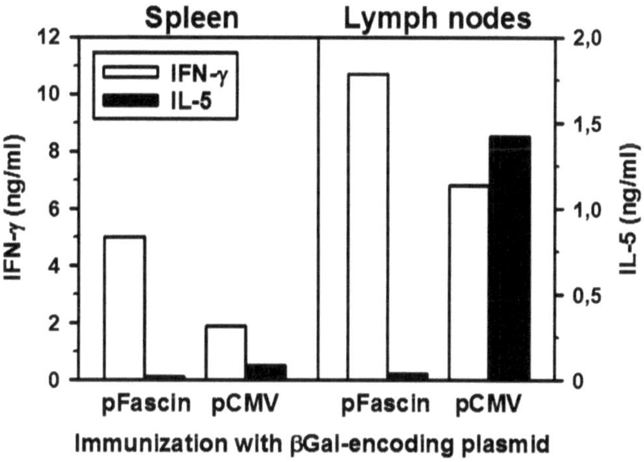

Fig. 2. Development of different Th cell subsets in draining lymph nodes following biolistic transfection with pFascin-βGal or pCMV-βGal. BALB/c mice ($n=4$) were immunized by five gene gun-mediated immunizations with pFascin-βGal and pCMV-βGal, respectively, in weekly intervals. Fifty-six days after the first immunization spleen and draining inguinal/axillary lymph nodes cells were prepared and cultured (5×10^6/well) in quadruplicates with or without 25 μg/mL βGal as antigen. After 72 h culture supernatants were pooled and the amount of IFN-γ (*open bars*) and IL-5 (*closed bars*) was determined.

tion of other cell types by plasmids harboring ubiquitously active viral promoters; and (3) since transgene expression is predominantly restricted to matured DCs when using the *fascin* gene promoter, tolerance induction known to be associated with antigen presentation by immature DCs is prevented. In conclusion, tran-

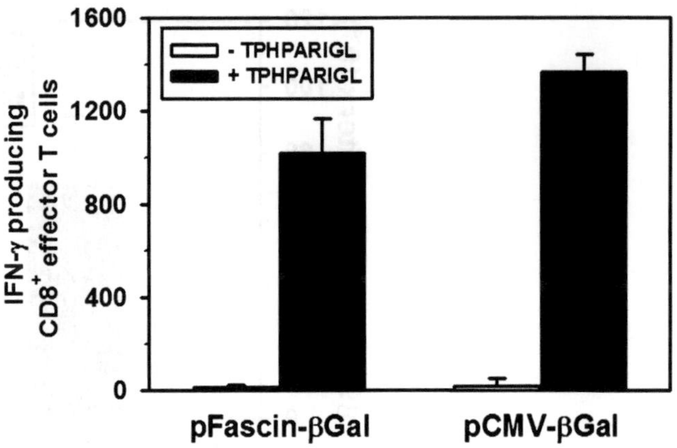

Fig. 3. Recruitment of IFN-γ-producing $CD8^+$ effector T cells following biolistic transfection with pFascin-βGal or pCMV-βGal. BALB/c mice ($n=4$) were immunized by three gene gun-mediated immunizations with pFascin-βGal and pCMV-βGal, respectively, in weekly intervals. Twenty-eight days after the first immunization numbers of IFN-γ-producing $CD8^+$ effector T cells among spleen cells were determined following incubation of splenocytes without (*open bars*) or with (*closed bars*) 0.1 μg/mL βGal-specific peptide TPHPARIGL for 22 h.

scriptional targeting of DCs using the promoter of the *fascin* gene for the purpose of focusing endogenous antigen production preferentially to primary stimulatory APCs represents an alternative strategy for immunomodulation and thus might provide new opportunities for immunotherapy and for improved vaccination strategies against cancer as well as infectious and allergic diseases.

2. Materials

Use ultrapure water for preparation of the solutions (Milli-Q-Integral System, Millipore) unless otherwise indicated, and reagents of analytical grade.

2.1. Plasmid DNA

1. Plasmid encoding β-galactosidase (βGal) under control of the ubiquitously active cytomegalovirus immediate early (CMV) promoter (pCMV-βGal) (see Note 1).
2. Plasmid encoding βGal under control of the DC-specific *fascin* gene promoter (pFascin-βGal) (see Note 1).

2.2. Cartridge Preparation

1. Tubing prep station, tefzel tubing, and tubing cutter (Bio-Rad, Hercules, CA, USA).
2. Nitrogen tank with attachments for tubing prep station.
3. Microfuge.

4. 15 mL polypropylene tubes.
5. 1.5 mL safe lock tubes.
6. 1.6 μm elemental gold particles (Bio-Rad).
7. 96% ethanol.
8. 0.05 M spermidine in endotoxin-free H_2O.
9. Polyvinylpyrrolidone (PVP, MW 360,000) (Bio-Rad)
   Stock solution (20 mg/mL): Dissolve 3.5 mg PVP in 175 μL 96% ethanol.
   Working solution (0.075 mg/mL): Dilute 30 μL of stock solution in 8 mL 96% ethanol.
10. 1 M $CaCl_2$ in endotoxin-free H_2O.
11. Ultrasonic bath.

2.3. Gene Gun Bombardment

1. Helios gene gun (Bio-Rad).
2. Compressed helium of grade 4.5 or higher.
3. Hearing protection device.
4. Electric shaver for trimming the abdominal fur of mice.
5. Cartridges loaded with gold/plasmid DNA.

2.4. Mice

Female BALB/c inbred mice (H-2^d), 6–10 weeks old at the beginning of the experiments (see Note 2).

2.5. ELISA and ELISPOT

1. 96-well ELISA microtiter plates (see Note 3) and ELISA plate seals.
2. 96-well ELISPOT microtiter plates (see Note 4).
3. ELISA washer.
4. Precision microplate reader.
5. ELISPOT reader system.
6. Phosphate buffered saline (PBS): 137.5 mM NaCl, 10 mM NaH_2PO_4, pH 7.2.
7. Recombinant βGal from *Escherichia coli* (Sigma-Aldrich, Deisenhofen, Germany) as antigen. Prepare 1 mg/mL βGal stock solution in PBS.
8. ELISA coating buffer: 0.1 M $NaHCO_3$, pH 8.2.
9. Diluent buffer: PBS with 1% bovine serum albumin.
10. Washing buffer: PBS with 0.05% Tween.
11. Recombinant mouse IL-5 (e.g., eBioscience, San Diego, CA, USA) and IFN-γ (e.g., BD Biosciences, San Diego, CA, USA) as standard.
12. Isotype-specific detection antibodies: biotin-labeled goat anti-mouse-IgG1 and goat anti-mouse-IgG2a, respectively (e.g., Biozol, Eching, Germany).

13. Cytokine-specific capture antibodies: rat anti-mouse-IL-5 (clone TRFK5) (BD Biosciences) and rat anti-mouse IFN-γ (clone R4-6A2) (20).
14. Cytokine-specific detection antibodies: biotin-labeled rat anti-mouse-IL-5 (clone TRFK4) (BD Biosciences) and biotin-labeled rat anti-mouse IFN-γ (clone AN18.17.24) (21).
15. Streptavidin-conjugated horseradish peroxidase (e.g., ExtrAvidin-Peroxidase, Sigma-Aldrich) as enzyme.
16. ELISA substrate buffer: 0.2 M NaH_2PO_4, 0.1 M sodium citrate, pH 5.0.
17. ELISA substrate: *o*-phenylenediamine (OPD) (Sigma-Aldrich).
18. ELISA stop solution: 1 M H_2SO_4.
19. ELISPOT substrate stock solution: 20 mg 3-amino-9-ethylcarbazole (AEC) (Sigma-Aldrich) dissolved in 2.5 mL dimethylformamide (DMF) (see Note 5).
20. ELISPOT substrate buffer: 0.1 M sodium acetate, pH 5.0.
21. 30% H_2O_2.
22. 0.45 μm filter.
23. Recombinant human IL-2 (rIL-2).
24. Peptide $\beta Gal_{876-884}$ (amino acid sequence: TPHPARIGL). Prepare a 2.5 μg/mL stock solution in PBS.

2.6. Antigen-Specific Restimulation of Lymphocytes In Vitro

1. Dissecting set.
2. Gey's lysis buffer: 10 mM $KHCO_3$, 155 mM NH_4Cl, 100 μM EDTA, pH 7.5; sterilize by filtration (0.2 μm).
3. Eagle's Minimum Essential Medium (EMEM) supplemented with 2 mM L-glutamine, 50 μM 2-Mercaptoethanol, 100 IU/mL penicillin, 100 μg/mL streptomycin, and 2% fetal calf serum (FCS).
4. Iscove's Modified Dulbecco's Medium (IMDM) supplemented with 2 mM L-glutamine, 50 μM 2-Mercaptoethanol, 100 IU/mL penicillin, 100 μg/mL streptomycin, and 10% FCS.
5. Recombinant βGal (Sigma-Aldrich) as antigen. Prepare a 5 mg/mL βGal stock solution in PBS.
6. 15 and 50 mL polypropylen tubes.
7. Sterile glass slides.
8. Cell strainer (Mesh size 40 μm).
9. Petri dishes (diameter 100 mm).
10. Centrifuge.
11. 24-well tissue culture plates.
12. Parafilm.

13. Hemocytometer.
14. Stereo microscope.
15. Trypan blue.
16. Humidified 10% CO_2 incubator for cell culture.

3. Methods

3.1. Coating Plasmid DNA onto Gold Particles

1. To process 30 cartridges, weigh 30 mg of gold particles into a safe lock tube.
2. Suspend gold particles in 100 μL of 0.05 M spermidine (see Note 6). To singularize gold particles, incubate tubes for 10 s in an ultrasonic bath.
3. Add 2 μg plasmid DNA per mg gold particles to the suspension and mix carefully. If the DNA volume exceeds 100 μL, match the volume of spermidine to that of the DNA.
4. Add 1 M $CaCl_2$ (use same volume as spermidine before) quickly and vortex the suspension immediately (see Note 7).
5. Incubate the suspension for 10 min at room temperature (RT) to allow the plasmid DNA to bind to the particles.
6. Microcentrifuge the tube for 10 min. Remove and discard the supernatant carefully. By flicking the tube, break up the pellet and add 1 mL of 96% ethanol to the pellet.
7. Repeat step 6 two additional times.
8. Carefully remove the supernatant completely. Solve the pellet in three steps in 1.75 mL PVP working solution and transfer the gold/DNA/PVP-suspension to a 15 mL polypropylen tube.
9. Seal the tube with parafilm and store at −20°C until cartridge preparation.

3.2. Preparation of Gene Gun Cartridges

1. Implement the tubing into the tubing prep station and dry the tubing by aeration with nitrogen.
2. Vortex the gold/DNA/PVP-suspension and immediately load the suspension slowly (see Note 8) into the nitrogen dried Tefzel tubing following the manufacturer's instructions.
3. Discard unevenly loaded tubing parts and cut the tubing with the tubing cutter into 1 cm pieces.
4. Store cartridges in a sealed container at 4°C up to 8 weeks.

3.3. Biolistic Transfection of Murine Skin Cells (see Note 9)

1. Shave the abdominal skin of the mouse selected for gene gun bombardment (see Note 10).
2. Attach the gene gun to the helium flask as written in the manufacturer's guidelines and set pressure at 400 psi (see Note 11).
3. Load the gene gun depot with the required cartridges. Two shots (see Note 12) per mouse are recommended to yield an amount of 4 μg DNA per mouse (see Note 13).
4. The distance piece (spacer) of the armed gene gun remains on the skin while the trigger is pulled (see Notes 14 and 15).

3.4. Determination of βGal-Specific IgG1 and IgG2a in Sera by ELISA

1. Take blood sample of a mouse by puncture of the retro-orbital plexus or by tail vein bleeding (see Notes 16 and 17). After 60 min at RT remove clot of blood and centrifuge the remaining cell suspension for 5 min at 10,000 × *g*. Carefully remove the supernatant as serum and store in safe lock tubes at −20°C.
2. Dilute βGal stock solution to 5 μg/mL in ELISA coating buffer. Pipet 100 μL of antigen solution into wells of a 96-well ELISA microtiter plate and incubate overnight at 4°C.
3. To avoid unspecific binding of antibodies, wash plate twice with washing buffer (see Note 18) and add 200 μL of diluent buffer to each well. Keep the plate for 1 h at RT.
4. Wash plate twice with washing buffer (see Note 18).
5. Add 100 μL of diluent buffer to each well. Serially dilute (1:2) the serum: pipet 100 μL of serum into the first well, mix carefully, and then transfer 100 μL of solution to the next well. Repeat seven times. To measure background reaction by the reagents, leave separate antigen-coated wells with diluent buffer only as blank control. Seal plate and incubate 2 h at RT or overnight at 4°C.
6. Wash plate as outlined in step 4. Add 100 μL of biotin-labeled goat anti-mouse-IgG1 or anti-mouse-IgG2a (1:5,000 in diluent buffer) to each well. Incubate 45 min at RT.
7. Wash plate as outlined in step 4. Add 100 μL of ExtrAvidin-Peroxidase (1:2,000 in diluent buffer) to each well (see Note 19). Incubate 45 min at RT.
8. Wash plate as outlined in step 4. Dissolve substrate OPD in ELISA substrate buffer (1 mg/mL) and add 1 μL/mL 30% H_2O_2 immediately before use. Add 100 μL of substrate solution to each well and watch for the reaction (appearance of yellow color) to occur.
9. Stop the enzymatic reaction by adding 100 μL of 1 M H_2SO_4 to each well (see Notes 20 and 21).
10. Measure optical density (OD) of the wells in a microplate reader at a wave length of 490 nm.

11. Subtract OD of blank wells from OD of sample wells. Calculate the antibody titer of the serum by linear regression analysis as the reciprocal serum dilution yielding an absorbance value of OD = 0.2.

3.5. Antigen-Specific Restimulation of Lymphocytes In Vitro

The following steps should be performed in a laminar flow under sterile conditions.

3.5.1. Preparation of Spleen and Lymph Node Cells

1. Kill mouse by cervical dislocation, remove the spleen and the draining inguinal/axillary lymph nodes and transfer the organs into a 50 mL polypropylen tube with 5 mL ice-cold supplemented EMEM (spleen) or a 15 mL polypropylen tube with 2 mL ice-cold supplemented EMEM (lymph nodes). Pool the organs of individual mice of one group. Store the tubes on ice.
2. Transfer organs in EMEM to 100 mm petri dishes. Generate single cell suspensions by mechanical disruption of the organs with sterile glass slides.
3. To remove cell debris, pass the cell suspension through a cell strainer into a 50 mL polypropylen tube (spleen) or a 15 mL polypropylen tube (lymph nodes).
4. Centrifuge the cells (10 min, $300 \times g$), discard the supernatant, and resuspend the cell pellet in 1 mL Gey's lysis buffer per spleen or 1 mL Gey's lysis buffer for all lymph nodes. Incubate 1 min at RT to remove erythrocytes.
5. Stop the reaction by adding 10 mL of supplemented EMEM per spleen and for all lymph nodes, respectively.
6. Centrifuge the cells for 10 min at $300 \times g$, discard the supernatant, and resuspend the cell pellet in 10 mL supplemented EMEM.
7. Repeat step 6, but resuspend the cell pellet in 10 mL supplemented IMDM.
8. Mix an aliquot of the cell suspension 1:10 with trypan blue and determine the living cell count under a stereo microscope using a hemocytometer (see Note 22).

3.5.2. Cell Culture

1. Adjust the cell number with supplemented IMDM to 1×10^7/mL. Transfer 500 μL of cell suspension (5×10^6 cells) into wells of a 24-well tissue culture plate.
2. Dilute βGal stock solution in supplemented IMDM to 50 μg/mL. Add 500 μL of antigen solution into a well for restimulation of cells (see Note 23) or 500 μL of supplemented IMDM without antigen in control wells (see Note 24).
3. Culture the cells for 72 h at 37°C in a humidified 10% CO_2 incubator.

4. Carefully harvest the culture supernatant in four aliquots of 200 μL each. Freeze at −20°C until thawed for determination of cytokines (see Note 25).

3.6. Determination of Cytokines (IFN-γ, IL-5) in Culture Supernatant by ELISA

1. Dilute capture antibodies in ELISA coating buffer (anti-IFN-γ: 2 μg/mL; anti-IL-5: 1 μg/mL). Pipet 50 μL of the appropriate antibody solution into wells of a 96-well ELISA microtiter plate and incubate overnight at 4°C.
2. Decant antibody solution by flicking the plate into a sink and firmly blotting the plate on paper towel.
3. To block residuary binding sites, add 200 μL of diluent buffer to each well. Keep the plate for 1 h at RT.
4. Wash plate twice with washing buffer (see Note 18).
5. Thaw aliquots of cell culture supernatants. Add 50 μL of diluent buffer to each well. Serially dilute (1:2) supernatant or cytokine standard: pipet 50 μL of supernatant or standard in appropriate dilution or concentration (see Note 26) into the first well, mix carefully, and then transfer 50 μL of solution to the next well. Repeat four times. To measure background reaction by the reagents, leave separate antibody-coated wells with diluent buffer only as blank control. Seal plate and incubate overnight at 4°C.
6. Wash plate as outlined in step 4. Add 50 μL of the appropriate cytokine-specific biotin-labeled detection antibody (1:5,000 in diluent buffer) to each well. Incubate 45 min at RT.
7. Wash plate as outlined in step 4. Add 50 μL of ExtrAvidin-Peroxidase (1:2,000 in diluent buffer) to each well (see Note 19). Incubate 45 min at RT.
8. Develop the ELISA as outlined in Subheading 3.4, steps 8–10, with the exception that the reagents are added to the wells in a volume of 50 μL.
9. Subtract OD of blank wells from OD of sample wells. Determine by linear regression analysis the reciprocal dilution of the supernatant yielding an absorbance value of OD = 0.2. Calculate the cytokine concentration in the supernatant using the linear regression curve of the standard.

3.7. ELISPOT Assay for Enumeration of IFN-γ-Producing CD8+ Effector T Cells (See Note 27)

The subsequent steps 1–7 should be performed in a laminar flow under sterile conditions.

1. Dilute anti-IFN-γ capture antibody in PBS to 10 μg/mL. Pipet 50 μL of the antibody solution into wells of a 96-well ELISPOT microtiter plate and incubate overnight at 4°C (see Note 28).
2. Wash plate twice with sterile PBS (see Note 18).
3. To block residuary binding sites, add 200 μL of supplemented IMDM to each well. Keep the plate for 2 h at 37°C.

4. Prepare spleen cells of mice as outlined in Subheading 3.5.1. Adjust the spleen cell number with supplemented IMDM to 1×10^7/mL.
5. Remove blocking solution by pipetting. Transfer 100 μL (1×10^6 cells) or 20 μL (2×10^5 cells) (see Note 29) of cell suspension into wells.
6. Dilute rIL-2 in supplemented IMDM to 1,000 U/mL. Add 50 μL of rIL-2 to each well (see Note 30).
7. Dilute peptide $\beta Gal_{876-884}$ stock solution in supplemented IMDM to 0.4 μg/mL. Pipet 50 μL of peptide solution into wells for stimulation (see Note 31) or 50 μL of supplemented IMDM without peptide in control wells (see Note 32).
8. Culture the cells for 22 h at 37°C in a humidified 10% CO_2 incubator.
9. Wash plate twice with washing buffer. Add 50 μL of biotin-labeled anti-IFN-γ detection antibody (1:5,000 in diluent buffer) to each well. Incubate 1 h at 37°C.
10. Wash plate as outlined in step 4. Add 50 μL of ExtrAvidin-Peroxidase (1:2,000 in diluent buffer) to each well. Incubate 1 h at 37°C.
11. Dilute 525 μL of ELISPOT substrate stock solution in 10 mL ELISPOT substrate buffer. Agitate thoroughly for 20 min. Filter the AEC solution using a 0.45 μm filter to remove precipitate.
12. Wash plate as outlined in step 4.
13. Add 1 μL/mL 30% H_2O_2 to the AEC substrate solution immediately before use. Add 50 μL of substrate to each well and watch for the reaction (appearance of red spots) to occur.
14. Stop the reaction when spots have reached the desired intensity by flicking the plate into a sink (see Note 33). Turn the plate upside down and allow the plate to air-dry overnight (see Note 34).
15. Store plate in the dark prior to analysis to prevent the fading of spots.
16. Count the spots using an EliSpot Reader System (see Note 35).

4. Notes

1. Plasmid preparations, dissolved and diluted in endotoxin-free water, should be free of endotoxin and protein contaminations.
2. Mice should be maintained under specific pathogen-free conditions on a standard diet. The principles of laboratory animal

care (NIH publication no. 85-23, revised 1985) should be followed. Elicit approval of the local ethics committee.

3. ELISA microtiter plates with high binding polystyrene surface (e.g., Microlon ELISA plates, Greiner bio-one, Frickenhausen, Germany) are recommended.
4. ELISPOT filter plates with nitrocellulose membrane (e.g., MultiScreen HA filter plates; Millipore, Bedford, MA, USA) are recommended.
5. Prepare ELISPOT substrate stock solution immediately before use. Solution should not be stored longer than 24 h.
6. Always use fresh spermidine solution.
7. DNA-coated gold particles should sink immediately to the bottom of the tube as they are chemically precipitated due to the addition of $CaCl_2$.
8. Ensure equal distribution of the gold suspension in the tubing and thus optimal coating of the tubing´s inner surface during the loading process.
9. For optimal induction of transgene-specific immune responses, biolistic transfection should be repeated up to five times in weekly intervals (16).
10. Take care not to injure the skin during the shaving process. Skin areas with scratches or open wounds should not be used for bombardment.
11. Depth of penetration of gold particles in the skin and consequently the efficiency of biolistic transfection is dependent on the helium pressure used for bombardment. Optionally try different pressures to optimize gene gun-mediated immunization.
12. Always use non-overlapping areas of abdominal skin for bombardment.
13. We have had best results in the induction of βGal-specific immune responses using a total amount of 4 μg βGal-encoding plasmid DNA for bombardment. However, to ensure optimal efficiency of gene gun-mediated immunization with other antigen-encoding plasmids, try different amounts of plasmid DNA.
14. A brown spot on the shaved skin indicates a successful application of particles.
15. To ensure even application of particles on the skin, softly touch the target area with the spacer so that the spacer is flush and the gene gun is perpendicular to the skin while pulling the trigger.
16. Usually immunoglobulins can be determined at the earliest 7 days after biolistic transfection. However, antibody produc-

tion in mice drastically increases with progressing time and with number of immunizations.

17. Minimum volume of blood sample should be 0.5 mL. Puncture of the retro-orbital plexus usually yields more blood than tail vein bleeding.
18. Remove the residuary washing buffer or PBS after each wash step by firmly blotting the plate on paper towel.
19. To minimize unspecific reaction of the substrate in the next step, add the enzyme without touching the walls of the well.
20. The reaction should be stopped at the latest, when the substrate solution in the blank control wells starts to change its color.
21. The color of the substrate solution in the wells changes from yellow into orange/brown.
22. Living cells exclude trypan blue and appear colorless under the microscope; dead cells incorporate trypan blue and appear blue.
23. The end concentration of βGal in the culture should account for 25 μg/mL.
24. We usually set up quadruplicates for stimulated wells and control wells, respectively.
25. It is recommended to freeze and store the aliquots in wells of 96-well culture plates. Use separate plates for the different aliquots in order to avoid several freeze–thaw cycles for the supernatant which is not required. Seal the plates with parafilm.
26. We usually use start concentrations of 5 ng/mL (IFN-γ) and 2 ng/mL (IL-5), respectively.
27. The frequency of βGal-specific $CD8^+$ effector T cells among splenocytes can be determined by stimulation of IFN-γ production by spleen cells using the $H\text{-}2L^d$-binding βGal-derived nonamer peptide TPHPARIGL ($\beta Gal_{876-884}$) (22).
28. Ensure that the nitrocellulose bottom of the well is completely covered with the antibody solution.
29. To equalize the volume, add another 80 μL of supplemented IMDM to the wells.
30. The end concentration of rIL-2 in the culture should account for 250 U/mL.
31. The end concentration of peptide $\beta Gal_{876-884}$ in the culture should account for 0.1 μg/mL.
32. Be aware that it is important to add first the rIL-2 and than the peptide $\beta Gal_{876-884}$.
33. The reaction can be stopped when discrete spots are visible, usually after 3–5 min.

34. The membrane tends to develop a more or less intensive background staining in the first hours after development of the ELISPOT. The background staining disappears after drying and the spots are now distinguishable.
35. Evaluate the spots by means of size, intensity, and shape. Always use (unstimulated) control wells to set thresholds for positive validation of spots.

Acknowledgment

This work was supported by the Stiftung Rheinland-Pfalz für Innovation (15212-386261/761).

References

1. Condon C et al (1996) DNA-based immunization by in vivo transfection of dendritic cells. Nat Med 2:1122–1128
2. Porgador A et al (1998) Predominant role for directly transfected dendritic cells in antigen presentation to CD8+ T cells after gene gun immunization. J Exp Med 188:1075–1082
3. Garg S et al (2003) Genetic tagging shows increased frequency and longevity of antigen-presenting, skin-derived dendritic cells in vivo. Nat Immunol 4:907–912
4. Chen D, Payne LG (2002) Targeting epidermal Langerhans cells by epidermal powder immunization. Cell Res 12:97–104
5. Stoecklinger A et al (2007) Epidermal Langerhans cells are dispensable for humoral and cell-mediated immunity elicited by gene gun immunization. J Immunol 179:886–893
6. Nagao K et al (2009) Murine epidermal Langerhans cells and langerin-expressing dermal dendritic cells are unrelated and exhibit distinct functions. Proc Natl Acad Sci USA 106:3312–3317
7. Stoecklinger A et al (2011) Langerin+ dermal dendritic cells are critical for CD8+ T cell activation and IgH γ-1 class switching in response to gene gun vaccines. J Immunol 186:1377–1383
8. Ross R et al (2003) Transcriptional targeting of dendritic cells for gene therapy using the promoter of the cytoskeletal protein fascin. Gene Ther 10:1035–1040
9. Mosialos G et al (1996) Circulating human dendritic cells differentially express high levels of a 55-kd actin-bundling protein. Am J Pathol 148:593–600
10. Ross R et al (1998) The actin-bundling protein fascin is involved in the formation of dendritic processes in maturing epidermal Langerhans cells. J Immunol 160:3776–3782
11. Ross R et al (2000) Expression of the actin-bundling protein fascin in cultured human dendritic cells correlates with dendritic morphology and cell differentiation. J Invest Dermatol 115:658–663
12. Bros M et al (2003) The human fascin gene promoter is highly active in mature dendritic cells due to a stage-specific enhancer. J Immunol 171:1825–1834
13. Mosialos G et al (1994) Epstein-Barr virus infection induces expression in B lymphocytes of a novel gene encoding an evolutionarily conserved 55-kd actin-bundling protein. J Virol 68:7320–7328
14. Edwards RA et al (1995) Cloning and expression of a murine fascin homolog from mouse brain. J Biol Chem 270:10764–10770
15. Jaffe R, DeVaughn D, Langhoff E (1998) Fascin and the differential diagnosis of childhood histiocytic lesions. Pediatr Dev Pathol 1:216–221
16. Sudowe S et al (2003) Transcriptional targeting of dendritic cells in gene gun-mediated DNA immunization favours the induction of type 1 immune responses. Mol Ther 8:567–575
17. Sudowe S et al (2006) Prophylactic and therapeutic intervention in IgE responses by biolistic DNA vaccination primarily targeting dendritic cells. J Allergy Clin Immunol 117: 196–203
18. Sudowe S et al (2008) Uptake and presentation of exogenous antigen and presentation of endogenously produced antigen by skin dendritic cells represent equivalent pathways for the priming of cellular immune responses fol-

lowing biolistic DNA immunization. Immunology 128:e193–e205

19. Ludwig-Portugall I et al (2004) Prevention of long-term IgE antibody production by gene gun-mediated DNA vaccination. J Allergy Clin Immunol 114:951–957
20. Spitalny GL, Havell EA (1984) Monoclonal antibody to murine gamma interferon inhibits lymphokine-induced antiviral and macrophage tumoricidal activities. J Exp Med 159: 1560–1565
21. Prat M et al (1984) Monoclonal antibodies against murine gamma interferon. Proc Natl Acad Sci USA 81:4515–4519
22. Gavin MA et al (1993) Alkali hydrolysis of recombinant proteins allows for the rapid identification of class I MHC-restricted CTL epitopes. J Immunol 151:3971–3980

Chapter 18

Particle-Mediated Administration of Plasmid DNA on Corneas of BALB/c Mice

Dirk Bauer, Susanne Wasmuth, Mengji Lu, and Arnd Heiligenhaus

Abstract

Gene gun administration of DNA is an invaluable technique for transfecting tissues with only 1 μg DNA/shot. Here, we describe a transfection technique of healthy corneas of BALB/c mice with a standard gene gun, using a technique that can avoid tissue destruction even when high pressure is used (e.g., 700 psi). The focal transfection of the cornea to improve corneal disease may be an advantage over other transfection methods in order to avoid unwanted bystander transfection in other compartments of the eye or body.

Key words: Biolistic, Gene gun, Transfection, Cornea, Mouse, BALB/c, Herpetic stromal keratitis

1. Introduction

Previously, new methods for the application of protective cytokine plasmid DNA for the transfection of mouse corneas has been described (1). The results showed that topical application of plasmid DNA onto the cornea may represent an attractive therapeutic approach with a great potential for clinical use. The drawbacks of these methods are that repetitive application and high dosages of DNA (up to 100 μg DNA/cornea) are required to achieve the effect. The high DNA doses can lead to undesired spread of plasmid DNA into the circulation and may result in unwanted bystander transfections in the iris, ciliar body, lymph nodes, and spleen (1, 2). Although the clinical effects have been attributed primarily to local cytokine expression, the spread of the plasmid DNA into other organs may significantly contribute to the therapeutic effect (1).

It has been shown that the gene gun treatment is a more effective vaccination method compared with the intramuscular injection of naked DNA. Most recently, the gene gun has been

Stephan Sudowe and Angelika B. Reske-Kunz (eds.), *Biolistic DNA Delivery: Methods and Protocols*, Methods in Molecular Biology, vol. 940, DOI 10.1007/978-1-62703-110-3_18, © Springer Science+Business Media, LLC 2013

successfully used to deliver DNA into the cornea of rabbits to induce expression of the related product (3–5). A single administration of IL-4 or IL-10 plasmid DNA onto the murine cornea before infection was useful as a protective application in experimental herpetic stromal keratitis (HSK), which is an experimental model of human herpetic keratitis that is a major cause of blindness world wide (6). Furthermore, the beneficial effect of IL-4 and CTLA4 as applied by gene gun technique was also shown in a murine corneal transplantation model (7–9).

Here, we demonstrate that this technique could be used to deliver plasmid loaded gold particles into the mouse corneas to induce transfection without inducing significant tissue destructions even with high helium pressure (up to 700 psi). However, we have shown previously that the transfection rate in the cornea was rather low with the gene gun method by using β-galactosidase as a reporter gene. β-galactosidase positive cells could be found on days 2 and 4 after gene gun delivery, but not later. Further, our results showed that β-galactosidase+ cells could be found in the epithelium, but no positively stained cells were found in the stromal cell layers, even when higher pressure was used (6).

The gene gun method described here has advantages as a research tool for in vivo studies on the influence of gene therapy for corneal diseases compared to other methods of DNA application.

2. Materials

1. Female BALB/c mice (6–8 weeks of age) (see Note 1).
2. Anesthetization: mix 1 mL ketamin hydrochloride (Ketamin 50), 1 mL mepivacaine hydrochloride (Scandicain 1%), 4 mL phosphate-buffered saline (PBS) (see Note 2).
3. 0.5% gentamicin ophthalmic solution (Merck, Germany).
4. Plasmids: B-gal (Clontech) and pCR 3.1 vector (Invitrogen). Use the Maxi plasmid purification kit (Qiagen) for preparing the plasmids.
5. Preparation of gold beads: coat gold beads according to the manufacturer's instructions (please find the instruction manual under: http://www.bio-rad.com/LifeScience/pdf/Bulletin_9541.pdf). Use 0.25 mg of gold/shot.
6. Plastic box for corneal transfection experiments: To protect the mouse eyes against unwanted tissue disruption due to the rapid helium release and to prevent unwanted strewing of gold particles to other tissues, use the cap of a plastic box (Cat-No: 975 502, Greiner, Germany) with a drilled hole of 3 mm in diameter. Use a 4 mm drill from the opposite site to avoid sharp edges. Remove the mounting parts of the cap before usage (Fig. 1).

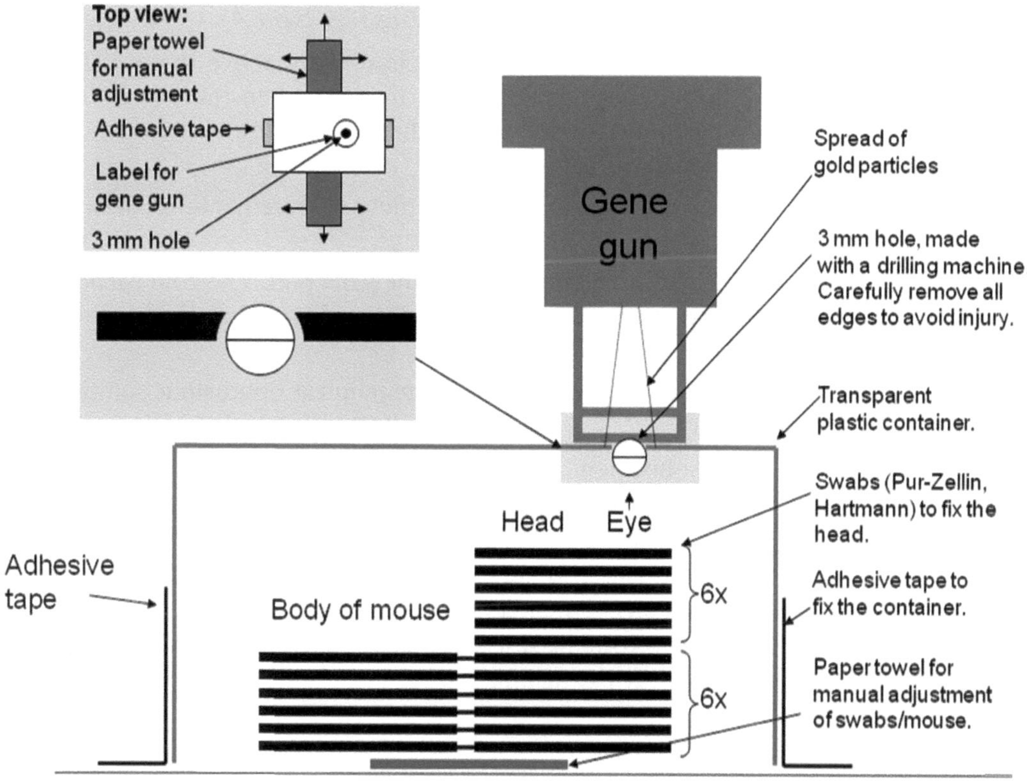

Fig. 1. Compositions for the gene gun treatment of healthy BALB/c mice. The head (but not the rest of the body) (see Note 5) of the mouse is fixed between swabs and plastic box. When the correct position of the mouse eye under the 3 mm hole is found, the box is fixed with the adhesive tape. The gene gun is then placed directly above the labeled hole to treat the cornea. After this the box is removed from the animal.

7. Helios gene gun (Bio-Rad).
8. Hematoxylin-eosin (Sigma-Aldrich).
9. β-galactosidase test kit (Roche).
10. Paper towels and swabs (Pur-Zellin, Hartmann).

3. Methods

Corneal administration of plasmid DNA:

1. Anesthetize mice by intraperitoneal injection of 0.1 mL of ketamine hydrochloride and mepivacaine hydrochloride in PBS per 10 g body weight (see Note 2).
2. Bend the paper towel and place the swabs on the towel (Fig. 1; see Note 3).

3. Place the mouse on the swabs (see Note 4).
4. Place the box with the hole directly above the eye. Use the paper towel to adjust the mouse. When the box is correctly positioned on the eye, fix the box with the adhesive tape on the desk (see Notes 5 and 6).
5. Take the gene gun and adjust it directly above the eye (see Note 7).
6. Release the pressure from the gene gun. The gold particles and the success of the treatment can be estimated after the treatment (see Notes 8–11).
7. Treat the mice with 0.5% gentamicin ophthalmic solution (see Note 12).
8. For histologic analysis, section paraffin-embedded tissue (5 μm thick) and stain according to a standard hematoxylin-eosin protocol (Fig. 2a–d) (10).
9. For β-galactosidase staining, prepare 5 μm-thick cryostat sections from treated and control eyes and stain by using the β-galactosidase test kit (Fig. 2e) according to the instructions given by the manufacturer.

4. Notes

1. The use of other mouse strains may also be possible. It may be necessary to adapt the protocol because of anatomic differences.
2. An optimal anesthesia is important for the success of the method. Use 0.1 mL/10 g mouse intraperitoneally. The duration of the anesthesia is 5–10 min. After repeated usage of the treatment, it may be necessary to increase the drug dosage.
3. Use 12 swabs to fix the head of the mice and 6 swabs for the body (Swabs: Pur-Zellin, Hartmann).
4. Cautiously open the eye to stretch the eye lid before procedure.
5. Compression of the mouse body should be avoided to prevent suffocation.
6. Always wear protection of the eye, ear, and disposable gloves during the procedure.
7. A marking on the box may be helpful to get the gene gun positioned above the eye.
8. The use of the gene gun may result in an acoustic shock, especially when higher pressures are used. Treatment of

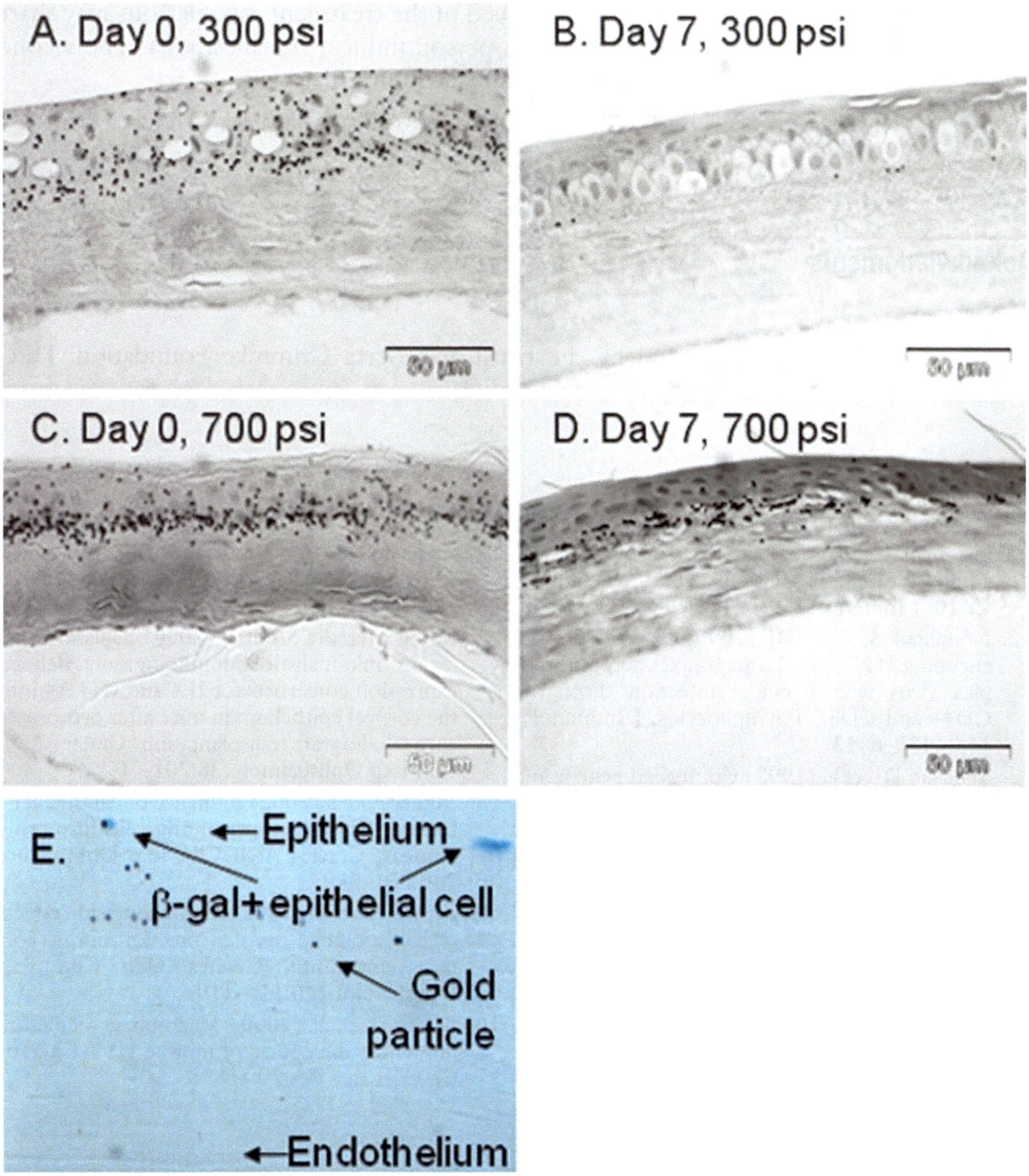

Fig. 2. Photomicrographs of corneas after gene gun treatment (300 or 700 psi). Central cornea section from 8-week-old BALB/c mice treated with 300 psi on day 0 (**a**) and after day 7 (**b**). Mouse corneas after treatment with 700 psi on day 0 (**c**) and after day 7 (**d**). One microgram of β-gal plasmid DNA was administered to the mouse cornea. (**e**) Representative section for the expression of β-gal in the corneas 2 days after gene gun treatment (300 psi). The results indicate that the transfection rate in the cornea was rather low.

mice in a separate room may prevent harmful stress for the other mice.

9. A portion of the gold particles are lost (on the box) during the treatment.
10. The gold particles that were administered on the cornea can be seen for 3–5 min.

11. To increase the speed of the treatment, two persons may share the work: the first person induces the anesthesia. The second person performs the gene gun treatment.
12. Use eye drops containing 0.5% gentamicin (Merck, Germany) to avoid bacterial infection.

Acknowledgments

Supported by the Ernst and Berta Grimmke Foundation, DFG Ba2248/1-2, He 1877/12-2.

References

1. Daheshia M et al (1997) Suppression of ongoing ocular inflammatory disease by topical administration of plasmid DNA encoding IL-10. J Immunol 159:1945–1952
2. Noisakran S, Carr DJ (2000) Plasmid DNA encoding IFN-alpha 1 antagonizes herpes simplex virus type 1 ocular infection through CD4+ and CD8+ T lymphocytes. J Immunol 164:6435–6443
3. Tanelian DL et al (1997) Controlled gene gun delivery and expression of DNA within the cornea. Biotechniques 23:484–488
4. Tanelian DL et al (1997) Semaphorin III can repulse and inhibit adult sensory afferents in vivo. Nat Med 3:1398–1401
5. Shiraishi A et al (1998) Identification of the cornea-specific keratin 12 promoter by in vivo particle-mediated gene transfer. Invest Ophthalmol Vis Sci 39:2554–2561
6. Bauer D et al (2006) Immunomodulation by topical particle-mediated administration of cytokine plasmid DNA suppresses herpetic stromal keratitis without impairment of antiviral defense. Graefes Arch Clin Exp Ophthalmol 244:216–225
7. König Merediz SA et al (2000) Ballistic transfer of minimalistic immunologically defined expression constructs for IL4 and CTLA4 into the corneal epithelium in mice after orthotopic corneal allograft transplantation. Graefes Arch Clin Exp Ophthalmol 238:701–707
8. Müller A et al (2002) Influence of ballistic gene transfer on antigen-presenting cells in murine corneas. Graefes Arch Clin Exp Ophthalmol 240:851–859
9. Zhang EP et al (2002) Minimizing side effects of ballistic gene transfer into the murine corneal epithelium. Graefes Arch Clin Exp Ophthalmol 240:114–119
10. Bauer D et al (2000) Macrophage-depletion influences the course of murine HSV-1 keratitis. Curr Eye Res 20:45–53

Part V

Biolistic DNA Vaccination in Animal Disease Models

Chapter 19

Optimizing Particle-Mediated Epidermal Delivery of an Influenza DNA Vaccine in Ferrets

Eric J. Yager, Cristy Stagnar, Ragisha Gopalakrishnan, James T. Fuller, and Deborah H. Fuller

Abstract

Particle-mediated DNA delivery technologies ("gene guns") have been shown in both animal and clinical studies to be an effective means of increasing the immunogenicity and protective efficacy of DNA vaccines. The primary goal in optimizing particle-mediated epidermal delivery (PMED) vaccination in different animal models is to achieve delivery of DNA-coated gold beads into the viable epidermis. Two key parameters that influence this outcome include the delivery pressure, which controls the penetrative force of the beads into the skin, and the anatomical site of DNA delivery. Although the ferret has been extensively used as an experimental model for influenza infection in humans, very few studies have investigated the capacity for PMED DNA vaccination to induce protective immune responses in ferrets. Here we describe methods to optimize DNA vaccine delivery using the PowderJect XR1 gene delivery in ferrets. We first assess the effects of firing pressure on both the delivery of DNA-coated gold beads into the desired epidermal layer and the degree of DNA vaccine reporter gene expression at the target site. Second, we evaluate the impact of vaccination site (skin or tongue) on DNA vaccine immunogenicity by measuring serum antibody responses to the model antigens influenza virus hemagglutinin and hepatitis B core antigen. Results from these studies support the use of the PowderJect XR1 device in ferrets for the study of prophylactic and therapeutic DNA vaccines against clinically important diseases such as influenza virus infection.

Key words: Influenza A virus, Hemagglutinin, DNA vaccine, Ferrets, PMED, Antibodies, Epidermis

1. Introduction

Particle-mediated epidermal delivery (PMED; also known as gene gun or biolistic delivery) of DNA vaccines has been successfully employed to induce immune responses in multiple animal models and against various infectious diseases, including influenza (1).

Stephan Sudowe and Angelika B. Reske-Kunz (eds.), *Biolistic DNA Delivery: Methods and Protocols*, Methods in Molecular Biology, vol. 940, DOI 10.1007/978-1-62703-110-3_19, © Springer Science+Business Media, LLC 2013

Additionally, PMED is, to date, one of the most effective strategies for inducing immune responses in humans (2, 3). The effectiveness of PMED is due to direct intracellular delivery of DNA into cells in vivo coupled to targeting the highly immune-competent epidermis of the skin (4, 5). However, PMED can be highly variable between species and successful delivery conditions for biolistic devices in mouse models are generally not optimal when applied to other species (2). Therefore, it is imperative that researchers using biolistic devices in different animal models optimize delivery parameters for each species.

Variables that influence successful PMED delivery in a particular species include the particular type of device employed, the delivery pressure, the site of delivery, and the vaccine dose (adjusted by increasing/decreasing the amount of DNA precipitated onto gold beads or the number of targets administered into the skin). Biolistic devices most commonly used among research include the Bio-Rad Helios device (6, 7), the PowderJect XR1 research device (later called PowderMed) (8), and the PowderMed ND series clinical devices (recently acquired by Pfizer) that were designed as disposable prototypes for human clinical trials (9–11). Devices are not equivalent and delivery parameters determined to be optimal for one device should not be used with a different device. For example, the Bio-Rad Helios biolistics device delivers a smaller, denser target into the skin than the XR1 or ND PMED delivery devices and requires a much higher dose (e.g., 50–100 targets) per immunization to induce a significant response in a larger animal model, such as a nonhuman primate, when compared to the XR1 or ND devices (e.g., 6–12 targets) (12–15).

Particle-mediated DNA vaccine delivery into the skin is not intradermal immunization. Instead, PMED uniquely delivers DNA vaccines directly into the cells of the epidermis, a skin layer that differs from the dermis in both cell composition and the capacity to express DNA vaccines. The goal for PMED is to deliver DNA-coated gold particles into the viable epidermis where the density of antigen presenting cells and keratinocytes capable of expressing DNA vaccines is highest. If gold bead penetration is too shallow, immunogenicity is reduced because the majority of the DNA vaccine dose is lost to the non-viable stratum corneum. On the other hand, if gold bead penetration is too deep, immunogenicity is also reduced because there are fewer cells in the dermis to express the DNA vaccine and penetration of the beads deeper into the skin compromises cell viability in the epidermis (16). For this reason, anatomical differences in the target skin site are the primary factors influencing optimal PMED delivery conditions in different animal models.

The translation of PMED influenza DNA vaccines from mice to swine to human clinical trials provides an example of how the device and delivery conditions were adjusted and optimized with each new species to achieve proof of concept in small and large animal models and to develop a device optimal for immunizing humans in clinical trials. Influenza DNA vaccine studies in mice provided the first evidence that PMED delivery into the epidermis provides superior immunogenicity and protection when compared to other routes of DNA delivery including intradermal and intramuscular immunization (17). The swine model provided proof of concept in a larger animal model of influenza (18) and was the animal model used to develop a disposable prototype device for the clinic because swine skin is anatomically similar to human skin (10).

Because they share a number of anatomical and physiological features with humans, ferrets have been used extensively in biomedical research in areas such as airway physiology, toxicology, reproductive physiology, endocrinology, and virology (19). Importantly, ferrets are widely accepted as the small animal model of choice for the study of influenza and to investigate the capacity of a vaccine to afford universal protection against diverse strains of influenza because they are susceptible to the same influenza viruses humans are susceptible to and develop many of the symptoms of influenza that are seen in humans (20, 21). Although a number of studies have examined the immunogenicity and protective efficacy of DNA vaccines in ferrets (3), very few studies have investigated the capacity for PMED DNA vaccination to elicit protective cellular and humoral responses against influenza in ferrets. As the first step towards critically evaluating PMED-based influenza DNA vaccines in ferrets, we conducted a study to optimize PMED DNA vaccination in ferrets by adjusting the delivery pressure and the site of delivery.

Here, we have detailed methods for optimizing PMED delivery conditions in ferrets as an example that could be used in other animal models. Our results show that, as expected, PMED immunogenicity in ferrets is influenced by delivery pressure. Furthermore, we show that particle-mediated delivery into the tongue of ferrets as an alternate route of DNA immunization was equally effective as skin for inducing serum antibody responses in ferrets, an outcome that is consistent with studies comparing skin and tongue PMED immunizations in swine (18). Additional studies are underway to determine if PMED administration into the tongue may offer a benefit for inducing mucosal responses and improved protection from heterosubtypic influenza virus challenges in ferrets.

2. Materials

2.1. Ferrets

Sixteen week-old outbreed male ferrets (Marshall Farms). Elicit approval for all experimental procedures by the Institutional Animal Care and Use Committee (IACUC).

2.2. Plasmids and Recombinant Proteins

1. Plasmid pPML7800 encoding the full-length hemagglutinin (HA) protein from influenza A/New Caledonia/20/99 (H1N1), constructed by inserting the codon-optimized HA gene into the pPJV7563 expression cassette (see Note 1).
2. Plasmid pCMV-βgal encoding the full-length β-galactosidase under the control of the human cytomegalovirus (CMV) immediate early promoter.
3. Plasmid pWRG7063 encoding the hepatitis B virus core antigen (HBcAg) under the control of the human CMV immediate early promoter.
4. Recombinant influenza A/New Caledonia/20/99 HA protein (Protein Sciences Corporation).
5. Recombinant HBcAg, amino acids 1–183 (Meridian Life Sciences, Inc.).

2.3. Particle-Mediated Epidermal Delivery of DNA Plasmids

1. Gene gun (XR1 research delivery device; Pfizer, Inc.).
2. Compressed helium gas (grade 4.5) with helium regulator.
3. DNA/Gold bead-coated cartridges (see Note 2).
4. Anesthetic ketamine hydrochloride (Ketaset®; Wyeth Animal Health) and xylazine hydrochloride (Xyla-Ject®; Phoenix Pharmaceutical, Inc.).
5. Electric animal clippers.

2.4. Measurement of β-Galactosidase Reporter Gene Expression in Tissue

1. Sterile surgical instruments: forceps, scalpel, and scissors.
2. Sterile phosphate buffered saline (PBS), 1×.
3. Sterile Dounce tissue homogenizers, 7 mL (BellCo Glass, Inc.).
4. cOmplete, Mini, EDTA-free protease inhibitor cocktail tablets (Roche Applied Science).
5. Galacto-Star™ mammalian β-galactosidase assay kit (Applied Biosystems).
6. Lyophilized β-galactosidase (Sigma-Aldrich) reconstituted to 1 mg/mL in 0.1 M sodium phosphate (pH 7.0), 0.1% BSA.
7. White luminescence-grade 96-well flat bottom microplate.
8. Microplate Luminometer.
9. DC™ Protein Assay Kit II (Bio-Rad).

2.5. X-Gal Staining of Intact Tissue for β-Galactosidase

1. Fixative: 2% paraformaldehyde in PBS.
2. X-gal staining solution (in PBS): 5 mM potassium ferricyanide, 5 mM ferrocyanide, 2 mM MgCl, 0.01% sodium deoxycholate, 0.02% Nonidet P-40 (NP-40).
3. X-gal stock solution (40×): 40 mg/mL X-gal (5-bromo-4-cholor-3-indolyl-β-D-galactopyranoside) in dimethylformamide; store at –20°C in small aliquots protected from light.

4. 6-well flat bottom tissue culture plate.
5. PBS, 1×.

2.6. Enzyme-Linked Immunosorbent Assays (ELISA)

1. BD Vacutainer® SST™ glass serum tubes, 13×75 mm×2.5 mL.
2. Goat anti-ferret IgG (H+L chain) antibody, horseradish peroxidase conjugated (Bethyl Laboratories, Inc.).
3. High protein-binding capacity polystyrene 96-well EIA plates (Corning-Costar).
4. Reagent reservoirs, disposable.
5. Bicarbonate/carbonate coating buffer (100 mM): 3.03 g Na_2CO_3, 6.0 g $NaHCO_3$, 1,000 mL distilled water; pH 9.5.
6. Blocking buffer: PBS containing 10% bovine serum albumin.
7. Wash buffer: PBS containing 1% bovine serum albumin and 0.05% Tween 20; PBS-T.
8. 3,3′,5,5′-Tetramethylbenzidine (TMB) liquid substrate.
9. 1 M Sulfuric acid in water.
10. Absorbance microplate reader.

2.7. Hemagglutination Inhibition Assay

1. Serum samples, negative control sera, and positive control sera (see Note 3).
2. Turkey red blood cells (TRBC; Lampire Biologics).
3. Sterile 1× Dulbecco's phosphate buffered saline (DPBS).
4. Influenza A/New Caledonia/20/99 virus (see Note 4).
5. Receptor-destroying enzyme II (RDE; Accurate Chemical).
6. 96-well round bottom plates.
7. Sterile micro-centrifuge tubes.
8. Multi-channel pipette (20–200 μL range) and tips.
9. Sterile reagent reservoirs, disposable.

3. Methods

3.1. Gene Gun Inoculation I: β-Galactosidase Reporter Gene Construct

1. Anesthetize ferrets by administering a cocktail of ketamine (10 mg/kg) and xylazine (2 mg/kg) via subcutaneous injection.
2. Shave abdomens using electric animal clippers.
3. Inoculate ferrets by gene gun using the XR1 device at 300, 350, or 400 psi compressed helium on their shaved abdomen. A total of 1 μg pCMV-βgal DNA/mg-coated 1–3 μm-sized gold particles and 1 mg of gold was delivered per actuation.

Gene gun inoculations at the three different pressure settings were performed in triplicate (see Note 5).

4. Euthanize ferrets as per IACUC guidelines 24 h post-vaccination for the collection of skin samples.

3.2. Measurement of β-Galactosidase Reporter Gene Expression in Tissue

1. Excise skin at each of the target sites (ca. 4 cm × 4 cm square) using sterilized surgical instruments (see Note 6).
2. Wash skin samples twice with PBS and then place individually in sterilized Dounce tissue homogenizers.
3. Homogenize skin samples in 2 mL of ice-cold Galacto-Star™ lysis buffer containing 1× cOmplete, Mini, EDTA-free protease inhibitor cocktail.
4. Transfer the extracts (supernatants) to a fresh tube and either analyze immediately or store at −70°C for future analyses.
5. Prepare standards by serially diluting β-galactosidase in Galacto-Star™ lysis buffer (2–20 ng of enzyme).
6. Prepare Reaction Buffer by diluting the Galacto-Star™ substrate 1:50 with the provided Reaction Buffer Diluent.
7. Transfer 10 μL of each extract and the diluted standards to wells of a microplate.
8. Add 100 μL of Reaction Buffer mix to each well, incubate at room temperature, in the dark, for 60 min.
9. Measure the signals in a luminometer (0.1–1 s/well).
10. Determine the protein concentration of the extracts using the DC™ Protein Assay microassay procedure following the manufacturer's instructions (see Note 7).
11. Calculate the ratio of β-galactosidase expression to the total amount of protein in each of the extracts (Fig. 1a).

3.3. X-Gal Staining of Intact Tissue for β-Galactosidase

1. Place collected skin tissue samples in separate wells of a 6-well flat bottom tissue culture plate containing 3 mL of ice-cold PBS (1×).
2. Remove the PBS by aspiration and replace with enough fixative (2% paraformaldehyde) to completely cover the skin sample. Incubate on ice (or at 4°C) for 30 min to several hours (to be determined empirically).
3. Remove the fixative by aspiration and rinse the samples 3–5× with 1× PBS.
4. Dilute the X-gal stock solution to 1× in the X-gal staining solution and incubate with the tissues at 37°C for 2–4 h (or until the desired level of staining is achieved).
5. Rinse the tissues many times in 1× PBS until the solution is no longer yellow in color (typically takes about five washes).

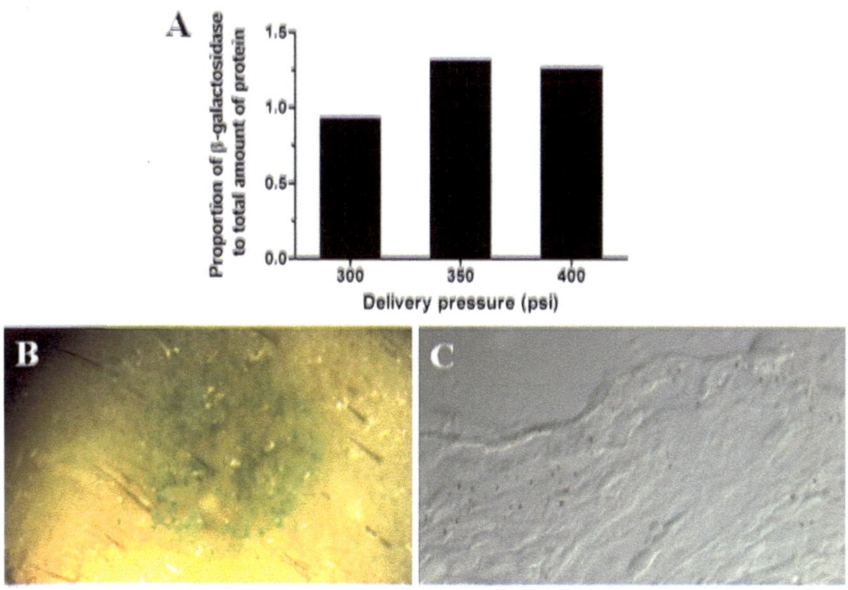

Fig. 1. Optimization of delivery pressure for effective particle-mediated epidermal delivery (PMED) DNA vaccination in ferrets. (**a**) Calculation of β-galactosidase reporter gene expression in skin samples taken from ferrets vaccinated with pCMV-βgal DNA using the XR1 gene gun device at the indicated delivery pressures. The highest level of β-galactosidase activity in skin homogenates at 24 h post-vaccination was detected at a delivery pressures of 350 and 400 psi. (**b**) Whole tissue X-gal histochemical staining revealed β-galactosidase reporter gene expression localized to the target site of PMED DNA vaccination (400 psi, 24 h post-vaccination). (**c**) Distribution of gold beads on ferret dermal tissue delivered by the XR1 gene gun device at a pressure setting of 400 psi. Notice how at this pressure setting the gold beads, seen as *black dots*, localize predominantly to the desired epidermal layer (magnification = ×400).

6. View the tissues under bright field microscopy to detect positive staining (light blue to dark blue coloring; see Note 6 and Fig. 1b).

3.4. Gene Gun Inoculation II: Influenza HA and HBcAg DNA Vaccines

1. Anesthetize ferrets by administering a cocktail of ketamine (10 mg/kg) and xylazine (2 mg/kg) via subcutaneous injection.
2. Gene gun inoculate ferrets at 400 psi compressed helium, one group on their shaved abdomen and the other group on the ventral and dorsal sides of their tongue, using a total of 2 μg DNA/mg-coated 1–3 μm-sized gold particles and 0.5 mg gold/shot giving 1 μg DNA/shot; 4 shots/ferret for a total dose of 4 μg of each plasmid DNA.
3. Gene gun immunize each group of ferrets again 4 weeks after priming.
4. Collect sera from ferrets prior to vaccination, at 4 weeks after vaccination, and at 2 weeks following booster immunization.

3.5. Intravenous Blood Sampling from Immunized Ferrets

1. Anesthetize ferrets by administrating a cocktail of ketamine and xylazine (see step 1 in Subheading 3.2).
2. Use electric animal clippers to remove hair from the neck area (ventral side) of anesthetized ferrets.
3. Working one animal at a time, situate anesthetized ferrets in a prone position on an examination table with their forelegs pulled over the end of the table. Hold the chin and head up in an extended position.
4. Palpitate the external jugular vein on one side of the ferret's neck. Apply digital pressure to the side of the vein as it arises from the chest to minimize the tendency of the vessel to roll (see Note 8).
5. Use 22 gauge needles and 3 cc syringes to collect blood from the jugular vein of the ferrets.
6. Place collected blood into BD Vacutainer® SST™ glass serum tubes and allow to clot.
7. Transfer separated sera into labeled micro-centrifuge tubes and store at –80°C until ready to analyze.

3.6. Enzyme-Linked Immunosorbent Assay

1. Dilute recombinant influenza A/New Caledonia/20/99 HA, or HBcAg, in coating buffer (bicarbonate/carbonate buffer) to a final concentration of 310 ng/mL.
2. Add 100 μL of diluted antigen to the appropriate number of wells of a 96-well EIA plate (31 ng antigen/well final); incubate plates overnight at 4°C.
3. Wash plates 1× with PBS (200 μL/well) and then block with blocking buffer (10% FBS in PBS) (250 μL/well) for 1 h at room temperature.
4. Wash plates 3× with wash buffer (PBS-T), 250 μL/well.
5. Prepare twofold serial dilutions, starting at 1/100, for each ferret serum sample. Add 100 μL of the prepared dilutions to designated wells of the coated and blocked plates; incubate the plates for 2–3 h at room temperature.
6. Wash plates 3× with PBS-T, 250 μL/well.
7. Add 100 μL of horseradish peroxidase conjugated goat anti-ferret IgG (diluted 1/5,000 in PBS-T) to the wells and incubate the plates for 1 h at room temperature.
8. Wash plates 3× with PBS-T, 250 μL/well.
9. Add 100 μL of TMB substrate to the wells and incubate the plates at room temperature for 30 min, or until maximum color development is reached.
10. Add 50 μL of 1 N H_2SO_4 to the wells to stop the reaction.
11. Measure the OD values at 450 nm wavelength using the microplate reader.

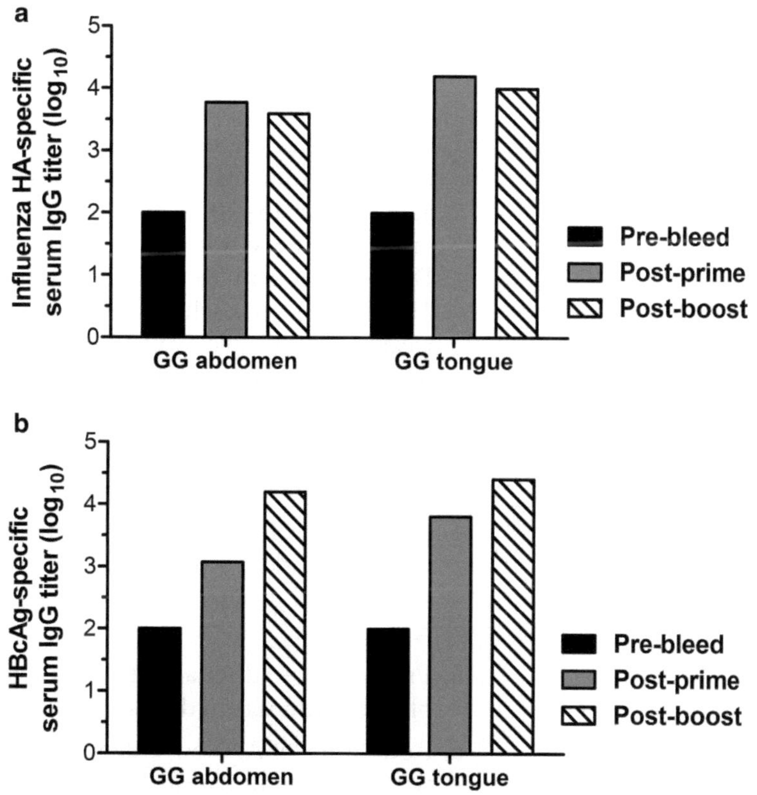

Fig. 2. Serum antibody titers in ferrets after PMED DNA vaccination at different anatomical sites. PMED was used to simultaneously deliver the DNA vaccine constructs pPML7800 and pWRG7063 to either the abdominal skin or tongue of ferrets at a delivery pressure of 400 psi. Serum antibodies specific for the hemagglutinin (HA) of influenza A/New Caledonia/20/99 (**a**) or Hepatitis B core antigen (HBcAg) (**b**) were measured by standard ELISA prior to vaccination, at 4 weeks post-prime, and at 2 weeks post-boost. Antibody titers are expressed as the geometric mean; $n = 3$.

12. Calculate reciprocal serum antibody titers specific for A/New Caledonia/20/99 HA (Fig. 2a; see Note 9) and HBcAg (Fig. 2b; see Note 10) as the highest dilution of sera giving an optical density (OD) reading two times greater than the OD reading of similarly diluted negative control sera samples.

3.7. Influenza A Virus Hemagglutination Inhibition Assay

1. *Preparation of erythrocytes*: Wash 2 mL of TRBC in 13 mL of sterile DPBS. Centrifuge the TRBC at $800 \times g$ for 10 min and then aspirate the supernatant from the tube without disturbing the pellet of erythrocytes. Wash the erythrocytes a total of three times in DPBS. Visually determine the volume of packed erythrocytes and add the appropriate volume of DPBS for a final erythrocyte concentration of 0.5% (e.g., 199 mL of DPBS to 1 mL of packed erythrocytes). Store the 0.5% erythrocyte solution at 4°C until ready to use (see Note 11).

2. *RDE treatment of sera samples*: Reconstitute a vial of lyophilized RDE with 20 mL of sterile DPBS. Aliquot 25–50 μL of each sera sample into individually labeled micro-centrifuge tubes. Add the appropriate volume of RDE equivalent to 3× the volume of the sera sample (i.e., 75–150 μL) to each micro-centrifuge tube. Mix the samples and then incubate the tubes at 37°C for 18–20 h. Inactivate the RDE by incubating the samples at 56°C for 30 min. Add the appropriate volume of DPBS equivalent to 6× the volume of the sera sample (i.e., 150–300 μL) to each micro-centrifuge tube and mix well; the treated sera samples will then be at a final dilution of 1:10. Samples may be used immediately in the assay or stored at -20°C for later use.
3. Add 25 μL of sterile DPBS to wells of a 96-well round bottom plate, leaving columns 1–11 of row A empty.
4. Add 50 μL of each RDE-treated sera sample to columns 1–9 of row A. Add 50 μL of negative control sera (normal ferret serum) and positive control sera to wells A10 and A11, respectively.
5. Prepare twofold serial dilutions of each sera sample down the plate to yield serum dilutions ranging from 1:10 to 1:1,280.
6. Add 25 μL of A/New Caledonia/20/99 viral stock (adjusted to 4 Hemagglutination Units (HAU)/25 μL) to each well in columns 1–11, and the first 4 wells in column 12 (virus control wells).
7. Cover the plates and incubate at room temperature for 30 min.
8. Add 50 μL of 0.5% TRBC to every well and incubate at room temperature for an additional 30 min.
9. Score plates after the last incubation for agglutination (clumped cells; cloudy well) or non-agglutination (tight pellet of TRBC).
10. Determine the HI titer for each sample based on the highest dilution of sera where agglutination is not observed. Samples exhibiting agglutination at the 1:10 dilution are assigned an arbitrary titer of 5 (Fig. 3; see Notes 12 and 13).

4. Notes

1. The HA coding sequence in pPML7800 was synthesized at GeneArt (Regensburg, Germany) from the full-length amino acid sequence of the influenza A/New Caledonia/20/99 hemagglutinin (HA) protein obtained from the Influenza

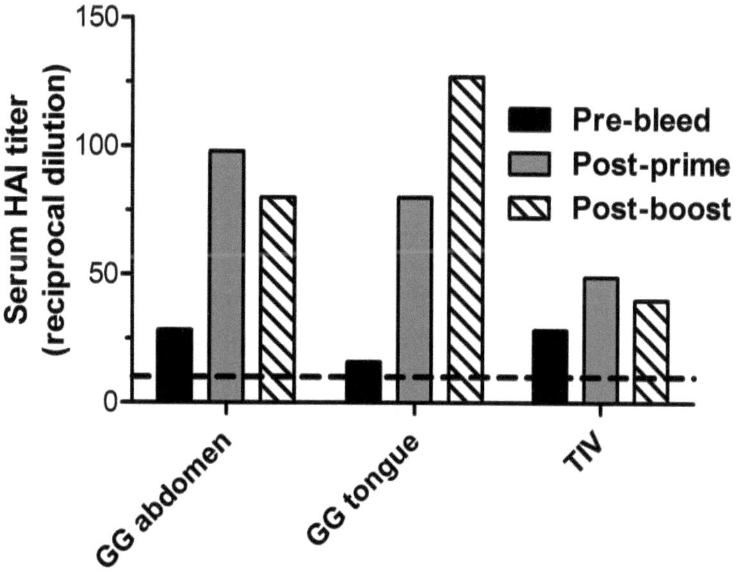

Fig. 3. PMED of DNA vaccines to the abdominal skin and tongue of ferrets elicits hemagglutination inhibition (HAI) antibodies. PMED was used to simultaneously deliver the DNA vaccine constructs pPML7800 and pWRG7063 to either the abdominal skin or tongue of ferrets at a delivery pressure of 400 psi. As a control, ferrets were injected i.m. with 0.5 mL of the human FluZone® inactivated flu vaccine (2005–2006 formulation; 0.5 mL = 15 μg HA). HAI antibodies against A/New Caledonia/20/99 were measured by standard HI assay prior to vaccination, at 4 weeks post-prime, and at 2 weeks post-boost. HI antibody titers are expressed as the geometric mean; $n = 3$ for gene gun inoculations, and $n = 2$ for TIV; *dotted line* = limit of detection.

Sequence Database (http://www.flu.lanl.gov). The expression cassette pPJV7563 uses the human CMV immediate early promoter. Additional sequences were included in the expression cassette to improve expression, specifically the HBV pre-S2 5′-UTR, CMV exon ½ (consisting of the first two CMV IE exons spliced together by deletion of the natural intron), rat insulin intron A, HBV env enhancer, and rabbit beta globin poly A (rGpA) (22).

2. Plasmids pPML7800 and pWRG7063 were delivered in concert at the same PMED target site. Plasmids pPML7800 and pWRG7063 were separately precipitated onto 1–3 μM gold beads at a rate of 2.0 mg DNA/0.5 mg of gold. The two DNA/gold bead preparations were combined together. Please refer to the following references for detailed protocols on the precipitation of plasmid DNA onto gold beads and the preparation of gene gun cartridges coated with plasmid DNA/gold beads (23, 24).

3. Sera from ferrets determined to be seronegative or seropositive to A/New Caledonia/20/99 was obtained from the Centers for Disease Control and Prevention (CDC), Atlanta, GA, and used as negative and positive control sera, respectively, in our assays.
4. Influenza A/New Caledonia/20/99 virus was produced in MDCK cells and a standard HA titration assay using TRBC was performed on infected culture supernatants to determine the HAU present in the viral preparation.
5. Ferret abdominal skin increases in thickness as you move caudally. Accordingly, we found it critical to administer multiple gene gun vaccinations on the same horizontal plain to ensure equal bead penetration/DNA delivery to the epidermis and consistent DNA vaccine expression.
6. For each of the three pressure settings, two of the target skin samples were assayed for β-galactosidase gene expression while the remaining sample was subjected to microscopic analysis to determine the impact of pressure on the distribution of gold beads in dermal tissue. While gene expression analyses (Fig. 1a) and whole tissue X-gal histochemical staining (Fig. 1b and data not shown) revealed the highest levels of β-galactosidase activity in skin homogenates at the 350 and 400 psi pressure settings (Fig. 1a), detailed microscopic analyses revealed a greater distribution of gold beads within dermal tissue vs. the stratum corneum at 400 psi than 350 psi (Fig. 1c and data not shown).
7. Due to the considerable amount of protein present in ferret skin extracts, we found it necessary to dilute the extracts 1:50 to 1:100 prior to performing the DC™ protein microassay to obtain accurate results.
8. Jugular venipuncture is generally performed with one person bleeding and another assisting with restraint. Additionally, venipuncture in ferrets is most successful if the neck of the animal is supported by the last three fingers of one hand while the vein is occluded with the thumb.
9. PMED pPML7800 DNA vaccination elicited influenza A/New Caledonia/20/99 HA-specific serum antibody responses in all recipients regardless of target site (abdomen or tongue). Peak HA-specific antibody responses were obtained 4 weeks after priming (Geometric Mean Titer (GMT) of 5,888 with gene gun administration to the abdomen; GMT of 15,488 with gene gun administration to the tongue), with little or no increase in titer observed during the 2-week interval following booster immunization. These results are comparable to analogous studies showing that influenza DNA vaccination induces

higher titers of HA-specific antibodies than conventional protein vaccination.

10. PMED pWRG7063 DNA vaccination elicited HBcAg-specific serum antibody responses in all recipients regardless of target site (abdomen or tongue). HBcAg-specific antibody responses were detectable 4 weeks after priming (GMT of 1,175 with gene gun administration to the abdomen; GMT of 6,456 with gene gun administration to the tongue), with additional increases in titer observed during the 2-week interval following booster immunization.
11. The 0.5% TRBC suspension should be mixed gently before each use to ensure a uniform distribution of cells in each of the test wells. Also, fresh TRBCs should be prepared regularly as the erythrocytes will begin to hemolyze following 5–7 days of storage.
12. The hemagglutination inhibition (HAI) assay is a standard measure of protective response to influenza. All vaccinated ferrets were found to develop HAI titers greater than or equal to 1:40, the level of antibody shown to be important for protection against influenza infection in clinical flu vaccine studies (25). However, we found that PMED DNA vaccination with pPML7800 elicited a rapid and robust rise in hemagglutination inhibition antibody titers in ferrets, as compared to the conventional trivalent protein influenza (TIV) vaccine, whether administered to the abdomen (approximately twofold increase in mean titer post-prime) or the tongue (approximately threefold increase in mean titer post-prime). We also observed that PMED vaccination on the tongue resulted in the highest HAI titers post-boost. Current belief in the field is that as HAI titers increase, the likelihood of flu infection occurring decreases. Experiments to compare the protective efficacy of our pPML7800 PMED DNA vaccine to conventional TIV vaccination in ferrets are currently underway.
13. Among the oral mucosa, the underside of the tongue (sublingual; s.l.) is considered a novel site for drug delivery since administered drugs are quickly absorbed and enter the bloodstream (26). Additionally, s.l. delivery of the conventional trivalent protein influenza vaccine was found to induce protective immune responses in mice, including the production of secretory IgA Abs in the respiratory tract (27). Though it is unclear from the current studies why PMED vaccination to the tongue elicited higher HAI and serum antibody titers in ferrets, the s.l. epithelium is known to contain a dense array of dendritic cells capable of quickly mobilizing to proximal lymph nodes (28).

Acknowledgments

This work was supported by U01 AI074509 from National Institute of Health/NIAID awarded to Deborah H. Fuller.

References

1. Fuller DH, Loudon P, Schmaljohn C (2006) Preclinical and clinical progress of particle-mediated DNA vaccines for infectious diseases. Methods 40:86–97
2. Payne LG, Fuller DH, Haynes JR (2002) Particle-mediated DNA vaccination of mice, monkeys and men: looking beyond the dogma. Curr Opin Mol Ther 4:459–466
3. Yager EJ, Dean HJ, Fuller DH (2009) Prospects for developing an effective particle-mediated DNA vaccine against influenza. Expert Rev Vaccines 8:1205–1220
4. Falo LD Jr (1999) Targeting the skin for genetic immunization. Proc Assoc Am Physicians 111:211–219
5. Falo LD Jr, Storkus WJ (1998) Giving DNA vaccines a helping hand. Nat Med 4: 1239–1240
6. Bennett AM et al (1999) Gene gun mediated vaccination is superior to manual delivery for immunisation with DNA vaccines expressing protective antigens from Yersinia pestis or Venezuelan Equine Encephalitis virus. Vaccine 18:588–596
7. Huber VC, Thomas PG, McCullers JA (2009) A multi-valent vaccine approach that elicits broad immunity within an influenza subtype. Vaccine 27:1192–1200
8. Pertmer TM et al (1995) Gene gun-based nucleic acid immunization: elicitation of humoral and cytotoxic T lymphocyte responses following epidermal delivery of nanogram quantities of DNA. Vaccine 13: 1427–1430
9. Roberts LK et al (2005) Clinical safety and efficacy of a powdered Hepatitis B nucleic acid vaccine delivered to the epidermis by a commercial prototype device. Vaccine 23: 4867–4878
10. Drape RJ et al (2006) Epidermal DNA vaccine for influenza is immunogenic in humans. Vaccine 24:4475–4481
11. Jones S et al (2009) DNA vaccination protects against an influenza challenge in a double-blind randomised placebo-controlled phase 1b clinical trial. Vaccine 27:2506–2512
12. Fuller DH et al (1997) Enhancement of immunodeficiency virus-specific immune responses in DNA-immunized rhesus macaques. Vaccine 15:924–926
13. Fuller DH et al (2002) Induction of mucosal protection against primary, heterologous simian immunodeficiency virus by a DNA vaccine. J Virol 76:3309–3317
14. Fuller DH et al (2006) DNA immunization in combination with effective antiretroviral drug therapy controls viral rebound and prevents simian AIDS after treatment is discontinued. Virology 348:200–215
15. Doria-Rose NA et al (2003) Multigene DNA priming-boosting vaccines protect macaques from acute CD4+-T-cell depletion after simian-human immunodeficiency virus SHIV89.6P mucosal challenge. J Virol 77:11563–11577
16. Eisenbraun MD, Fuller DH, Haynes JR (1993) Examination of parameters affecting the elicitation of humoral immune responses by particle bombardment-mediated genetic immunization. DNA Cell Biol 12:791–797
17. Fynan EF, Robinson HL, Webster RG (1993) Use of DNA encoding influenza hemagglutinin as an avian influenza vaccine. DNA Cell Biol 12:785–789
18. Macklin MD et al (1998) Immunization of pigs with a particle-mediated DNA vaccine to influenza A virus protects against challenge with homologous virus. J Virol 72:1491–1496
19. Marini RP et al (2002) Biology and diseases of ferrets. In: Fox JG, Anderson LC, Loew FM, Quimby FW (eds) Laboratory animal medicine. Academic, San Diego, pp 483–517
20. Maher JA, DeStefano J (2004) The ferret: an animal model to study influenza virus. Lab Anim (NY) 33:50–53
21. Matsuoka Y, Lamirande EW, Subbarao K (2009) The ferret model for influenza. Curr Protoc Microbiol Chapter 15:Unit 15G 12
22. Loudon PT et al (2010) GM-CSF increases mucosal and systemic immunogenicity of an H1N1 influenza DNA vaccine administered into the epidermis of non-human primates. PLoS One 5:e11021

23. Roy MJ et al (2000) Induction of antigen-specific CD8+ T cells, T helper cells, and protective levels of antibody in humans by particle-mediated administration of a hepatitis B virus DNA vaccine. Vaccine 19:764–778
24. Wang S, Joshi S, Lu S (2004) Delivery of DNA to skin by particle bombardment. Methods Mol Biol 245:185–196
25. Couch RB, Kasel JA (1983) Immunity to influenza in man. Annu Rev Microbiol 37: 529–549
26. Kildsgaard J et al (2007) Sublingual immunotherapy in sensitized mice. Ann Allergy Asthma Immunol 98:366–372
27. Song JH et al (2008) Sublingual vaccination with influenza virus protects mice against lethal viral infection. Proc Natl Acad Sci USA 105:1644–1649
28. Cuburu N et al (2007) Sublingual immunization induces broad-based systemic and mucosal immune responses in mice. Vaccine 25: 8598–8610

Chapter 20

Methods for Monitoring Gene Gun-Induced HBV- and HCV-Specific Immune Responses in Mouse Models

Gustaf Ahlén, Matti Sällberg, and Lars Frelin

Abstract

The hepatitis B and C viruses (HBV/HCV) are major causes for chronic liver disease globally. For HBV new antiviral compounds can suppress the viral replication for years, but off-therapy responses are rare. Current therapies based on interferon and ribavirin can cure 45–85% of the treated HCV-infected patients largely depending on the viral genotype. New regimens including protease inhibitors where introduced during 2011 and have increased the cure rates for the hardest to treat HCV genotype 1 from 45 to 65%. Here a major need is to replace the immunomodulatory effects of interferon and/or ribavirin. Thus, therapeutic vaccines have a place in both chronic HBV and HCV infection. Unfortunately, none of these viruses can infect mice whereby substitute models are needed. We have used several types of murine models to predict the clinical efficacy of therapeutic vaccines for chronic HBV and HCV infections. In this chapter we describe transdermal delivery of genetic vaccines using the Helios Gene Gun device. A central role is that the model should have generally functional immune response, but with selective defects towards HBV and/or HCV. Thus, mice with stable integrated transgenes are useful. However, as a simple model to study the hepatic entry and functionality of a HBV- and/or HCV-specific immune responses other models are needed, where a killed transgenic hepatocyte is replaced by a healthy non-transgenic hepatocyte. Here we can effectively apply a technique termed hydrodynamic injection, which makes 10–30% of hepatocytes transiently transgenic for any plasmid. Within this chapter the methods used to characterize transiently transgenic mice are described. The main methods are the hydrodynamic injection technique, detection of transgene expression by immuno-precipitation, western blot, and immunohistochemistry. Finally, the in vivo functionality of T cells can be determined by using stably transfected syngeneic tumor cell lines expressing HBV and/or HCV proteins. The tumor challenge model enables studies of in vivo T cell function, whereas the cytotoxicity assay is used to determine T cell function in vitro. Overall, these models effectively reveal the efficiency by which various vaccine technologies, including biolistic DNA vaccination can kill the "infected" hepatocyte.

Key words: Gene gun, Genetic vaccination, Hepatitis, HBV, HCV, Transiently transgenic mice, Tumor challenge model

Stephan Sudowe and Angelika B. Reske-Kunz (eds.), *Biolistic DNA Delivery: Methods and Protocols*, Methods in Molecular Biology, vol. 940, DOI 10.1007/978-1-62703-110-3_20, © Springer Science+Business Media, LLC 2013

1. Introduction

1.1. Characteristic Features of Hepatitis B and C Viruses

The hepatitis B and C viruses (HBV/HCV) are major causes for chronic liver disease globally (1, 2). Every third person on the planet has been in contact with HBV or HCV. These viruses are the major cause for liver transplantation in the western world due to its ability to cause liver failure and liver cancer. Thus, these two viruses cause a significant disease burden. Available treatment for HBV consists of interferon (IFN)-α and nucleoside/nucleotide inhibitors, which effectively reduces the viral replication, but when treatment is stopped the viral load most often rebounds (3). Today's treatment of HCV consists of a combination therapy of IFN-α and ribavirin. The treatment is successful in 45–85% of the treated patients, widely dependent on the viral genotype (4). During the year of 2011 new antiviral compounds, protease inhibitors, become available. These protease inhibitors are used in combination with IFN-α and ribavirin, which increases the clearance rate in patients with approximately 10–20% (5). Even though more effective treatments for HCV are being available, there are still a high proportion of patients that do not respond to the current therapy. Therefore, therapeutic vaccines are desirable for treatment of both HBV and HCV infections. Patients that clear or control acute or chronic HBV and HCV infections have strong multispecific CD4+ and CD8+ T cell responses to HBV and HCV proteins, whereas patients that are unable to clear the infections lack these strong T cell responses (6, 7). One way to activate strong CD4+ and CD8+ T cell responses in chronic HBV and HCV patients is to utilize DNA-based therapeutic vaccines since they activate antiviral CD4+ and CD8+ T cells responses (8, 9).

1.2. Animal Models of Hepatitis B and C Virus Infection

One of the major problems with HBV and HCV is that the viruses only infect humans and chimpanzees, thereby making testing of new HCV therapeutics cumbersome. We and others have therefore developed substitute animal models, mainly mouse models to enable testing of new antiviral drugs and vaccines for HBV and HCV prior to clinical evaluation. To fully evaluate new HBV and HCV therapeutics, the animal model used needs to have a functional immune response. Many studies have been performed in mice with liver-specific stably integrated transgenes for HBV and HCV (10–14). These stable transgenic mouse lineages have a functional immune response but with selective defects towards HBV and/or HCV, enabling immunogenicity studies of therapeutic vaccines. One problem with the stably transgenic mice is that a killed liver cell is always replaced with a new HBV- or HCV-transgenic liver cell. We have therefore established a "transiently transgenic mouse model" (15). In this model, any wild-type mouse is made

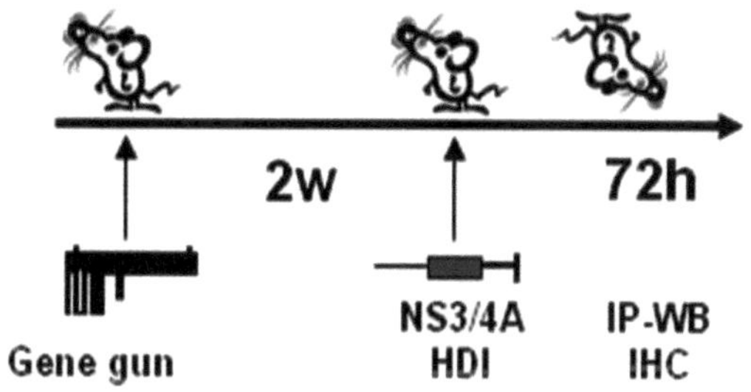

Fig. 1. Schematic outline of the transiently transgenic mouse model. Mice are immunized transdermally with a codon optimized HCV NS3/4A DNA vaccine using the Helios Gene Gun device. Two weeks later the same mice receive a hydrodynamic injection (HDI) of 100 μg codon optimized HCV NS3/4A DNA to generate a transient hepatic expression of the transgene. At 72 h post-HDI the mice are sacrificed and liver specimen are analyzed for presence of HCV NS3-protein by immuno-precipitation and western blot (IP-WB) and by immunohistochemistry (IHC).

transiently transgenic for HBV or HCV proteins (or other proteins) through a hydrodynamic injection, where 1.8 mL of Ringer's solution containing 100 μg HBV or HCV plasmid DNA is injected rapidly (<5 s) intravenously in the tail vein. This causes, due to the increased pressure, a perfusion of the liver with the Ringer's solution containing the HBV or HCV plasmid DNA. The DNA is taken up by the liver cells and is subsequently expressed into protein. Approximately 10–30% of the liver cells express the injected gene-product already 24 h postinjection and the protein expression persist for around 2 weeks depending of the method of detection (15). The transiently transgenic mouse model allows for studies where one can evaluate how effectively vaccine-primed HBV- and HCV-specific T cells can enter the liver and clear HBV- and/or HCV-expressing liver cells. The general procedure is outlined in Fig. 1. One effective way to prime HCV-specific T cells is by particle bombardment using the Helios Gene Gun device, which transdermally delivers the DNA directly into the cytoplasm of the target area. We have vast experience using the Helios Gene Gun device to prime humoral and cellular HBV and HCV immune responses (16–18). The Helios Gene Gun device has shown some success in humans where a hepatitis B virus surface antigen (HBsAg)-expressing plasmid delivered using the Helios Gene Gun device was found to raise protective levels of anti-HBs (19). Thus, this may well be an attractive way of DNA vaccination in humans. One unique feature of the transiently transgenic mouse model is that any mouse strain (wild-type, knock-out or transgenic) may be used and this largely helps to explain which components are important for viral clearance. Importantly, every killed transiently transgenic liver cell is replaced by a non-transgenic liver cell, making this model more "infection-like." To characterize the ability of vaccine-primed T cells to enter the liver and clear "infected or protein-expressing liver cells" one may utilize different approaches. We have mainly utilized two techniques to monitor clearance of HBV- and/or HCV-expressing liver cells, (1) immuno-precipitation and western blot, and (2) immunohistochemistry. Immuno-precipitation and western blot allows for sensitive measurement of HBV or HCV protein content in the liver of the transiently transgenic mice. In addition, immunohistochemistry allows us to determine the number of cells that are expressing the HBV or HCV proteins as well as monitor liver inflammation. This enables us to compare mice that have received a previous vaccination with naïve or non-immunized mice. In summary, both immuno-precipitation and western blot, and immunohistochemistry allow us to determine how effectively the vaccine-primed T cells can eradicate HBV- or HCV-expressing liver cells in mice with a functional immunity (see Fig. 2). Another substitute model is the "tumor challenge model" (16, 17, 20). In this model we may determine the in vivo functionality of HBV- and HCV-primed T cells by

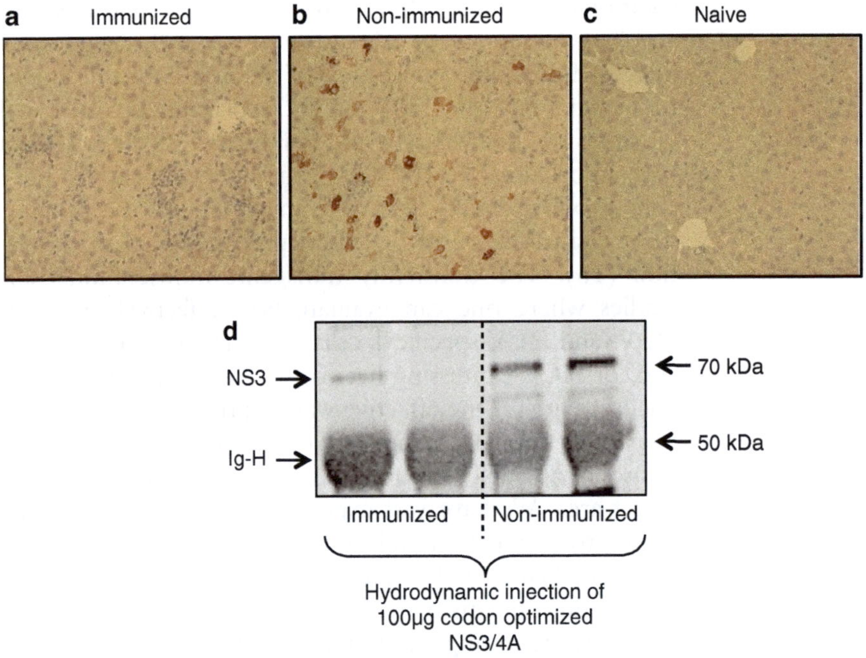

Fig. 2. Detection of transiently transgenic hepatocytes generated by a hydrodynamic injection of 100 μg of codon optimized HCV NS3/4A DNA plasmid (**a**) and (**b**). In vivo clearance of HCV NS3-protein (cells stained in *brown*) in a mouse immunized once with 2 μg codon optimized HCV NS3/4A DNA plasmid using the Helios Gene Gun device 2 weeks prior the hydrodynamic injection (**a**) and in a non-immunized mouse (**b**). In (**c**) a naïve control mouse is shown. The immuno-histochemical detection of the HCV NS3-protein is visualized 72 h post the hydrodynamic injection. In (**d**) detection of HCV NS3-protein in liver homogenates from immunized and non-immunized mice using immuno-precipitation and western blot. The mouse Ig heavy chain (50 kDa) was used as a loading control.

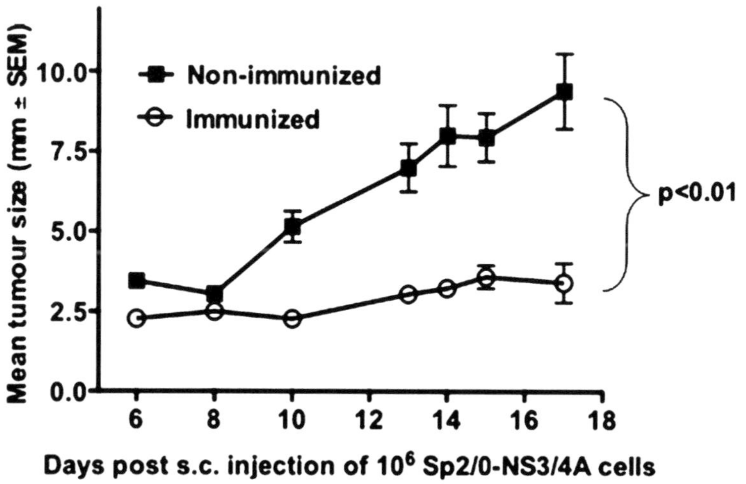

Fig. 3. In vivo protection against tumor growth in BALB/c (H-2[d]) mice after 2 monthly Gene Gun immunizations. Groups of ten mice were either immunized (2 μg codon optimized NS3/4A DNA) or left untreated. Two weeks after the last immunization, both groups were sub cutaneously inoculated with 1×10^6 Sp2/0 cells expressing the HCV NS3/4A protein. Tumor sizes were measured from day 6 to 18 using a sliding caliper. The area under the curve (AUC) for the two curves was statistically different (ANOVA; $p<0.01$). Statistical comparisons were performed using GraphPad Instat 3 for Macintosh (version 3.0b; GraphPad Software, San Diego, CA) and Microsoft Excel 2008 for Macintosh (version 12.2.8; Microsoft, Redmond, WA).

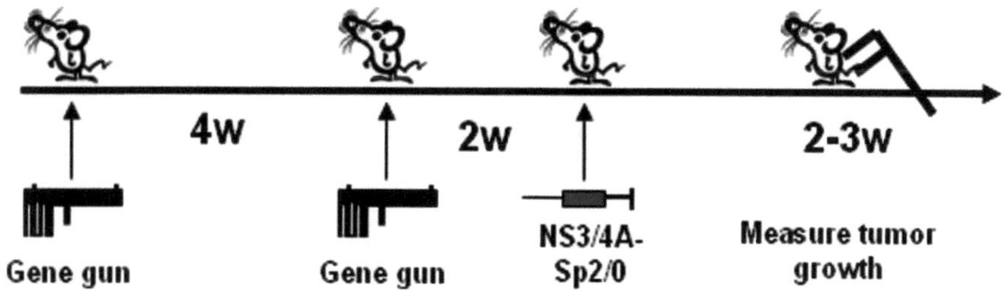

Fig. 4. Schematic outline of the tumor mouse model. Mice are immunized transdermally with a codon optimized HCV NS3/4A DNA vaccine using the Helios Gene Gun device. Two weeks later the same mice are inoculated with 1×10^6 Sp2/0 cells expressing the HCV NS3/4A protein. Tumor sizes/volumes are measured during a 2–3 week period using a sliding caliper.

transplanting stably transfected syngeneic tumor cell lines expressing HBV or HCV proteins into laboratory mice. Here we may study how efficient HBV- or HCV-specific T cells can localize and eradicate the sub cutaneous tumors expressing HBV or HCV proteins (see Fig. 3). By measuring the tumor size/volume using a sliding caliper we may follow the kinetics of tumor growth and how efficient primed T cells can eradicate the solid tumor. The general procedure is outlined in Fig. 4. Moreover, to monitor the lytic activity of our vaccine-primed T cells we are utilizing a classical cytotoxicity assay, namely the ^{51}Cr release assay. This functional in vitro assay enables us to monitor the lytic activity of Helios Gene Gun immunized mice vs. non-immunized mice (see Fig. 5). This

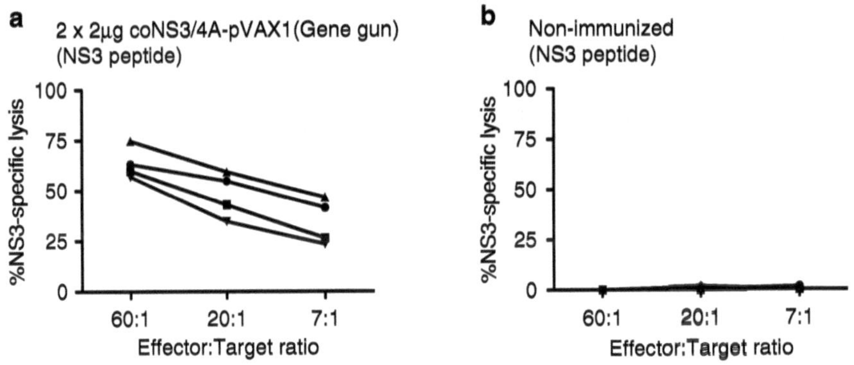

Fig. 5. Priming of in vitro detectable CTLs in C57BL/6 (H-2^b) mice after 2 monthly Gene Gun immunizations. Groups of four to five mice were either immunized ((**a**); 2 µg codon optimized NS3/4A DNA) or left untreated (**b**). Two weeks after immunization the ^{51}Cr release assay was setup. The percent specific lysis corresponds to the percent lysis obtained with NS3-peptide-coated RMA-S cells minus the percent lysis obtained with unloaded cells. Values are given for effector to target (E:T) cell ratios of 60:1, 20:1, and 7:1. Each *line* indicates an individual mouse. The area under the curve (AUC) for the two groups was statistically different (ANOVA; $p<0.01$). Statistical comparisons were performed using GraphPad Instat 3 for Macintosh (version 3.0b; GraphPad Software, San Diego, CA) and Microsoft Excel 2008 for Macintosh (version 12.2.8; Microsoft, Redmond, WA).

assay provides us with information whether the activated T cells can recognize and kill target cells.

In summary, the herein described materials and methodology provide detailed protocols highlighting some of the available animal models to study and monitor HBV and HCV infections.

2. Materials

2.1. Preparation of DNA/Microcarrier Cartridges and Delivery of DNA/Microcarrier Complexes to Mice Using the Helios Gene Gun Device

2.1.1. Preparation of DNA/Microcarrier Cartridges

1. Marking pen.
2. Lab timer.
3. Adjustable pipettors and tips (0.5–1,000 µL).
4. Purified plasmid DNA (1 mg/mL) resuspended in phosphate buffered saline (PBS). This protocol is optimized for a codon optimized HCV NS3/4A gene inserted into the pVAX1 vector.
5. Compressed nitrogen gas of highest quality (99.998%).
6. Nitrogen regulator.
7. Absolute ethanol (>99.5%). Always use an unopened bottle of absolute ethanol. A high percentage of ethanol is essential for optimal DNA/microcarrier preparation.
8. Spermidine. Stock solution 0.5 M (10×) diluted in sterile water, single use aliquots stored at −80°C.

9. Calcium chloride ($CaCl_2$), stock solution 1 M diluted in sterile water. Store the $CaCl_2$ stock solution wrapped in parafilm at +4°C.
10. Polyvinylpyrrolidone (PVP, MW 360,000, Bio-Rad), stock solution 20 mg/mL diluted in absolute ethanol. Store the PVP stock solution wrapped in parafilm at +4°C.
11. Gold microcarriers, 0.6, 1.0, 1.6 μm in diameters (Bio-Rad). In this protocol we are using gold microcarriers with 1.0 μm diameter in combination with the codon optimized HCV NS3/4A DNA.
12. Gold-Coat tubing (Bio-Rad). Approximately 75 cm may be inserted into the tubing prep station at a time.
13. Tubing cutter (Bio-Rad). Used to cut ca. 1.25 cm cartridges of the coated sample tubing.
14. Cartridge collection/storage vials (one desiccant pellet per vial).
15. Tubing prep station (Bio-Rad).
16. Ultrasonic bath.
17. Vortex.
18. Analytical balance.
19. 1.5 mL Sarstedt vials with screw-cap lids.
20. 15 mL Falcon tubes with screw-cap lids.
21. 5 mL syringe with connected elastic plastic tubing (approximately 20 cm in length).

2.1.2. Delivery of DNA/ Microcarrier Complexes to Mice Using the Helios Gene Gun Device

1. Ethical permission for the proposed animal experimentation.
2. Required documentation that allows you to work with laboratory mice.
3. Compressed helium gas of highest quality.
4. Helium regulator (Bio-Rad).
5. Helios Gene Gun device (Bio-Rad).
6. Battery (9 V, LR61).
7. Shaver (for laboratory mice).
8. Laboratory mice (strain of your choice, we have used $H\text{-}2^b$ (C57BL/6) and $H\text{-}2^d$ (BALB/c)).
9. Ear protection (discharging the Helios Gene Gun generates a sound of approximately 108 db at 400 psi).
10. Cartridge holder for delivery. Cartridge holder may be loaded with 12 cartridges for delivery.
11. Cartridges coated with plasmid DNA, e.g., codon optimized HCV NS3/4A gene (each cartridge contains 1 μg plasmid DNA).

2.2. Transiently Transgenic Mouse Model

1. Ethical permission for the proposed animal experimentation.
2. Required documentation that allows you to work with laboratory mice.
3. Laboratory mice (strain of your choice, we have used H-2^b (C57BL/6)).
4. Hypnorm: 0.315 mg/mL fentanyl citrate, 10 mg/mL fluanisone (VetaPharm Ltd), or other neuroleptanalgesic drug.
5. Mouse restrainer (for 25 g mice).
6. Infrared "heat" lamp.
7. Ethanol (70%).
8. Purified plasmid DNA (1 mg/mL) resuspended in PBS. We have used a codon optimized HCV NS3/4A gene inserted into the pVAX1 vector.
9. 1 and 2 mL syringe.
10. 27G, 19 mm needle.
11. Ringer's solution.

2.2.1. Immuno-Precipitation and Western Blot (IP-WB)

1. Marking pen.
2. Lab timer.
3. Wet ice.
4. Scissor and forceps.
5. Ripa buffer pH 7.4: 150 mM NaCl, 50 mM Tris–HCl, 1% Triton-X 100, 1% Na-deoxycholate, 1% SDS. Add 0.2 mM PMSF, 0.5 mM DTT and 1 mM Na_3VO_4 before use and keep the solution on ice.
6. Protein-A Sepharose.
7. PBS.
8. Mouse-anti-NS3 IgG antibody (produced in-house, titer of 1:77.760).
9. Non-denaturating buffer: 1% Triton-X, 50 mM Tris–HCl pH 7.4, 300 mM NaCl, 5 mM EDTA, 0.02% Sodium azide, 1 mM PMSF diluted in sterile water.
10. Bovine Serum Albumin (BSA).
11. Wash buffer: 0.1% Triton-X, 50 mM Tris–HCl pH 7.4, 300 mM NaCl, 5 mM EDTA, 0.02% Sodium azide diluted in sterile water.
12. 2× loading dye (10 mL contains: 2.5 mL 0.5 M Tris–HCl, pH 6.8, 0.4 g SDS, 2 mL Glycerol, 20 mg Bromophenol blue, 400 μL 2-Mercaptoethanol diluted in sterile water).
13. NuPAGE 4–12% Bis-Tris gels (Invitrogen) or similar.
14. Gel-loading tips.

15. Running buffer, MOPS SDS running buffer (20×) (Invitrogen) dilute in sterile water or similar.
16. Molecular weight marker (MagicMark XP western protein standard, Invitrogen) or similar.
17. iBlot transfer stacks, mini or regular (Invitrogen) or similar.
18. WesternBreeze chemiluminescent anti-mouse kit (Invitrogen) or similar. The solutions are diluted as indicated by the following proportions:
    - Blocking solution, 2/10 Blocker A, 1/10 Blocker B, and 7/10 sterile water.
    - Wash buffer, 1/16 wash solution, 15/16 sterile water.
    - Secondary antibody, alkaline phosphatase (AP), ready to use.
    - Chemiluminescent substrate, 19/20 CDP-Star substrate and 1/20 Nitro-Block-II enhancing reagent.
19. Scale (0.1–1 g).
20. Sonicator.
21. Gel system, XCell SureLock Mini-Cell (Invitrogen) or similar.
22. Power supply (200 V).
23. Heating block for 1.5 mL Sarstedt tubes (100°C).
24. Western blot transfer system, iBlot Gel transfer device (Invitrogen) or similar.
25. iBlot gel transfer stacks Nitrocellulose, mini or regular (Invitrogen) or similar.
26. Rotating table
27. Chemiluminescent detection system (Gene Gnome, Syngene Bio Imaging) or similar.

2.2.2. Immunohistochemistry

1. Xylene.
2. Ethanol (99%, 95%, 70%).
3. Sterile water.
4. Hydrogen peroxide (H_2O_2).
5. Antigen retriever solution: 0.01 M Citrate buffer.
6. Normal horse serum (NSH) 2.5% in PBS.
7. Rabbit-anti-NS3 IgG antibody (produced in-house, titer of 1:156.250).
8. ImmPRESS reagent kit peroxidase, anti-rabbit-Ig (Vector laboratories).
9. DAB peroxidase substrate kit (Vector laboratories). (For 2.5 mL solution: 2.5 mL sterile water, 1 drop buffer, 2 drop DAB, and 1 drop H_2O_2. Mix after each addition.) One mouse liver section requires approximately 30 μL solution.

10. Mayers HTX.
11. Mounting media, Pertex or similar.
12. Imm Edge Pen or similar.
13. Adjustable pipettors and tips (0.5–1,000 μL).
14. Pressure cooker, 2100 retriever (Prestige medical) or similar.
15. Microscope (×100 magnification).
16. Cuvette (slide chamber) for microscope slides.
17. Microscope slides (Thermo Scientific) or similar.
18. Humidified chamber.
19. Coverslips (In vitro diagnostics) or similar.

2.3. Tumor Challenge Model

1. Ethical permission for the proposed animal experimentation.
2. Required documentation that allows you to work with laboratory mice.
3. Adjustable pipettors and tips (0.5–1,000 μL).
4. Tumor cells. We have used Sp2/0 myeloma cells ($H\text{-}2^d$) stably expressing HCV NS3/4A.
5. Trypan blue stain 0.4%.
6. Bürkner chamber and cover glass.
7. Laboratory mice (strain of your choice, we have used $H\text{-}2^d$ (BALB/c)).
8. PBS.
9. 1 mL syringe.
10. 25G, 19 mm needle.
11. 27G, 19 mm needle.
12. Hypnorm (0.315 mg/mL fentanyl citrate and 10 mg/mL fluanisone) (VetaPharm Ltd.) or other neuroleptanalgesic drug.

2.4. Cytotoxicity Assay

1. Incomplete medium: RPMI 1640 plus 100 U/mL penicillin, 100 μg/mL streptomycin.
2. Complete medium: RPMI 1640 medium plus 10% fetal bovine serum (FBS, prior to use the FBS is heat inactivated for 30 min at +56°C), 100 U/mL penicillin, 100 μg/mL streptomycin, 10 mM HEPES buffer, 2 mM L-glutamine, 1 mM sodium pyruvate, 1 mM nonessential amino acids, 50 μM 2-mercaptoethanol.
3. 15 mL Falcon tubes with screw-cap lids.
4. 50 mL Falcon tubes with screw-cap lids.
5. Adjustable pipettors and tips (0.5–1,000 μL).
6. Red Blood Cell Lysing Buffer (Sigma-Aldrich).

7. Cell culture flask, 25 cm^2 (Nunc).
8. Cell strainer, 70 μm pore size (BD Biosciences).
9. Petri dishes, 10 mm.
10. PBS.
11. 96-Well microtiter plate (Corning Inc. Costar).
12. HCV NS3 major histocompatibility complex (MHC) class I peptide (HCV NS3 GAVQNEVTL, H-$2D^b$). The peptide is used for in vitro stimulation and the ^{51}Cr-release assay.
13. $Chromium^{51}$ (5 mCi, 185 MBq).
14. V-bottom plates.
15. Isoplate 96-well sample plate (PerkinElmer).
16. Optiphase Super Mix (PerkinElmer).
17. Sealing tape (PerkinElmer).
18. Sterile syringe filters 0.22 μm.
19. Incubator 37°C, 5% CO_2.
20. Centrifuge (12,000 × *g*, +4°C).
21. 1450 Microbeta Trilux liquid scintillation and luminescence counter (Wallac) or similar instrument for detection gamma radiation.
22. Irradiator (Gamma Cell 2000, Cs-137) or similar instrument.

3. Methods

The methodology section outlines the following procedures: (1) preparation of DNA/microcarrier samples, (2) delivery of DNA/microcarrier samples to laboratory mice using the Helios Gene Gun device, (3) monitoring primed immune responses using (a) transiently transgenic mice, (b) tumor challenge model, and (c) measurement of T cell lytic activity.

3.1. Preparation of DNA/Microcarrier Samples and Delivery of DNA/Microcarrier Samples to Mice Using the Helios Gene Gun Device

The following protocol is used to prepare approximately 120 Gene Gun samples/cartridges. Each cartridge contains 1 μg of plasmid DNA on microcarriers. The procedure requires laboratory work involving two persons for approximately 3 h.

3.1.1. Preparation of HCV NS3/4A-DNA/Microcarrier Cartridges

1. Prior to starting preparation of DNA/microcarrier cartridges, the amount of needed DNA and gold has to be determined. The amount of DNA loaded per mg of microcarriers is called the DNA loading ratio (DLR). The DLR typically ranges from 1 to 5 μg DNA per mg gold. The amount of microcarriers delivered per cartridge/shot is called the microcarrier loading

quantity (MLQ). The MLQ typically ranges from 0.25 to 0.5 mg gold per cartridge. The DLR and MLQ has to be optimized and tested to see which amounts that work the best in vivo. We have used a MLQ of 0.5 mg gold per shot and a DLR of 2 μg DNA per mg gold, which gives 1 μg DNA per cartridge/shot.

2. Weigh 37.5 mg gold microcarriers (1.0 μm in diameters, see Note 1) and put in each of the two 1.5 mL Sarstedt tubes.
3. Prepare a 15 mL Falcon tube with 0.01 mg/mL PVP working solution (see Note 2). Mix 5.3 μL of PVP (stock solution 20 mg/mL) in 10.5 mL of absolute ethanol (2,000× dilution, see Note 3) in the 15 mL Falcon tube.
4. Add 10 μL Spermidine (0.5 M, 10×) + 90 μL of absolute ethanol to each gold microcarrier Sarstedt tube. This generates a gold microcarrier solution containing 0.05 M spermidine.
5. Vortex the tubes vigorously for 10 s.
6. Add directly 75 μL of plasmid DNA (1 mg/mL) per Sarstedt tube (see Note 4).
7. Vortex the tubes vigorously for 10 s and directly hold the tubes in the ultrasonic bath for 10 s (70% effect) (see Note 5).
8. Unscrew the lid and start vortexing immediately. Add 100 μL $CaCl_2$ (1 M) per tube drop-wise (see Note 6).
9. Let the mixture precipitate for 10 min at room temperature.
10. Centrifuge the Sartstedt tubes for 1 min at 2,500 × *g* at room temperature and discard the supernatant (see Note 7).
11. Wash the pellet with 950 μL absolute ethanol. Carefully resuspend the pellet by flicking the Sarstedt tubes prior to centrifugation. Repeat the washing step tree times.
12. After the last wash, repeatedly add 500 μL of the 0.01 M PVP solution to the Sarstedt tubes and transfer the mixture to a 15 mL Falcon tube. Continue until all mixture is transferred to the 15 mL Falcon tube.
13. Split the 10.5 mL solution containing the plasmid DNA/microcarrier complexes into three new 15 mL Falcon tubes. Mix extensively to make sure that each of the tree new Falcon tubes gets similar amounts of complexes. At this step, each Falcon tube contains approximately 3.5 mL, which equals 25 mg of gold microcarriers (see Notes 8 and 9).
14. Prepare the Tubing prep station by open the nitrogen flow into the tubing prep station for >15 min before start and then turn off the nitrogen. Adjust the nitrogen flow to 0.3–0.4 L/min (LMP) on the valve on the flow-meter.
15. Cut three pieces of 75 cm Gold-Coat tubing.

16. Attach one piece of 75 cm Gold-Coat tubing to the elastic plastic tubing that is connected to a 5 mL syringe.
17. Vortex one of three Falcon tubes containing 3.5 mL of DNA/microcarrier complexes for 10 s, directly put the tube in the ultrasonic bath for 10 s (70% effect) (see Note 10), and then immediately bring up 3.5 mL of the mixture in the 75 cm length tubing by using a syringe (see step 15). Keep the 75 cm Gold-Coat tubing horizontally during the whole procedure (see Note 11).
18. Directly put the tubing into the tubing prep station, and let stand for 5 min to allow the gold to be separated from the solution (see Note 12).
19. Remove the solution from the tubing by using the syringe. The solution should be removed within 20 s (use a constant speed during the removal) (see Note 13).
20. Immediately turn on the rotation on the tubing prep station by pressing button 1, and let rotate for 30 s and then turn on the nitrogen (0.3–0.4 LMP) and let dry for 3.5 min (see Note 14).
21. Turn off the motor on the tubing prep station (button 0). Turn off nitrogen by closing the valve on the flow-meter.
22. Remove tubing.
23. Repeat steps 16–22 for the two remaining Falcon tubes containing the mixture.
24. Turn off the nitrogen cylinder and open the valve on the flow-meter to empty the tubing prep station from the remaining nitrogen gas.
25. Inspect the coated sample tubing and remove parts that have not been coated evenly (see Note 15).
26. Cut the coated sample tubing into ca. 1.25 cm cartridges using the tubing cutter.
27. Store the cartridges in cartridge collection/storage vials, each containing one desiccant pellet.
28. The cartridges may be stored at +4°C for later use (see Note 16).

3.1.2. Delivery of HCV NS3/4A-DNA/Microcarrier Complexes to Mice Using the Helios Gene Gun Device

The following protocol describes the procedure for immunizing laboratory mice using the Helios Gene Gun device. The procedure requires laboratory work involving two persons. The time needed depends on the number of laboratory mice to be immunized. Roughly, the procedure takes 2 h for shaving and immunization of 20 laboratory mice.

1. Carefully shave the abdominal area of the mouse that is being used for delivery, see Fig. 6a, b (see Note 17).
2. Insert the 9 V battery in the Helios Gene Gun device.

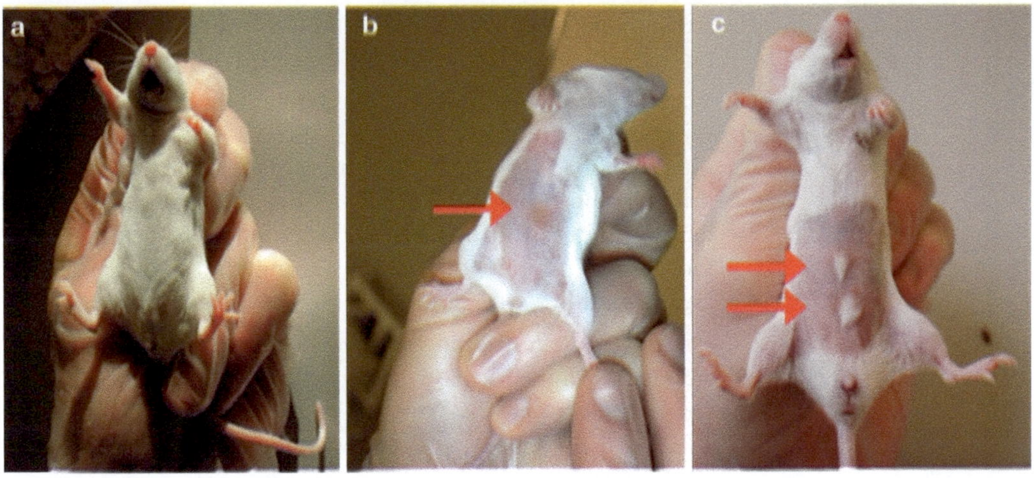

Fig. 6. Illustrative pictures of BALB/c mice prior to shaving and immunization (**a**), after shaving and one immunization (**b**), and 2 weeks postimmunization (**c**). The *red arrow* in (**b**) shows the target area of the Helios Gene Gun immunization (*gold colored mark*). The *arrows* in (**c**) indicate increased fur growth at the place for delivery.

3. Load up to 12 DNA-coated gold cartridges into the holder (see Note 18).
4. Insert the holder into the Helios Gene Gun device.
5. Put on the ear protection.
6. Open the helium gas on the regulator and adjust the pressure to 500 psi (see Note 19).
7. Restrain the mouse by holding it in your hand (person 1), see Fig. 6a, b.
8. The other person (person 2) then puts the Helios Gene Gun device to the upper part of the shaved abdominal area of the mouse and fire the first shot, see Fig. 6b (see Note 20).
9. Move the Helios Gene Gun device to the lower part of the shaved abdominal area, reload and fire the second shot (see Note 21).
10. After the last immunization, close the helium cylinder and fire the Helios Gene Gun until the pressure falls down to zero psi (see Note 22).
11. Remove the 9 V battery from the Helios Gene Gun device.
12. Store the Helios Gene Gun device at room temperature.

3.2. Transiently Transgenic Mouse Model

The following protocol describes the technique for generating transient expression of HCV NS3-protein in a mouse liver. The protocol can easily be used for expression of any gene of interest in the liver. The procedure requires laboratory work involving two persons and approximately 2.5 h for generation of 20 mice with transient hepatic gene expression.

1. Prepare 100 μg DNA in 1.8 mL Ringer's solution (for a 25 g mouse) (see Note 23).
2. Restrain the mouse in your hand and inject the mouse intra peritoneal with a low dose of Hypnorm using a 1 mL syringe and a 27G needle 5–10 min prior to the hydrodynamic injection. This will anesthetize the mouse during the procedure (see Note 24).
3. Put the mouse in a restrainer (see Note 25).
4. Place the restrainer with the mouse tail facing the heating lamp for 1 min (see Note 26).
5. Wet the tail with 70% ethanol (see Note 27).
6. Inject the mouse with 1.8 mL intravenously within 5 s. During the procedure, one person holds the restrainer and the other person performs the hydrodynamic tail vein injection (see Notes 28–31).
7. Quickly remove the mouse from the restrainer and place the mouse in a cage located under the heating lamp.
8. Check the status of the mouse until fully recovered from anesthesia.

3.3. Immuno-Precipitation and Western Blot (IP-WB)

The herein described protocol is optimized for detection of liver-specific expression of HCV NS3-protein after hydrodynamic injection of a codon optimized HCV NS3/4A DNA plasmid. The protocol can be adjusted for detection of any protein of interest. The procedure requires laboratory work involving one person for approximately 12 h separated over 2 days.

3.3.1. Preparation of Liver Homogenates

1. Sacrifice the mouse through cervical dislocation and excise the liver. Weigh 100 mg of liver and cut the specimen into small pieces using a scissor (see Note 32).
2. Immediately put the liver pieces into a 15 mL Falcon tube containing 1 mL RIPA buffer. Keep the samples on ice.
3. Homogenize the liver tissue by sonication in turns of maximum 30 s using amplitude 60 on the Cell Vibra unit. Keep the sample tube on ice during the whole procedure (see Note 33).
4. Incubate the sample on ice for 15 min, vortex and transfer to a 2 mL Sarstedt tube.
5. Centrifuge at 12,000 ×*g* for 2 min at +4°C.
6. Transfer the supernatant to a new 1.5 mL Sarstedt tube and keep on ice (or store at −80°C for later use).

3.3.2. Preparation of Protein-A Sepharose

1. Weigh 125 mg Protein-A Sepharose in a 1.5 mL Sarstedt tube. Add 0.5 mL PBS (100% Protein-A Sepharose) and mix carefully (see Note 34).

2. Add another 0.5 mL PBS (gives you a 50% Protein-A Sepharose solution) and let the Protein-A Sepharose swell for at least 30 min at room temperature.
3. Wash the Protein-A Sepharose five times in PBS: centrifuge at 500 × *g* for 1 min. Remove 500 μL of the supernatant and add new fresh 500 μL PBS. Mix gently between each washing step.

3.3.3. Preparation of Antibody-Conjugated Beads

1. Perform all the following steps on ice.
2. Aliquot 5 μL of a mouse HCV NS3 IgG antibody into individual Sarstedt tubes (see Note 35).
3. Directly add 0.5 mL ice-cold PBS and 30 μL 50% Protein-A Sepharose. Mix the Protein-A Sepharose between every sample addition (see Note 36).
4. Put the tube on rotation at 12 × *g* for 1 h at +4°C (see Note 37).
5. Centrifuge the sample for 1 min at 12,000 × *g* at +4°C.
6. Discard the supernatant by flicking the tube.
7. Add 1 mL of Non-denaturing lysis buffer and invert the tube five times until the pellet has dissolved (see Notes 38 and 39).
8. Repeat steps 5–7 twice.
9. After the last centrifugation, use a 1 mL pipette to remove the supernatant (see Note 40).

3.3.4. Immuno-Precipitation (IP)

1. Add 10 μL 10% BSA to the antibody-Protein-A Sepharose complex (see Note 41).
2. Add maximum 1 mL of homogenate/lysate (e.g., protein).
3. Rotate at 12 × *g* overnight at +4°C.
4. Centrifuge the Sarstedt tubes for 1 min, 12,000 × *g* at +4°C.
5. Discard the supernatant (containing unbound proteins).
6. Add 1 mL ice-cold wash buffer and resuspend the Protein-A Sepharose-beads by inverting the tube four times.
7. Repeat steps 4–6 four times (see Notes 42 and 43).
8. Wash beads once more using 1 mL ice-cold PBS and after centrifugation aspirate supernatant completely using a 1 mL pipette.
9. Add 15 μL 2× loading dye. Continue with SDS-PAGE or store samples at −20°C for later use.

3.3.5. SDS-PAGE

1. Prepare a NuPAGE 4–12% Bis-Tris gel in running buffer in a gel chamber and connect to a power unit.
2. Denaturate the samples for 5 min at 100°C and immediately place samples back on ice.

3. Load the samples to the gel (15 μL/well) using a gel-loading tip (see Note 44).
4. Include a molecular weight marker (7.5 μL of MagicMark XP protein western standard) that you load to the NuPAGE 4–12% Bis-Tris Gel. There is no need to heat or reduce the standard prior to loading.
5. Run gel at 200 V for approximately 60 min (see Note 45).

3.3.6. Western Blot (WB)

1. Transfer the proteins to a nitrocellulose membrane using the iBlot Dry Blotting System.
2. Remove the sealing of the iBlot anode stack (bottom) and place it in the iBlot device, in the blotting surface, with the membrane facing up (see Note 46).
3. Place the gel on the membrane and use the roller to remove any bubbles between the gel and the membrane (see Note 47).
4. Wet a Whatman paper in sterile water and place it on the gel.
5. Open the cathode stack and place the cathode on the Whatman paper with the copper electrode facing up.
6. Place a disposable sponge with the metal contact in the upper right corner of the lid.
7. Close the iBlot device lid and run a program for 10 min at 20 V.
8. Transfer the membrane to a dish (with the size of the membrane) and wash the membrane twice in 10 mL sterile water for 5 min on a rotating table (50 rpm).
9. Block the nitrocellulose membrane with 10 mL blocking solution for 30 min on a rotating table (50 rpm).
10. Discard the blocking solution and wash the membrane twice in 10 mL sterile water for 5 min on a rotating table (50 rpm).
11. Incubate the membrane with 10 mL of primary antibody solution (e.g. mouse polyclonal NS3 IgG antibody) diluted 1:100 in blocking solution for 1 h on a rotating table (50 rpm, see Note 48).
12. Wash the membrane four times in 10 mL antibody wash solution for 5 min on a rotating table (50 rpm).
13. Incubate the membrane with 10 mL of secondary antibody solution (anti-mouse alkaline phosphatase (AP)-conjugated antibody) for 30 min on a rotating table (50 rpm).
14. Wash the membrane four times in 10 mL antibody wash solution for 5 min on a rotating table (50 rpm).
15. Wash the membrane three times in 10 mL of sterile water for 5 min on a rotating table (50 rpm).

3.3.7. Chemiluminescence Detection

1. Place the membrane on a clean sheet of transparency plastic. Mix 2.375 mL CDP-Star substrate and 0.125 mL Nitro-Block-II enhancing reagent for one membrane. Apply the 2.5 mL chemiluminescent substrate to the membrane surface. Let the reaction develop for 5 min (see Note 49).
2. Gently discard the excess of chemiluminescent substrate solution and expose the membrane to light for 1 min.
3. Develop the membrane in the Gene Gnome instrument (or similar) for approximately 50 s (see Note 50).

3.4. Immuno-histochemistry

Formalin fixed, paraffin-embedded sections of liver, sectioned at 3.5 μm are used for immuno-histochemical staining of the HCV NS3-protein. This protocol can be adjusted for detection of other proteins. The procedure requires laboratory work involving one person for approximately 8 h separated over 2 days.

1. Incubate the microslides in a cuvette if other not stated.
2. Deparaffinize and hydrate the sections by incubating the tissue slides 3 × 5 min in xylene, 2 × 5 min in 99% ethanol, 5 min in 95% ethanol and 2 min in 70% ethanol.
3. Wash the slides 10 min in sterile water.
4. Incubate the slides for 10 min in 0.3% hydrogen peroxide (H_2O_2) in sterile water to inhibit endogenous peroxidase activity.
5. Wash the slides 2 × 5 min in sterile water.
6. Using a pressure cooker, boil the slides in antigen retriever solution (0.01 M citrate buffer) for 5 min.
7. Leave the slides in the antigen retriever solution to cool down for at least 45 min before proceeding to the next step.
8. Incubate the slides 5 min in PBS.
9. To prevent nonspecific protein binding, block the tissue using 2.5% NHS diluted in PBS for 30 min in a humidified chamber (see Note 51).
10. Apply the primary antibody, rabbit-anti-NS3 IgG, diluted 1:1,000 in PBS containing 2.5% NHS and incubate over night at +4°C in a humidified chamber (see Note 52).
11. Wash the slides 3 × 10 min in PBS.
12. Incubate the tissue with ImmPRESS anti-rabbit-Ig-peroxidase reagent for 30 min in a humid chamber.
13. Wash the slides 2 × 3 min in PBS.
14. Incubate the tissue in DAB (peroxidase substrate) until adequate staining can be visualized (approximately 2–6 min). Use a microscope to monitor the enzymatic activity (see Note 53).
15. Stop the reaction by washing the slides 3 min in sterile water.

16. Incubate the tissue in Mayers HTX for 3 min.
17. Incubate in sterile water for 5 min to visualize the blue coloring of the cell nuclei (see Note 54).
18. Dehydrate the sections by incubating the tissue slides in 70% ethanol for 3 min, 95% ethanol for 3 min, 99% ethanol for 2 × 3 min and xylene 2 × 3 min.
19. Mount the tissue by adding four drops of Pertex on the microscope slide and place a coverslip on top (see Note 55). Let dry for at least 1 h.
20. Analyze the results in a microscope.

3.5. Tumor Challenge Model

This section describes the in vivo tumor challenge model. The procedure includes preparation of tumor cells and the rationale for sub cutaneous transplantation of tumor cells in laboratory mice. This protocol is based on inoculation of 1×10^6 tumor cells. The procedure requires laboratory work involving one person. The time needed depends on the number of tumor cells needed for transplantation and the number of laboratory mice to be inoculated.

1. Harvest the number of tumor cells needed (see Note 56).
2. Centrifuge (450 × *g*, 5 min, +4°C) the harvested tumor cells and discard the cell culture medium supernatant.
3. Resuspend the cell pellet in 50 mL sterile PBS.
4. Centrifuge (450 × *g*, 5 min, +4°C) the resuspended tumor cells and discard the supernatant (see Note 57).
5. Repeat step 3 and 4 once.
6. Resuspend cells in an appropriate volume (1–5 mL) of PBS depending of the size of the cell pellet.
7. Prepare cells for counting by mixing 10 μL cell suspension + 90 μL Trypan blue stain 0.4%.
8. Count cells in a Bürkner chamber.
9. Dilute cells to 1×10^6 cells/200 μL in sterile PBS (see Note 58).
10. Keep cells on ice until cell transplantation.
11. Restrain the mouse in your hand and inject the mouse intra peritoneal with 20–30 μL of Hypnorm using a 1 mL syringe and a 27G needle 5–10 min prior to the tumor cell transplantation. This will anesthetize the mouse during the procedure (see Note 59).
12. Mix the tumor cell suspension several times prior to transplantation to obtain a single cell suspension (see Note 60).
13. When the laboratory mouse is anesthetized after Hypnorm treatment, restrain the mouse in your hand and sub cutaneously inoculate 200 μL of the tumor cell suspension under the skin of the right flank of the mouse using a 1 mL syringe and a 25G needle (see Note 61).

14. Palpate and measure tumor growth using a sliding caliper at appropriate time points, see Fig. 3 (see Note 62).
15. Inspect the tumor bearing laboratory mouse every day for signs of illness due to the tumor growth (see Note 63).
16. Sacrifice the laboratory mouse when you reach the experimental end-point.

3.6. Cytotoxicity Assay

This section describes the procedure to measure in vitro cytotoxic T lymphocyte (CTL) activity in spleen cells from DNA-immunized mice. The procedure requires laboratory work preferable performed by two persons for approximately 7 h the first day and another 9 h 5 days later. This is estimated on an experiment based on 20 laboratory mice. The protocol also requires good knowledge and permission to work with radioactive material.

3.6.1. Preparation of Spleen Cells (Setup of Re-stimulation Cultures)

1. Sacrifice the laboratory mice and excise the spleen using scissor and forceps.
2. Immediately put the spleen in a 15 mL Falcon tube containing 2 mL of incomplete medium (see Subheading 2.4, item 1).
3. The following procedures should be performed under sterile conditions in a laminar airflow bench.
4. Make single cell suspension by pushing the spleen, with the upper end of a 1 mL syringe, on the net of a cell strainer placed in a Petri dish.
5. Transfer the cell suspension to a new 15 mL Falcon tube, wash the net with 5 mL of incomplete medium and collect the wash in the same tube.
6. Centrifuge for 5 min at $450 \times g$, +4°C.
7. Discard the supernatant and resuspend the pellet in the remaining medium.
8. Add 1 mL of Red Blood Cell Lysing buffer and incubate at room temperature for 1 min (see Note 64).
9. Add 12 mL of PBS to the Falcon tube to inactivate the Red Blood Cell Lysing buffer and proceed immediately with centrifugation for 5 min at $450 \times g$, +4°C.
10. Discard the supernatant and resuspend the cell pellet in 2 mL of complete medium (see Subheading 2.4, item 2).
11. Prepare cells for counting by mixing 10 μL cell suspension + 90 μL Trypan blue stain 0.4%.
12. Count cells in a Bürkner chamber.
13. Prepare a cell suspension of 25×10^6 cells in a cell culture flask (25 cm^2) with a total volume of 12 mL complete medium (see Note 65).

14. Co-culture the spleen cells from immunized mice with 25×10^6 irradiated (2,000 rad = 20 Gy (gray)) spleen cells from naïve mice (see Note 66).
15. Dilute the MHC class I HCV NS3 peptide $H2D^b$ (amino acid sequence: GAVQNEVTL) to a concentration of 50 μM in complete medium.
16. Sterile filter the dissolved peptide solution through a 0.22 μm filter.
17. Add 120 μL of the sterile HCV NS3 peptide solution (50 μM) into the 12 mL of complete medium of the cell culture flask. The peptide is diluted 100× to a final concentration of 0.5 μM (see Note 67).
18. Incubate the cell culture flask for 5 days in a humidified incubator (+37°C and 5% CO_2) (see Note 68).
19. Subheadings 3.6.2 and 3.6.3 are performed in parallel.

3.6.2. Preparation of Target Cells

1. Harvest and prepare 1×10^6 target cells (e.g., RMA-S, see Note 69) in 500 μL complete medium, with and without 50 μM of the HCV $NS3_{(GAVQNEVTL)}$ peptide (see Notes 70 and 71). Use 15 mL Falcon tubes for this procedure.
2. Incubate the RMA-S cells for 90 min at +37°C and 5% CO_2. Mix carefully every 15 min.
3. Centrifuge the cells for 5 min at $450 \times g$ at +4°C and thereafter discard the supernatant forcefully by hand turning the tube upside down (see Note 72).
4. The following procedure has to be performed in a designated room for work with radioactive material.
5. Add 25 μL Chromium 51 (^{51}Cr, 5 mCi/mL) to each target cell tube (1×10^6 RMA-S cell pellet) (see Notes 73 and 74).
6. Incubate for 1 h at 37°C and 5% CO_2. This allow for passive uptake of ^{51}Cr by the RMA-S target cells.
7. Wash the RMA-S target cells three times in PBS, centrifuge for 5 min at $450 \times g$, +4°C (see Note 75).
8. Dilute the RMA-S target cells to a concentration of 50,000 cells/mL in complete medium.

3.6.3. Preparation of Effector Cells

1. Harvest the effector cells (e.g., CTLs) in a 15 mL Falcon tube using a 10 mL pipette (see Note 76).
2. Centrifuge the effector cells for 5 min at $450 \times g$, +4°C and resuspend the pellet in complete medium at a concentration of 3×10^6 cells/mL.
3. Add effector cells to a 96-well V-bottom plate, according to effector:target ratios shown in Tables 1 and 2. Prepare triplicates of all samples (see Note 77).

Table 1
Volumes of effector/target cell suspensions used in the cytotoxicity assay

Ratio	60:1 (μL)	20:1 (μL)	7:1 (μL)
Effector cells	100	33.3	11.1
Complete medium	–	66.7	88.9
Target cells	100	100	100

Target cells: 5,000/well (50,000 cells/mL)
Effector cells: 300,000 cells/100 μL (3×10^6/mL)

Table 2
Numbers of effector/target cells used in the cytotoxicity assay

Ratio	Effector cells (number of cells per well)	Target cells (number of cells per well)	Effector cells (volume in μL when conc. is 3×10^6/mL)
60:1	300,000	5,000	100
20:1	100,000	5,000	33.3
7:1	35,000	5,000	11.1

4. Add 100 μL/well of the ^{51}Cr-labeled RMA-S target cells (5,000 target cells/well) to the effector cells.
5. Also prepare wells with target cells in medium to be used as positive (maximum release) and negative (spontaneous release) controls.
6. Incubate for 4 h at 37°C and 5% CO_2 (see Note 78).
7. After the 4-h incubation, transfer 25 μL of supernatant from each well to an Isoplate 96-well sample plate containing 200 μL OptiPhase Super Mix (see Subheading 3.6.4, step 1). As negative control (spontaneous release), transfer 25 μL supernatant from target cells incubated in medium alone (see Note 79). For the positive control (maximum release), mix the supernatant and cells and transfer 25 μL to the Isoplate 96-well sample plate.

3.6.4. Measure ^{51}Cr Release in the Supernatants

1. Add 200 μL OptiPhase Super Mix to an Isoplate 96-well sample plate.
2. Add 25 μL supernatant to the Isoplate 96-well sample plate containing 200 μL OptiPhase Super Mix (see Subheading 3.6.3, step 7).

3. Seal the plate by attaching a plastic cover (Sealing tape).
4. Mix the plate vigorously by hand for a few seconds to ensure a homogenous solution.
5. Clean the plate cover with 70% ethanol and dry the cover using a paper towel, if needed.
6. Turn on the 1450 MicroBeta TriLux scintillation counter, with the MicroBeta workstation software 4.0.
7. Put the plate in a rack (plate cassette) and inset into the counter.
8. Setup program (use 1 min counting/well) and run program to count the plate.
9. Calculate the specific lysis as follows:

% specific lysis = (sample counts per minute/spontaneous release)/(max release/spontaneous release) × 100.

4. Notes

1. The diameter of the gold microcarriers will affect the depth of penetration when administered to the tissue of interest. This has to be optimized for different nucleic acids, target tissue/organ etcetera. We have used gold with a diameter of 1.0 μm in this protocol.
2. PVP serves as an adhesive during the cartridge preparation. The optimal amount of PVP to be used must be determined empirically, but typical concentrations of PVP ranges from 0.01 to 0.1 mg/mL. We have used a concentration of 0.01 mg/mL.
3. Opened bottles of ethanol absorb water and the presence of water in the tubing while drying the cartridges will lead to streaking, clumping, and uneven coating of the microcarrier over the inner surface of the Gold-Coating tubing, resulting in poor or unusable cartridges.
4. Addition of 75 μL of DNA (1 mg/mL) in this step generates Gene Gun cartridges that contain 1 μg DNA.
5. The solution in the Sarstedt tubes should vibrate during the 10 s to allow for proper gold microcarrier separation. It is important to avoid gold microcarrier aggregates for proper coating.
6. One person holds the tube in the vortexer and the other person adds the $CaCl_2$ drop-wise to the mixture. When adding $CaCl_2$ the DNA and microcarriers will associate.
7. Carefully discard the supernatant since parts of the pellet may detach.

8. Proper mixing of the DNA/microcarrier complexes is crucial to obtain similar amounts of DNA coated on the tubing.
9. At this step the Falcon tubes containing the DNA/gold microcarriers may be stored at +4°C for a shorter time period or at -20°C for up to 2 months. Remember to wrap the tubes in parafilm.
10. The mixture/solution should vibrate when applied in the ultrasonic bath to make sure that no gold microcarrier aggregates are present.
11. It is essential to bring up the 3.5 mL solution as fast as possible to get the gold microcarrier evenly distributed in the 75 cm Gold-Coat tubing. Avoid bubbles when bringing up the mixture and keep the tubing horizontally during the whole procedure.
12. Check that you can see a nice line with gold microcarrier on the bottom of the tubing in the tubing prep station.
13. Keep the speed constant when removing the solution to avoid disrupting the gold microcarrier separated from the solution. Too fast removal or changing the speed of removal may affect/move the gold microcarrier layer. If the solution is removed too slowly, the gold microcarrier will start to dry in the tubing prior to start of tubing rotation. This will cause a non-evenly distributed coating, which is unusable.
14. Start of rotation is critical to obtain an evenly distributed coating. When the nitrogen is turned on the gold microcarrier will dry inside the tubing. This is clearly visible. When the nitrogen is turned on, look at the tubing, close to the flow-meter, and you will see the gold microcarrier drying, this will continue all the 75 cm towards the end of the tubing.
15. Only use coated cartridges with evenly distributed gold microcarrier layer. This is to secure that each cartridge contain the same amount of DNA.
16. We have used cartridges for at least 1 month without seeing any differences in priming immune responses in laboratory mice. Although, this is probably depending on multiple parameters, such as quality of reagents, solutions, purity of DNA, diameters of the gold microcarriers, storage temperature etcetera.
17. The abdominal area has to be completely free from hair to allow for proper delivery of the DNA coated on gold microcarriers. Make sure that you do not cause any wounds in the shaved skin area. Wounds in the skin may cause ruptures in the skin when delivery of the gold microcarriers are performed using high helium pressure.
18. When loading the cartridge holder with cartridges check that they have similar sizes and that the coated cartridges have gold

microcarriers evenly distributed. Discard cartridges that do not fulfill these criteria's.

19. Fire the Helios Gene Gun device two times to make sure that the pressure is 500 psi before starting the immunization on mouse abdomens.
20. During the delivery procedure the mouse should be restrained with the same force towards the Helios Gene Gun device. The Helios Gene Gun device target area is 2 cm^2.
21. Directly after firing the two shots, two round-shaped gold colored marks will be visible on the abdomen of the mouse, see Fig. 6b. After approximately 2 weeks you may see an increased fur growth at the place for delivery, see Fig. 6c.
22. Do not disconnect the Helios Gene Gun from the helium gas cylinder without first empty the equipment from gas.
23. The injection volume and the weight of the mouse are important for efficient uptake and expression of the plasmid DNA. Volumes below 1.8 mL drastically reduce the expression efficiency. Mice weighing more than 25 g would need larger injection volumes for efficient perfusion of the liver.
24. The volume of Hypnorm to be used has to be adjusted depending on mouse strain and weight of the animals. The Hypnorm will anesthetize the laboratory mice for approximately 30–60 min depending on the dose.
25. It is important that the mouse is restrained properly to avoid movement during the injection.
26. Warming of the mouse tail vein makes it more visible since it expands when heated.
27. The ethanol makes the veins more visible. Avoid ethanol at the place where you will hold the tail with your fingers. The ethanol makes the tail slippery and may result in losing the grip during injection.
28. Hold the mouse tail strained during the whole injection. Any movement of the mouse will increase the risk of slipping out of the vein.
29. Injection speed longer than 5 s will decrease the perfusion efficiency.
30. If the needle goes out of the vein during injection the resistance will prevent any further injection.
31. The whole volume has to be injected at once. If injection stops the mouse has to be excluded from the experiment.
32. If protein expression from several laboratory mice will be analyzed, excise representative pieces from the different liver lobes.

33. The sonication will generate heat, it is therefore recommended to use several shorter turns of sonication with the tube placed on ice, to avoid protein degradation.
34. Mix until no clumps can be seen in the mixture.
35. The choice of antibody depends on the protein to be detected.
36. To get the same amount of Protein-A Sepharose in each sample it is absolutely necessary to mix the Protein-A Sepharose vigorously between every pipetting.
37. Make sure that the Protein-A Sepharose antibody solution flows in the tube during rotation. If not, flick the tube until the solution flows neatly.
38. This step is important to remove unbound or weakly bound antibody.
39. If the pellet does not dissolve properly when adding the Non-denaturing lysis buffer, carefully flick the tube until the pellet is completely dissolved.
40. Remove as much as possible of the supernatant without removing any of antibody-Protein-A Sepharose pellet.
41. This step is important to block unspecific protein binding to the NS3-antibody-Protein-A Sepharose complex.
42. Several washing steps are required to remove unbound protein in the antibody-Protein-A Sepharose solution.
43. The pellet will become more distinct after each washing step.
44. The gel-loading tip is preferable to avoid pipetting the Protein-A Sepharose-beads into the gel.
45. Stop the gel when the loading dye starts to run out in the running buffer.
46. Keep the stack in the plastic tray.
47. Keep the roller wet to avoid breaking the gel.
48. Since we are using the same antibody in the immuno-precipitation and in the protein detection step, we will also detect the mouse heavy (50 kDa) and light chain (25 kDa) on the developed membrane. If this is not appropriate, one can use two different antibodies with different species origin, recognizing the same protein.
49. Make sure that the whole membrane is covered with the substrate during the whole incubation period to avoid insufficient signal.
50. The 50 s exposure time is chosen for this protocol where we detect the HCV NS3-protein. This parameter has to be optimized for detection of other proteins.
51. Use an Imm Edge Pen to mark the area around the tissue. This will reduce the volume needed for each incubation and keep the solution in place.

52. As a negative control, omit the primary antibody from this procedure.
53. Positive cells will stain brown.
54. If no or weak nuclei staining appear go back and repeat step 16.
55. Hold the microscope slide tilted when placing the coverslide on top, which reduces the risk of bubbles in the Pertex.
56. It is important to obtain a single cell suspension regardless if the cells are grown adherent or in suspension.
57. The washing of the tumor cells is important to get rid of medium components and supplements to avoid toxic effects in the laboratory mice.
58. The number of tumor cells needed to establish tumor growth in vivo has to be evaluated for each cell line and mouse strain. In general, $1–5 \times 10^6$ tumor cells per mouse is a good starting point.
59. The volume of Hypnorm to be used has to be adjusted depending on mouse strain and weight of the animals. The Hypnorm will anesthetize the laboratory mice for approximately 30–60 min depending on the dose. This is important to avoid leakage of the transplanted tumor cells when the mice are moving in the cage.
60. Insufficient mixing of the tumor cell suspension prior to each transplantation may cause that each mouse is inoculated with an incorrect number of cells.
61. After penetrating the skin, push the needle approximately 1–1.5 cm towards the right front leg by moving the needle just under the skin. This is important to avoid any leakage from the needle hole.
62. Tumors are usually palpable and measurable around day 3–5 post inoculation. We suggest measuring the tumors every 1–3 days. Each experiment should have an end-point, such as a maximum allowed tumor size or volume when the mice should be terminated. This is regulated by the local and national guidelines for animal welfare.
63. Some laboratory mice may get wounds on and around the solid tumor (a lot depending on the tumor cell line used) and these situations should be handled according to the local and national guidelines for animal welfare.
64. Shorter incubation will result in insufficient lysis of the red blood cells. Longer incubation will affect the condition of the splenocytes and may results in low cell yield.
65. The use of 25×10^6 spleen cells is optimized for detection of HCV NS3 CTLs in vitro. This parameter may need to be evaluated for detection of CTLs specific for other antigens.

66. The spleen cells from naïve mice are used as feeder cells. After irradiation they may only divide once, but will produce and secrete necessary growth factors, cytokines and be a source for professional antigen presenting cells (APCs). These properties are important for optimal proliferation of the specific CTLs obtained from the immunized mice.
67. The peptide concentration has to be optimized for every peptide. The amount of peptide needed for stimulation may vary a lot.
68. Our experience is that 5-days of incubation is needed to allow for optimal proliferation of HCV NS3-specific CTLs.
69. The RMA-S cells were kindly provided by professor K. Kärre, Karolinska Institutet, Stockholm, Sweden. The RMA-S cells are TAP (transporter associated with antigen processing) deficient, which allows exogenous loading of peptides on MHC class I without disturbance of endogenous peptides.
70. This incubation is necessary to load the target cell MHC class I molecules with the HCV NS3-specific CTL peptide.
71. The RMA-S target cells have to be passaged 1 day prior to the ^{51}Cr-release assay.
72. It is important to remove as much possible of the supernatant to allow proper labeling of the cells with radioactive ^{51}Cr.
73. The volume depends on the activity of the ^{51}Cr. Our experience is to add another 5 μL to the total volume for each week after the date of isotope production. We are using batches of ^{51}Cr for no more than 4 weeks from production.
74. The radioactive work should be performed with necessary safety precautions. Try to work as much as possible with the radioactive source at distance from your body. Work behind protective shielding and use led coat and eye protection.
75. Washing of the labeled RMA-S cells is necessary to wash away excess of ^{51}Cr.
76. Pipette the medium up and down several times to collect all cells that have attached to the bottom of the cell culture flask.
77. The effector:target ratio is optimized for detection of NS3-specific CTLs. The optimal effector:target ratio has to evaluated for other peptides/antigens.
78. This incubation allow the HCV NS3 peptide-specific CTLs to recognize and lyse peptide pulsed RMA-S target cells. Lysis of RMA-S target cells will result in release of ^{51}Cr into the supernatant. The amount of released ^{51}Cr can be measured and correlated to the lytic activity of the CTLs.
79. Carefully transfer the supernatant without touching the cell pellet at the bottom of the wells.

Acknowledgment

The following work was supported by grants from the Swedish Research Council (K2012-99X-22017-01-3), the Swedish Cancer Society, Stockholm County Council. Dr. Lars Frelin was supported by grants from the Swedish Society of Medical Research, the Swedish Society of Medicine, Goljes Memorial Fund, the Åke Wiberg Foundation, the Royal Swedish Academy of Sciences, and from Karolinska Institutet. Dr. Gustaf Ahlén was supported by grants from Karolinska Institutet/Södertörn University (postdoctoral grant), Lars Hiertas Memorial fund, Goljes Memorial Fund, the Royal Swedish Academy of Sciences, and from Magnus Bergvalls foundation.

References

1. Alter MJ (1997) Epidemiology of hepatitis C. Hepatology 26:62S–65S
2. Milich DR (1997) Pathobiology of acute and chronic hepatitis B virus infection: an introduction. J Viral Hepat 4(suppl 2):25–30
3. Zoulim F, Locarnini S (2009) Hepatitis B virus resistance to nucleos(t)ide analogues. Gastroenterology 137:1593–1608
4. Manns MP et al (2001) Peginterferon alfa-2b plus ribavirin compared with interferon alfa-2b plus ribavirin for initial treatment of chronic hepatitis C: a randomised trial. Lancet 358:958–965
5. McHutchison JG et al (2009) Telaprevir with peginterferon and ribavirin for chronic HCV genotype 1 infection. N Engl J Med 360:1827–1838
6. Ferrari C et al (1990) Cellular immune response to hepatitis B virus-encoded antigens in acute and chronic hepatitis B virus infection. J Immunol 145:3442–3449
7. Bowen DG, Walker CM (2005) Adaptive immune responses in acute and chronic hepatitis C virus infection. Nature 436:946–952
8. Yang NS et al (1990) In vivo and in vitro gene transfer to mammalian somatic cells by particle bombardment. Proc Natl Acad Sci USA 87:9568–9572
9. Tang DC, DeVit M, Johnston SA (1992) Genetic immunization is a simple method for eliciting an immune response. Nature 356:152–154
10. Chisari FV (1996) Hepatitis B virus transgenic mice: models of viral immunobiology and pathogenesis. Curr Top Microbiol Immunol 206:149–173
11. Pasquinelli C et al (1997) Hepatitis C virus core and E2 protein expression in transgenic mice. Hepatology 25:719–727
12. Frelin L et al (2006) The hepatitis C virus and immune evasion: non-structural 3/4A transgenic mice are resistant to lethal tumour necrosis factor alpha mediated liver disease. Gut 55:1475–1483
13. Kriegs M et al (2009) The hepatitis C virus non-structural NS5A protein impairs both the innate and adaptive hepatic immune response in vivo. J Biol Chem 284:28343–28351
14. Lerat H et al (2002) Steatosis and liver cancer in transgenic mice expressing the structural and nonstructural proteins of hepatitis C virus. Gastroenterology 122:352–365
15. Ahlen G et al (2005) In vivo clearance of hepatitis C virus nonstructural 3/4A-expressing hepatocytes by DNA vaccine-primed cytotoxic T lymphocytes. J Infect Dis 192:2112–2116
16. Frelin L et al (2003) Low dose and gene gun immunization with a hepatitis C virus nonstructural (NS) 3 DNA-based vaccine containing NS4A inhibit NS3/4A-expressing tumors in vivo. Gene Ther 10:686–699
17. Frelin L et al (2004) Codon optimization and mRNA amplification effectively enhances the immunogenicity of the hepatitis C virus nonstructural 3/4A gene. Gene Ther 11:522–533
18. Nystrom J et al (2010) Improving on the ability of endogenous hepatitis B core antigen to prime cytotoxic T lymphocytes. J Infect Dis 201:1867–1879
19. Roy MJ et al (2000) Induction of antigen-specific CD8+ T cells, T helper cells, and protective levels of antibody in humans by particle-mediated administration of a hepatitis B virus DNA vaccine. Vaccine 19:764–778
20. Frelin L et al (2009) A mechanism to explain the selection of the hepatitis e antigen-negative mutant during chronic hepatitis B virus infection. J Virol 83:1379–1392

Chapter 21

Gene Gun Immunization to Combat Malaria

Elke S. Bergmann-Leitner and Wolfgang W. Leitner

Abstract

DNA immunization by gene gun against a variety of infectious diseases has yielded promising results in animal models. Skin-based DNA vaccination against these diseases is not only an attractive option for the clinic but can aid in the discovery and optimization of vaccine candidates. Vaccination against the protozoan parasite *Plasmodium* presents unique challenges: (a) most parasite-associated antigens are stage-specific; (b) antibodies capable of neutralizing the parasite during the probing of the mosquitoes have to be available at high titers in order to prevent infection of the liver; (c) immunity to liver-stage infection needs to be absolute in order to prevent subsequent blood-stage parasitemia. Gene gun vaccination has successfully been used to prevent the infection of mice with the rodent malaria strain *P. berghei* and has been employed in a macaque model of human *P. falciparum*. DNA plasmid delivery by gene gun offers the opportunity to economically and efficiently test novel malaria vaccine candidates and vaccination strategies, which include the evaluation of novel molecular adjuvant strategies. Here we describe the procedures involved in making and delivering a pre-clinical malaria DNA vaccine by gene gun as well as the correct approach for the in vivo evaluation of the vaccine. Furthermore, we discuss various approaches that either have already been tested or could be employed to improve DNA vaccines against malaria.

Key words: Malaria, *Plasmodium*, Molecular adjuvant, Genetic adjuvant, Plasmid co-delivery

1. Introduction

Malaria is an infectious disease caused by the protozoan parasite of the genus *Plasmodium* (phylum: Apicomplexa) and transmitted by various genera of mosquitoes. Several species selectively infect humans (of which *P. falciparum* is responsible for the majority of the reported 1–2 million fatalities every year) while other species selectively infect rodents (*P. berghei*, *P. yoelii*) or birds (e.g., *P. gallinaceium*). Currently, no effective vaccine exists for any of these parasite species and infection is prevented by and treated with drugs (such as Quinine (and its derivatives), Chloroquine or

Stephan Sudowe and Angelika B. Reske-Kunz (eds.), *Biolistic DNA Delivery: Methods and Protocols*,
Methods in Molecular Biology, vol. 940, DOI 10.1007/978-1-62703-110-3_21, © Springer Science+Business Media, LLC 2013

Artemisinin). However, antimalarial resistance is very common, which emphasizes the urgent need for an effective vaccine. Despite the differences between rodent and human malaria species, rodent models are still a valuable tool for the identification of the elusive correlates of protection and for the pre-selection of promising vaccine candidates and vaccination strategies (1). Moreover, some of the vaccine candidates with high protective efficacy have been evaluated in non-human primate models (2).

Malaria is among the many diseases in which DNA vaccination was shown to be capable of protecting experimental animals from infection. At least in the rodent model system of *P. berghei*, the skin was a superior target site for the vaccine compared to the muscle (3) and—when targeting this vaccination site—gene gun delivery induced better protection than the intradermal or intramuscular injection of plasmid (see Table 1). Even in the absence of a clear path to the clinic for the biolistic delivery of a malaria vaccine, gene gun-based vaccination studies in model systems of malaria (rodent or primate) are highly attractive and allow researchers to address the following issues.

Table 1
Effect of the immunization route on various aspect of the resulting immune response

Modality[a]	Intramuscular injection	Epidermal delivery	Intradermal injection
Dose[b]	100 μg	2 μg	100 μg
Th type[c]	Strong Th1	Strong Th2	Strong Th1
Predominant cytokine[d]	IFN-γ	IL-4, IFN-γ	IFN-γ
Induction of cytolytic T cells[e]	+/−	++	+
Magnitude of antibody response[f]	+/−	+++	+
IgG1:IgG2a ratio[f]	<0.5	2–3	0.5
IgE[f]	ND	Yes	No
IgM[f]	ND	Yes	Late
Vaccine efficacy[g]	22–45%	>95%	40%

[a]DNA was injected either by needle (for intramuscular and intradermal) or by gene gun (epidermal immunization)
[b]Dose delivered per immunization. Three immunizations in 4-week intervals were given. Analysis of the immune response was done 2 weeks after the last (third) immunization
[c]Determined by the isotype of antigen-specific antibodies and the cytokine profile of antigen-specific T cells
[d]Measured in ex vivo EliSpot assays; cells were stimulated with 3 μg/mL recombinant circumsporozoite protein (CSP) or GST (as negative control antigen)
[e]Cells were stimulated in vitro and then used for ^{51}Cr release assays using peptide pulsed target cells
[f]Measured by quantitative ELISA using recombinant CSP as plate antigen
[g]Efficacy = $[1-[(I/n \text{ vaccine})/(I/n \text{ control})]]\times 100$. I = number of infected animals; n = total of mice in the group. Control animals were immunized with empty plasmid

1.1. The Genome of Protozoan Parasites Is Relatively Large

Compared to other microorganisms, the genome of protozoan parasites is relatively large and accommodates a vast number of genes encoding many proteins which could be targeted by vaccines. The choice of target antigens is further complicated by the selective expression of stage-specific antigens throughout the parasite's life cycle. Only a very limited number of malaria antigens have so far been studied in detail, most notably CSP and TRAP on the surface of the sporozoite and AMA-1 and MSP-1 on the blood-stage parasite (reviewed in ref. (4)). Genomic studies have revealed a treasure trough of potential vaccine targets waiting to be tested, such as the recently described CelTOS antigen, which is highly conserved among *Plasmodium* species and capable of inducing cross-strain protection (5). DNA vaccination by gene gun enables large-scale, economical screening experiments of such novel vaccine targets requiring minute amounts of rapidly generated mammalian expression plasmids.

1.2. Recombinant Protein Vaccines as Well as Recombinant Viral Vectors Suffer from Several Shortcomings, Which Complicate Their Use

1.2.1. Potentially Inadequate Expression Levels

Potentially inadequate expression levels due to differences in the codon usage between protozoa and organisms used as expression platforms some malaria antigens are inherently difficult to produce. This has (partially) been remedied by synchronizing the codon usage in the expression plasmid with that of the expression platform using either codon optimization or codon harmonization (6). A major shortcoming of codon optimization (i.e., translating the *Plasmodium* preferred codons into the expression vector's codons for which the tRNAs are most abundant) is that the resulting proteins are often truncated, misfolded, or insoluble. Codon harmonization is a unique concept in which the codon frequency of the expression system is adjusted to the *relative* codon frequency of *Plasmodium* resulting in a translation rate (or speed) closely mimicking that observed in the natural host. This approach assures that the growing polypeptide chain has sufficient time to assume the correct folding.

1.2.2. Forced, High-Level Expression in Recombinant Bacteria

Forced, high-level expression in recombinant bacteria frequently yields insoluble recombinant protein that requires denaturation and refolding to produce a soluble protein. This represents an empirical process, which requires extensive expertise and has a success rate that depends on the biochemical nature of the protein of interest.

1.2.3. Aberrant Glycosylation

Aberrant glycosylation due to either inherent glycosylation patterns in the expression systems (in case of recombinant proteins) or due to the misinterpretation of the protozoan sequences in mammalian cells (in case of viral vectors) result in immunologically altered antigens and, therefore, antibodies, which may not recognize the pathogen-associated antigen during infection.

1.3. Immunogenicity of recombinant and highly pure vaccines

To become immunogenic, recombinant and highly pure vaccines are entirely dependent on adjuvants, which provide the necessary innate immune stimulation. In addition to "traditional" adjuvants, i.e., molecules which activate innate immune responses, numerous "molecular" adjuvants have been described (7). Such molecules are designed to modulate or trigger highly defined immune pathways or route antigens into distinct antigen processing pathways in distinct target cells (most often dendritic cell subsets, B cells). Molecular adjuvants previously used in the context of malaria DNA vaccines include the costimulatory molecule B7.1 and B7.2 (8), the complement fragment C3d (9, 10), and pro- as well as anti-apoptotic molecules (particularly Bax or Bcl (11)). The selection of any immunomodulatory strategy in a disease such as malaria is complicated by the lack of known immune correlates of protection thus requiring the laborious screening of a variety of approaches. Many of these molecules can be encoded on a plasmid (and are therefore sometimes referred to as "genetic adjuvants") and delivered alongside the DNA vaccine, but some require co-delivery to the same target cells (e.g., costimulatory molecules, pro/anti-apoptotic molecules) in order to be efficacious. Gene gun delivery eliminates the need for co-expression of the two molecules from the same plasmid since a mixture of plasmids co-precipitated onto the same gold particle will also be co-delivered to the same host cell thus drastically facilitating the pairing of target antigens and immunomodulatory molecules.

1.4. Considerations Preceding Immunization

1.4.1. Cloning of Malaria Antigens

Initially, the protein sequence of the (novel) malaria antigen selected as the vaccine candidate should be screened for at least the following motifs: (a) Glycosylation sites; when produced in mammalian cells, these sites will be *N*-glycosylated and this non-native glycosylation pattern often results in altered antigenic structure and thus altered recognition of crucial B cell epitopes (12). (b) GPI-anchor sequences that result in the retention of the translated protein in the ER of transfected cells. Therefore, the antigen remains "invisible" for B cells (13). (c) Motifs associated with complement binding as some malarial antigens are known to bind complement factors as part of immune escape mechanisms (14).

Consider cloning multiple constructs with truncations or deletions to address the issues listed above, which affect the efficacy of DNA immunization in ways that can be difficult to predict.

1.4.2. Co-Immunization

(a) DNA vaccine delivery by gene gun enables the co-delivery of plasmids encoding the target antigen (=vaccine) and plasmids encoding immunomodulatory molecules by simple co-precipitation on gold particles thus eliminating the need for engineering bi-cistronic expression vectors (15). (b) Co-delivery of multiple malaria antigens; most vaccines currently used in the clinic contain more than one antigen. It would be surprising if an efficacious human malaria vaccine, which needs to protect against multiple malaria strains and species of *Plasmodium* would be based on a single antigen. Thus, it is likely that a clinical malaria vaccine will have to include several antigens targeting e.g. different developmental stages such as pre-erythrocytic, asexual, and sexual erythrocytic stages or will contain several antigens from one stage. In cases of a multiantigen immunization regimen into the same target site it is imperative to exclude potential antigenic competition within the transfected cell. Studies using recombinant poxviruses expressing several malaria antigens have revealed antigenic interference and the loss of immunity directed to some of the antigens in the mixture (16, 17). Therefore, immunogenicity data with individual constructs have to be obtained before testing any vaccines containing multiple antigens.

1.4.3. Immunization Regimen

As for any vaccine, the ultimate goal is to achieve protective immunity through a single immunization. However, essentially every experimental malaria vaccine tested so far required multiple doses to induce significant levels of protection but in most studies little attention had been paid to the importance of the interval between vaccinations. The duration of antigen expression after DNA vaccination (and particularly after gene gun delivery) is poorly understood and depends on multiple factors including the vaccine construct and the immunogenicity of the encoded antigen. While it is difficult to detect plasmid-encoded antigen a few days after vaccination, some studies suggest that small, but immunologically relevant amounts of antigen continue to be produced for extended periods of time resulting in poor boosting of the still ongoing immune response if the booster immunizations are delivered too soon. As shown in one study, extending the time between gene gun deliveries of a malaria antigen drastically increased the immunogenicity and protective efficacy of the vaccine (1). Therefore, multiple immunization regimens should be considered and tested.

The number of shots used for each immunization varies between protocols. While there is no significant increase in immunogenicity between two and three shots, three shots per immunization are recommended to assure optimal immunization and reduce mouse-to-mouse variations (caused e.g., by variations in the coating of individual bullets or the ability of the investigator to reliably deliver a shot).

1.4.4. Expression Levels of Plasmids

In most studies with malaria DNA vaccines (as well as DNA vaccines for other diseases) the pathogen-derived antigen is under the control of a strong promoter (most commonly the cytomegalovirus immediate early promoter) to assure high-level antigen expression in virtually all types of mammalian target cells. Expression levels are also influenced by additional factors such as the polyadenylation sequence (provided either by the target gene or the expression vector) or the presence of introns (two forms of the CMV promoter are commonly being employed; with and without intron A). The nature of the antigen used also affects the expression level: (a) toxicity of the antigen: some malaria antigens exhibit biological functions in mammalian cells (see below), (b) use of the native parasite DNA sequence or use of re-coded (harmonized) DNA sequences, (c) use of full-length sequences or removal of sequences encoding e.g., the protozoan GPI anchor, which is responsible for the cytoplasmic retention of the antigen. Expression levels from DNA vaccines are frequently cited as a major determinant of vaccine immunogenicity (and vaccine efficacy); however, in the presence of (additional) immunostimulatory signals antigen levels are secondary and no longer (reliably or predictably) correlate with immunogenicity. Such signals can originate from conventional adjuvants (not routinely used for gene gun-delivered plasmids because of the limitations of co-delivery, but shown to be efficacious when used to pre-treat the gene gun-targeted skin areas) or molecular adjuvants (in case of DNA vaccines often delivered as "genetic" adjuvants in the form of plasmid-encoded immunostimulators/ immunomodulators such as cytokines, costimulatory molecules or pro/anti-apoptotic molecules). Furthermore, certain plasmid vectors deliver additional immunostimulatory signals such as alphaviral replicase-based plasmids, which trigger powerful anti-viral innate immune pathways (18) and result in immunostimulatory host cell apoptosis (19). Antigen expression levels from such vectors do not necessarily correlate with the vector's immunogenicity and have successfully been used in a variety of pathogen models, but to date not for malaria. Their usefulness for antigens such as the CSP may actually be limited when the antigen itself is pro-apoptotic and, therefore, cell death in the transfected cell may be induced too rapidly thus preventing the production of immunologically relevant amounts of antigen as is the case when co-delivering too much of a plasmid encoding a pro-apoptotic protein (11). The optimal choice of plasmid vectors or immunostimulatory molecules for malaria antigens is entirely empirical and depends on the parasite antigen and whether (as well as how) the antigen sequence has been modified to remove certain biologically active sequences (see Subheading 1.4).

1.4.5. Vaccine Testing

The ultimate test for a malaria vaccine is the challenge of the immunized host with life parasites. For clinical trials, infected mosquitoes are generally used while in preclinical studies are routinely infected by the intravenous injection of parasites isolated from the salivary glands of infected mosquitoes. Although logistically more attractive than the laborious challenge by mosquito bite, this approach should be avoided! Not only are the challenge results highly variable (20) but this extremely artificial route of infection prevents anti-sporozoite antibodies to exert their effect during the parasite's migration from the skin to the liver (21) thus leading to wrong conclusions about the vaccine's efficacy (22). The subcutaneous injection of parasites represents a compromise, which is technically simple but still targets the relevant host tissue (20).

2. Materials

The reagents required for attaching malaria antigen-encoding plasmids onto gold particles are essentially not different from those used for gene gun vaccination with other types of plasmids. Thus the following protocols are appropriate for any gene gun-based immunization protocol in preclinical models. Prepare all solutions using ultrapure water (prepared by purifying deionized water to attain a sensitivity of 18 MΩ cm at 25°C). The use of high-quality (e.g., molecular-grade) reagent is essential since contaminants can act as adjuvants and thus introduce artifacts and unacceptable variability in the experiments. The main contaminant most frequently found in inadequately purified plasmid preparations is endotoxin (lipopolysaccharide, LPS), which is a potent innate immune stimulator even at very small concentrations. LPS contamination is predominantly a concern when delivering larger amounts of DNA by injection and the effect the contaminant may have when only minute amounts of DNA are delivered by gene gun is unclear. Nevertheless, this represents an undesirable variable, which can easily be avoided.

1. Gold particles: Micron-sized gold particles can be obtained from BioRad or directly from the manufacturer (DeGussa Corporation Metal Group, Ferro Electronic Material Systems). Particles have an average diameter of approximately 1.4–1.6 μm (see Note 1).
2. Eppendorf® tubes: 1.5 mL.
3. Eppendorf® centrifuge.
4. Spermidine: (1,8-Diamino-4-azaoctane, *N*-(3-Aminopropyl)-1,4-diaminobutane). Prepare stock solutions (0.05 M) with tissue-culture grade deionized water (see Note 2).
5. Calcium Chloride ($CaCl_2$): Prepare 1 M stock solution of $CaCl_2$ with deionized water (see Note 3).
6. Anhydrous (200 proof) ethanol (see Note 4).

7. ETFE (Tefzel®) tubing: pre-treated Ethylene tetrafluoroethylene tubing (BioRad) (see Note 5).
8. Compressed Nitrogen (see Note 6).
9. Plasmids: highly purified plasmids for immunizations (see Note 7).
10. Compressed helium: Purity grade of >4.5 (i.e., 99.995%) with a maximum pressure of 2,600 psi.
11. Parafilm®.
12. Airtight containers (such as 20 mL-scintillation vials).
13. 3 cc syringes, short piece of silicone tubing that can be attached to the tubing (to load tubing and to remove ethanol from tubing).
14. Dessicant pack (dessication pellets); to be added to the storage container for bullets.
15. Dessicant (e.g., silica gel) and Dessicator.
16. Razor blades, one sided for cutting tubing to size.
17. Sonicating waterbath.
18. Electric clipper (e.g., Oster®) with clipper blade size 40.
19. Tris-EDTA (optional, for resuspending or diluting plasmid preparations). Commercially obtained Tris-EDTA should be explicitly endotoxin-free!
20. Tubing prep station ("Bullet maker"; BioRad).
21. Siliconized microfuge tubes (e.g., from Costar Inc.): 0.5 and 1.5 mL.
22. Dissection medium: RPMI-1640 with 5% mouse serum (see Note 8).
23. Glass wool.
24. Small glass plate or Petri dish for dissection.
25. Scalpel.
26. 20-gauge hypodermic needle.
27. Hemocytometer.
28. Phase contrast microscope.
29. Syringes: 1 cc with 0.1 cc increments.
30. 27-gauge needles.
31. Microscopic (glass) slides.
32. Methanol.
33. Giemsa staining solution.
34. Microscopic slide carrier and Glass trough for slide staining.
35. Surgical scissors.

3. Methods

3.1. Preparing Gold Slurry

1. Weigh 30 mg gold into an Eppendorf® tube and add 100 μL of spermidine (see Note 1).
2. Vortex vigorously, then briefly (for several seconds) incubate in a sonicating waterbath to break up any clumps.
3. Add 60 μg of plasmid DNA (see Note 7). (DNA concentration should be ca. 1 mg/mL or higher. Avoid adding more than 100 μL of plasmid per tube! If the DNA concentration is too low, precipitate plasmid and resuspend in a smaller volume of Tris-EDTA buffer). When co-delivering multiple plasmids, keep the total plasmid loading rate below approximately 5 μg DNA/mg gold to avoid aggregation of gold particles (see Note 9).
4. Wait until the gold particles have settled; then, spin the Eppendorf® tube for 20 s at maximum speed (Eppendorf® centrifuge).
5. Carefully remove (by pipetting, not decanting!) the supernatant.
6. Break up the gold pellet and add 0.5 mL Ethanol (see Note 10).
7. Spin for 30 s and repeat washing procedure two more times.
8. Resuspend gold in a total of 3 mL of 200 proof ethanol in a 15 mL polypropylene tube after the third wash. This slurry can be stored at −20°C for extended periods of time. When using banked slurry, allow it to warm to room temperature before opening the tube to prevent condensation, which interferes with the process of coating the gold beads in the Tefzel® cartridges. For extended storage tubes should be capped tightly and caps should be sealed with Parafilm®.

3.2. Preparing Gene Gun Cartridges ("Bullets")

1. Purge the empty tubing with nitrogen gas for at least 15 min (to remove moisture) at a pressure of 1–2 psi and a flow rate through the tubing of 0.4–0.5 L/min before loading the gold slurry into the Tefzel® tubing.
2. Turn off nitrogen gas.
3. Cut a section of the purged Tefzel® tube (ca. 30 in/76 cm) and attach to a 3 mL-syringe through a piece of silicone tubing.
4. *Quickly* draw the freshly resuspended slurry into the tubing and immediately insert into the tubing station. Speed is of the essence as the gold particles quickly settle resulting in an uneven distribution of gold in the tubing.
5. Allow the gold to settle for several minutes with the syringe still attached.

6. Remove the ethanol with the syringe or peristaltic pump at a rate of 0.5–1 in. (approx. 1.3–2.5 cm) per second making sure not to disturb the settled gold particles, and then remove the syringe or peristaltic pump.
7. Quickly turn the tubing 90° inside the tubing prep station and wait for a few seconds, then turn again 90° and wait for a few seconds before turning on the tube turner (this initiates the breaking up of the thick gold slurry).
8. Rotate the tubing for ca. 15 s, then initiate the flow of nitrogen (ca. 0.4 L/min) and continue to rotate for another 3–4 min to completely evaporate the ethanol.
9. Examine the Tefzel® tubing and remove any section that is not evenly coated with gold.
10. Cut the tubing into 0.5 in. (1.27 cm) sections (also referred to as bullets or cartridges). Frequently change razor blades used for cutting the Tefzel® tubing to avoid squeezing the tubing and damaging the gold coating.
11. Store cut tubing in airtight containers (e.g., glass scintillation vials) with a desiccant pack. Further seal the cap with Parafilm®. Ideally, the storage vials are kept at 4°C in a desiccator.

3.3. Immunization

Anesthesia of small animals for gene gun immunization is neither necessary nor recommended.

1. Remove abdominal hair with an electric clipper (chemical depilation (e.g., with Nair®) should be avoided because of the unknown or poorly studied effect, which depilation agents may have on the immune status of the skin).
2. Apply three non-overlapping shots on the abdomen of each mouse for each immunization using a helium pressure of 300 psi (see Notes 11 and 12).

3.4. Malaria Challenge of Immunized Mice

1. Prepare Ozaki tubes (23) as follows: puncture the bottom of a siliconized 0.5 mL microfuge tube with a hot 20 gauge hypodermic needle; plug the hole with balled-up glass wool (to capture mosquito debris); place the Ozaki tube into a 1.5 mL siliconized tube.
2. Obtain mosquitoes 18–20 days post feeding on *P. berghei* infected mice or hamsters. The rating of the mosquitoes is ideally done based on the oocyte count, which is an indication of the mosquitoes' infectivity rate.
3. Kill mosquitoes by quickly submersing them in 70% Ethanol (see Note 13); then transfer mosquitoes to dissection medium.
4. Remove mosquitoes from the liquid by pouring them onto a glass plate or Petri dish. Pull head from thorax using a scalpel. Collect heads and thoraces in Ozaki tubes (Notes 14 and 15).

5. Spin Ozaki tubes at 8,000 ×*g* for 1 min at RT.
6. Remove Ozaki tube from receptacle tube. Resuspend the pellet (i.e., isolated sporozoites) with the liquid in the receptacle tube and transfer the suspension into a fresh siliconized microfuge tube (collection tube).
7. Return Ozaki tube to receptacle tube, add 100 μL dissection medium and repeat steps 5 and 6 two more times.
8. Take collection containing the suspension of all three centrifugations and gently mix. Remove aliquot for cell counting.
9. Load hemocytometer with aliquot of sporozoite suspension and wait 5 min until the parasites have settled before counting.
10. Count sporozoites at ×200–400 magnification using a phase contrast objective.
11. Adjust the concentration of sporozoites so that 100 μL of RPMI-1640 with 10% mouse serum contain the sporozoites required for one mouse (see Note 16).
12. Inject sporozoites subcutaneously with a 27 gauge needle into the left and right inguinal region of the mouse dispensing 50 μL per side.
13. Prepare blood smears by cutting the very tip of the mouse's tail with surgical scissors on day 7 and day 14 after challenge. Mice without blood parasitemia on day 14 are scored as sterilely protected.
14. Spot blood onto microscopic slide and prepare a thin blood smear.
15. Air-dry slides, then fix smears by submerging the slides in methanol.
16. Transfer slides into a 10% Giemsa solution and stain for 15 min at RT.
17. Remove slides from the glass trough and differentiate the staining by immersing the slides in water.
18. Air-dry slides, and then evaluate blood smears microscopically at ×1,000 magnification (see Note 17).
19. Calculate vaccine efficacy (see Note 18).

4. Notes

1. Aliquots (30 mg/Eppendorf® tube) can be stored frozen together with spermidine (100 μL/tube).
2. Spermidine deaminates with time. Therefore, always store solutions frozen and avoid repeated thawing (i.e., store small aliquots in Eppendorf® tubes).

3. $CaCl_2$ is used to precipitate plasmid onto gold particles. Aliquots of stock solution (1 M) can be stored at room temperature or frozen. The stock solution should be prepared with tissue-culture grade deionized water. Calcium chloride is an irritant and eye protection should be worn.
4. Anhydrous (100%; 200 proof) ethanol is used to wash the plasmid-coated gold particles. Contaminating water interferes with plasmid binding and coating of the Tefzel® tubing with the gold particles. Therefore, special attention must be paid to maintaining the ethanol water-free: The ethanol should not be cooled (or frozen) for use. Although this enhances its ability to precipitate DNA it leads to condensation. Ethanol bottles should only be opened for brief periods of time and ethanol from bottles, which had previously been used multiple times, should not be used to prepare the final gold slurry, but only for washing of the formulated gold (see protocol Subheading 3.1).
5. ETFE (Tefzel®) tubing is specifically pre-treated Ethylene tetrafluoroethylene tubing and can be purchased from BioRad. Untreated tubing may not allow coating of the tubing with the gold film.
6. Compressed Nitrogen should have a purity grade of >4.5 (i.e., 99.995%) and a maximum pressure of 2,600 psi, using a single-stage, low-pressure nitrogen tank regulator (final pressure between 30 and 50 psi).
7. High-quality plasmid is isolated most effectively using the Endo-free plasmid kit from Qiagen (Maxi-prep). This will assure efficient removal of endotoxin derived from the recombinant bacteria used for plasmid production. If cesium chloride gradient centrifugation is used for plasmid purification, an additional (potential) contaminant is CsCl, which needs to be removed from the final plasmid preparation. Various commercially available mammalian expression vectors have successfully been used to deliver malaria antigens such as pCI (Promega) and pcDNA3 (Invitrogen). Prior to using newly generated constructs for immunizations it is essential to perform in vitro transfections (e.g., using BHK cells (11, 24)) to determine not only the quality of the resulting protein product (i.e., appropriate length and absence of truncated protein; recognition by specific antibodies) but also the effect of the protein on the viability of the transfected cells. Some malaria antigens such as the CSP antigen have been shown to contain a ribosome binding motif thus interfering with protein biosynthesis in the experimentally transfected or naturally infected cell and resulting in host cell apoptosis. Analyzing both, transfected cells and culture supernatant of transfected cells (by Western Blotting) will show whether the plasmid-encoded protein is secreted or retained in the transfected cell (either because of

the lack of appropriate secretion sequences or because of unique properties of the protein such as the presence of protozoan GPI-anchor sequences, which result in cytoplasmic retention (13, 25)).

8. The quality of the mouse serum used for the parasite resuspension medium is crucial when performing subcutaneous (s.c.) or intradermal (i.d.) challenges. While sporozoites delivered by intravenous injection (i.v.) do not have to be fully functional as they will be delivered into the liver by the blood stream, the sporozoites that are injected s.c. or i.d. need to find their way to the liver by migrating through the skin into the blood stream. We have noticed that using medium with freshly (i.e., same day) obtained serum (harvested through cardiac puncture of designated donor mice, ideally litter mates of immunized mice) yielded sporozoites that had the highest functional activity as measured by sporozoite motility assays as well as number of successful challenges. Avoid using previously frozen mouse serum or serum that had been stored in the refrigerator for extended periods of time.
9. Reported bead loading rates range from 0.1 to 5 μg DNA per mg of gold. The "standard" bead loading rate (routinely used for malaria studies) is 2 μg plasmid DNA/mg gold resulting in Tefzel® cartridges, which deliver a calculated amounts of 0.5 mg gold coated with 1 μg DNA. At this bead loading rate, the surface of the gold particles is only partially coated with DNA and therefore the bead loading rate can be increased up to tenfold. However, increasing the bead loading rate also increases clumping of the gold, which results in poor coating of the Tefzel® cartridges and thus variable and inadequate delivery of gold particles during immunization. Increasing the amount of plasmid encoding the vaccine (antigen of interest) is not advisable since it does not appear to increase vaccine efficacy. This leaves sufficient capacity for co-delivered plasmids (i.e., plasmids encoding additional pathogen-derived antigens or molecular adjuvants). Co-delivered plasmids should not by default be delivered at a 1:1 ratio but titration experiments should be conducted to determine the most effective ratio. Higher doses of co-delivered pro-apoptotic molecules will result in host cell death before the antigen of interest is produced in sufficient amounts and the co-delivery of large amounts of helper-antigens (designed to trigger a "bystander" CD4 helper response) can lead to immunodominance of the helper antigen thus not providing the desired adjuvant effect (15). However, gene gun vaccination permits the rapid and simple comparison of multiple ratios of antigen:molecular adjuvant without the need for cloning different plasmids. The same is true for the co-delivery of multiple plasmids encoding different malaria antigens (see Subheading 3.1).

10. Adding polyvinylpyrrolidone (PVP) to the gold slurry as described in other gene gun protocols is not recommended. PVP is an adhesive used to facilitate the binding of gold particles to the Tefzel® tubing, but particularly at lower (i.e., 300 psi) helium pressure, it can cause retention of some gold and therefore inadequate immunization.
11. Note that *Plasmodium* is highly sensitive to innate immune responses (26). The duration of the innate immune responses following vaccination is determined by the type of vaccine, the delivery method, and adjuvant used. Adjuvanticity of a vaccine is not only determined by exogenously added adjuvants (such as Alum or Freund's adjuvant) but also the ability of the vaccine itself to stimulate innate immune responses (e.g., CpG motifs on DNA plasmids, which is a more relevant consideration when delivering large amounts of plasmids by injection with needle and syringe). Thus it is imperative to include relevant vector controls in the immunization experiment to control for innate protection against infection. If animals immunized only with vector controls cannot be reliably infected with *Plasmodium* the interval between booster immunization and challenge has to be extended. For some malaria antigens the interval between booster immunization and challenge also determines the durability of the protective immune response with short intervals resulting in only transient protection. Therefore, when testing the protective efficacy of any malaria vaccine, it is advisable to (a) explore different intervals between the last immunization and the parasite challenge and (b) rechallenge protected animals to determine the durability of the immune protection since the first exposure to the parasites may have resulted in the editing of the protective response and thus loss of protection (9, 10).
12. Before discarding used bullets, check for residual gold. Retention of some gold is generally not a problem when no PVP had been added to the ethanol used to resuspend the gold. The problem can be remedied by simply increasing the helium pressure used to deliver the gold particles. However, consider that changing the helium pressure will alter the depth of tissue penetration of the gold particles, which is determined by both, gas pressure and size of gold particles. If using gold particles of different sizes than recommended, it is advisable to determine the location of the gene gun-delivered gold (using several pressure settings) by conventional histology of the targeted skin to assure their presence in the epidermis.

    Before first use of the gene gun consult the manufacturer's manual to assure that the helium pressure used does not exceed the maximum pressure for the gun.

13. The time the mosquitoes spend in the 70% Ethanol should be minimized as the diffusion of the alcohol into the tissue ultimately kills the sporozoites. Therefore, it is imperative to work quickly at this stage.
14. Assure that the mosquito parts do not dry up as this will greatly affect the viability of the sporozoites. One Ozaki tube can be filled with material derived from as many as 100 mosquitoes.
15. Alternatively, sporozoites can be obtained by carefully removing the heads from the mosquitoes and slowly pulling out the attached salivary glands. These salivary glands are then spun down in siliconized microfuge tubes. This method will yield the highest purity of sporozoites, but requires a significant amount of practice and skill. The Ozaki method is more popular because of its ease of use and high yields.
16. The dose of sporozoites required for a reliable infection of ≥90% of control animals depends on the mouse strain (20). BALB/c show inherent resistance to *P. berghei* and therefore require 3,000–4,000 sporozoites while C57BL6 mice can be reliably challenged with 300 sporozoites. Outbred mice such as the CD-1 mice or AJ-mice require the highest doses (12,000 sporozoites/mouse).
17. To determine the presence of blood-stage parasites or to conclude that a mouse is sterilely protected at least 20 microscopic fields have to be evaluated.
18. Formula to calculate vaccine efficacy:

    % Vaccine efficacy = [1 – [{(number of infected animals in experimental group)/(number of total animals in experimental group)}/{(number of infected animals in control group)/(number of total animals in control group)}]]*100. Statistical analysis of the results is performed using a Fisher's Exact Test.

References

1. Leitner WW et al (1997) Immune responses induced by intramuscular or gene gun injection of protective deoxyribonucleic acid vaccines that express the circumsporozoite protein from *Plasmodium berghei* malaria parasites. J Immunol 159:6112–6119
2. Walsh DS et al (2006) Heterologous prime-boost immunization in rhesus macaques by two, optimally spaced particle-mediated epidermal deliveries of *Plasmodium falciparum* circumsporozoite protein-encoding DNA, followed by intramuscular RTS,S/AS02A. Vaccine 24:4167–4178
3. Weiss R et al (2000) Genetic vaccination against malaria infection by intradermal and epidermal injections of a plasmid containing the gene encoding the *Plasmodium berghei* circumsporozoite protein. Infect Immun 68:5914–5919
4. Sharma S, Pahtak S (2008) Malaria vaccine: a current perspective. J Vector Borne Dis 45:1–20
5. Bergmann-Leitner ES et al (2010) Immunization with pre-erythrocytic antigen CelTOS from *Plasmodium falciparum* elicits cross-species protection against heterologous challenge with *Plasmodium berghei*. PLoS One 5:e12294
6. Angov E, Legler PM, Mease RM (2010) Adjustment of codon usage frequencies by codon harmonization improves protein expression and folding. Methods Mol Biol 705:1–13

7. Sin JI et al (2000) Engineering of DNA vaccines using molecular adjuvant plasmids. Dev Biol (Basel) 104:187–198
8. Maue AC (2004) CD80 and CD86, but not CD154, augment DNA vaccine-induced protection in experimental bovine tuberculosis. Vaccine 23:769–779
9. Bergmann-Leitner ES et al (2007) C3d-defined complement receptor-binding peptide p28 conjugated to circumsporozoite protein provides protection against *Plasmodium berghei*. Vaccine 25:7732–7736
10. Bergmann-Leitner ES et al (2005) C3d binding to the circumsporozoite protein carboxy-terminus deviates immunity against malaria. Int Immunol 17:245–255
11. Bergmann-Leitner ES et al (2009) Molecular adjuvants for malaria DNA vaccines based on the modulation of host-cell apoptosis. Vaccine 27:5700–5708
12. Kang Y, Calvo PA (1998) Comparison of humoral immune responses elicited by DNA and protein vaccines based on merozoite surface protein-1 from *Plasmodium yoelii*, a rodent malaria parasite. J Immunol 161:4211–4219
13. Scheiblhofer S et al (2001) Removal of the circumsporozoite protein (CSP) glycosylphosphatidylinositol signal sequence from a CSP DNA vaccine enhances induction of CSP-specific Th2 type immune responses and improves protection against malaria infection. Eur J Immunol 31:692–698
14. Bergmann-Leitner ES, Leitner WW, Tsokos GC (2006) Complement 3d: from molecular adjuvant to target of immune escape mechanisms. Clin Immunol 121:177–185
15. Leitner WW et al (2009) Enhancement of DNA tumor vaccine efficacy by gene gun-mediated codelivery of threshold amounts of plasmid-encoded helper antigen. Blood 113:37–45
16. Ockenhouse CF et al (1998) Phase I/II a safety, immunogenicity and efficacy trial of NYVAC-Pf17, a pox-vectored, multiantigen, multistage vaccine candidate for *Plasmodium falciparum* malaria. J Infect Dis 177:1664–1673
17. Tine JA et al (1996) NYVAC-Pf17: a poxvirus-vectored, multiantigen, multistage vaccine candidate for *Plasmodium falciparum* malaria. Infect Immun 64:3833–3844
18. Leitner WW et al (2003) Alphavirus-based DNA vaccine breaks immunological tolerance by activating innate antiviral pathways. Nat Med 9:33–39
19. Leitner WW et al (2004) Apoptosis is essential for the increased efficacy of alphaviral replicase-based DNA vaccines. Vaccine 22:1537–1544
20. Leitner WW, Bergmann-Leitner ES, Angov E (2010) Comparison of *Plasmodium berghei* challenge models for the evaluation of pre-erythrocytic malaria vaccines and their effect on perceived vaccine efficacy. Malar J 9:145
21. Kebaier C, Voza T, Vanderberg J (2009) Kinetics of mosquito-injected *Plasmodium sporozoites* in mice: fewer sporozoites are injected into sporozoite-immunized mice. PLoS Pathog 5:e1000399
22. Vanderberg J et al (2007) Assessment of antibody protection against malaria sporozoites must be done by mosquito injection of sporozoites. Am J Pathol 171:1405–1406
23. Ozaki LS, Gwadz RW, Godson GN (1984) Simple centrifugation method for rapid separation of sporozoites from mosquitoes. J Parasitol 70:831–833
24. Leitner WW et al (2000) Enhancement of tumor-specific immune response with plasmid DNA replicon vectors. Cancer Res 60:51–55
25. Bergmann-Leitner ES et al (2011) Cellular and humoral immune effector mechanisms required for sterile protection against sporozoite challenge induced with the novel malaria vaccine candidate CelTOS. Vaccine 29:5940–5949
26. Smith TG et al (2002) Innate immunity to malaria caused by *Plasmodium falciparum*. Clin Invest Med 25:262–272

Chapter 22

Identification of T Cell Epitopes of *Mycobacterium tuberculosis* with Biolistic DNA Vaccination

Toshi Nagata and Yukio Koide

Abstract

Tuberculosis (TB) has been listed as one of the most prevalent and serious infectious diseases worldwide. The etiological pathogen of TB is *Mycobacterium tuberculosis* (Mtb), a facultative intracellular bacterium. *Mycobacterium bovis* bacillus Calmette-Guérin (BCG) is the only approved vaccine against TB to date. BCG has been widely used, but the efficacy is questionable, especially in adult pulmonary TB. Therefore, more effective, safe and reliable TB vaccines have been urgently needed. T cell-mediated cellular immune response is a key immune response for effective protective immunity against TB. DNA vaccines using Mtb antigens have been studied as promising future TB vaccines. Most TB DNA vaccine studies so far reported used intramuscular or intradermal injection with needles, as these methods tend to induce a type 1 helper T lymphocyte (Th1)-type immune response that is critical for the protective immunity. We have been using DNA vaccines with gene gun bombardment for T cell epitope identification of various Mtb antigens. We show here our strategy to identify precise Mtb T cell epitopes using DNA vaccines with gene gun bombardment.

Key words: *Mycobacterium tuberculosis*, Codon usage, T cell epitope, Cytotoxic T lymphocyte, Type 1 helper T lymphocyte, Interferon-γ

1. Introduction

1.1. Tuberculosis and the Vaccine

According to the global burden of disease caused by tuberculosis (TB) in 2009, there were 9.4 million incident cases of TB with approximately one third of the world total population being infected (1). *Mycobacterium bovis* bacillus Calmette-Guérin (BCG) is the only approved attenuated live vaccine to date against TB (2, 3). Despite the fact that BCG is among the most widely used vaccines throughout the world, TB still poses a serious global health threat. Whereas BCG is believed to protect newborns and

Stephan Sudowe and Angelika B. Reske-Kunz (eds.), *Biolistic DNA Delivery: Methods and Protocols*, Methods in Molecular Biology, vol. 940, DOI 10.1007/978-1-62703-110-3_22, © Springer Science+Business Media, LLC 2013

young children against early manifestations of TB, its efficacy against pulmonary TB in adults is still a subject of debate (4) and was reported to wane with time since vaccination (5). Variable levels of the protective efficacy ranging from 0 to 80% have been reported in different studies (2, 4). Moreover, the viable nature of BCG makes it partly unsafe in case of immunocompromised individuals. This highlights the need to develop more effective, safe and reliable vaccines against TB, and several TB vaccine candidates have now entered clinical trials (6).

The T cell-mediated immune response is critical for the development of resistance against mycobacterial infection (7, 8). It has been well established that major histocompatibility complex (MHC) class II-restricted $CD4^+$ type 1 helper T lymphocytes (Th1) are important mediators of host defense against TB. In addition, MHC class I-restricted $CD8^+$ cytotoxic T lymphocytes (CTL) have also been reported to be required for the optimum control of mycobacterial infection (9, 10).

1.2. Codon Usage

In DNA vaccines against pathogens such as bacteria, protozoa, and viruses, interspecies difference of codon usage is one of the major obstacles for the effective induction of specific immune responses. We evaluated the codon optimization effect on CTL induction using the DNA vaccine against *Listeria monocytogenes* and malaria parasite (11). Using mammalian culture cells, we analyzed the translation efficiency of several genes composed of different levels of optimization to mammalian cells, but encoding an identical CTL epitope, and showed that the codon optimization level of the genes is not precisely proportional to, but correlates well with the translation efficiency in mammalian cells. These results also correlated well with the induction level of specific CTL response in vivo (12). For evaluation of the codon optimization level, the relative synonymous codon usage (RSCU) value has been used (13). The RSCU values of codons used in *L. monocytogenes* showed the opposite relationship to the RSCU values of codons used in mice and humans, indicating that native codons frequently used in *L. monocytogenes* are rarely used in mice and humans (see Table 1, Note 1). However, such a relationship is not necessarily applicable in *Mycobacterium tuberculosis* (Mtb). The RSCU values of codons used at high frequency in Mtb genes are quite similar to those in mouse and human genes (see Table 1). The Mtb genome has a similar G + C content as mammalian genomes. Therefore, the effect of codon difference on expression efficiency of Mtb DNA vaccines would be minimal.

1.3. DNA Vaccination Against TB

Many papers on DNA vaccination against Mtb have been published since 1996 (14, 15). So far, a variety of Mtb antigen genes have been used for DNA vaccines, which include heat shock protein

Table 1
RSCU values of codons in genes derived from *L. monocytogenes*, *M. tuberculosis*, *Mus musculus*, and *Homo sapiens*

RSCU						RSCU					
		L. monocytogenes	*M. tuberculosis*	*Mus musculus*	*Homo sapiens*			*L. monocytogenes*	*M. tuberculosis*	*Mus musculus*	*Homo sapiens*
Phe	UUU	1.304	0.419	0.845	0.771	Ser	UCU	1.225	0.251	1.148	1.091
	UUC	0.696	1.581	1.155	1.130		UCC	0.624	1.260	1.334	1.364
Leu	UUA	2.420	0.101	0.365	0.409		UCA	1.153	0.410	0.825	0.851
	UUG	0.943	1.087	0.767	0.727		UCG	0.569	2.106	0.336	0.338
Leu	CUU	1.231	0.350	0.756	0.741	Pro	CCU	1.206	0.240	1.206	1.122
	CUC	0.285	1.084	1.215	1.215		CCC	0.229	1.175	1.229	1.334
	CUA	0.885	0.295	0.457	0.406		CCA	2.017	0.430	1.119	1.082
	CUG	0.236	3.083	2.441	2.502		CCG	0.548	2.155	0.446	0.462
Ile	AUU	1.736	0.463	0.996	1.030	Thr	ACU	1.259	0.278	0.970	0.910
	AUC	0.769	2.380	1.558	1.521		ACC	0.390	2.621	1.449	1.483
	AUA	0.495	0.157	0.447	0.449		ACA	1.691	0.353	1.137	1.055
Met	AUG	1.000	1.000	1.000	1.000		ACG	0.659	1.175	0.445	0.471
Val	GUU	1.468	0.382	0.562	0.685	Ala	GCU	1.427	0.337	1.155	1.049
	GUC	0.450	1.532	0.879	0.985		GCC	0.364	1.789	1.544	1.644
	GUA	1.392	0.229	0.386	0.425		GCA	1.513	0.398	0.884	0.876
	GUG	0.690	1.858	1.614	1.905		GCG	0.697	1.476	0.417	0.431
Tyr	UAU	1.401	0.460	0.822	0.840	Cys	UGU	1.384	0.512	0.915	0.860
	UAC	0.599	1.113	1.178	1.160		UGC	0.616	1.488	1.085	1.140

(continued)

Table 1
(continued)

RSCU						RSCU					
ter	UAA	–	–	–	–	ter	UGA	–	–	–	–
ter	UAG	–	–	–	–	Trp	UGG	1.000	1.000	1.000	1.000
His	CAU	1.380	0.578	0.786	0.795	Arg	CGU	2.094	0.695	0.531	0.510
	CAC	0.620	1.422	1.214	1.205		CGC	0.839	2.321	1.106	1.203
Gln	CAA	1.711	0.525	0.513	0.507		CGA	0.788	0.598	0.718	0.654
	CAG	0.289	1.475	1.486	1.493		CGG	0.487	2.004	1.118	1.243
Asn	AAU	1.365	0.408	0.835	0.892	Ser	AGU	1.474	0.389	0.887	0.743
	AAC	0.636	1.592	1.165	1.108		AGC	0.954	1.583	1.469	1.257
Lys	AAA	1.710	0.504	0.767	0.820	Arg	AGA	1.575	0.111	1.268	1.195
	AAG	0.290	1.496	1.233	1.180		AGG	0.218	0.272	1.259	1.195
Asp	GAU	1.483	0.546	0.878	0.904	Gly	GGU	1.475	0.774	0.697	0.646
	GAC	0.516	1.454	1.122	1.096		GGC	0.785	2.031	1.342	1.392
Glu	GAA	1.653	0.689	0.807	0.819		GGA	1.264	0.415	1.036	0.981
	GAG	0.347	1.310	1.193	1.181		GGG	0.477	0.785	0.925	0.980

(Hsp) 65, Hsp 70, Antigen (Ag) 85A, Ag85B, and ESAT6. DNA immunization with naked DNA has been shown to efficiently induce cellular as well as humoral immune responses. DNA vaccines in most of these reports used needle injection via intramuscular or intradermal routes, although some studies used gene gun (16). The DNA immunization with needle injection tends to raise predominant Th1 responses, which is indispensable for induction of the protective immunity. On the other hand, gene gun DNA immunization is apt to produce "mixed type" (Th1 and Th2; producing interferon (IFN)-γ) and interleukin (IL)-4) T cell responses, which is not necessarily adequate for induction of the protective immunity (17). The difference is considered to be mainly due to the difference in the amount of antigen produced from the plasmids (high amounts in needle injection and low amounts in gene gun bombardment). Therefore, DNA vaccination with gene gun will need additional factors such as adjuvants for eliciting protective immunity against Mtb.

We realized that DNA immunization with gene gun bombardment is an excellent method for identification of Mtb T cell epitopes, as it is highly reproducible and efficiently induces T cell responses, especially $CD8^+$ CTL (18). Identification of T cell epitopes in Mtb antigens is indispensable for accurate analysis of T cell responses against Mtb antigens by analyses with specific MHC tetramers or intracellular cytokine staining. A variety of T cell epitopes of Mtb antigens have been reported. Some of them are listed in Table 2. Huygen and colleagues have reported identification of T cell epitopes of Ag 85 family proteins (Ag85A, Ag85B, and Ag85C) (19, 20) using intramuscular DNA immunization. We have used gene gun DNA immunization method for identification of $CD8^+$ and $CD4^+$ T cell epitopes of Mtb antigens including MPT51 (21–23), MDP1 (24), and low-molecular-mass secretory antigens (CFP11, CFP17, and TB18.5) (25). After immunization, immune spleen cells were examined for their IFN-γ responses to overlapping peptides covering full-length proteins by measuring IFN-γ levels by enzyme-linked immunosorbent assay (ELISA) or by counting the numbers of IFN-γ-secreting cells by enzyme-linked immunospot assay (ELISPOT). We combined these methods with computer algorithms to predict T cell epitopes (Fig. 1). These programs are helpful for narrowing down the amino acid region of the bona fide T cell epitope. However, the algorithms are still not perfect for accurate identification of T cell epitopes at this time. A peptide that shows the highest score in these algorithms is not necessarily the best T cell epitope. Experimental validation is definitely necessary to determine actual T cell epitopes.

Table 2
T cell epitopes of Mtb antigens (examples)

Antigen	Epitope peptide	MHC restriction	Reactive T cells	References
Ag85A	p60–68 (9-mer)	K^d	Mouse CD8	(19)
	p144–152 (9-mer)	K^d	Mouse CD8	(19)
	p101–120 (20-mer)	E^d	Mouse CD4	(32)
	p241–260 (20-mer)	A^b	Mouse CD4	(32)
	p261–280 (20-mer)	A^b	Mouse CD4	(20)
Ag85B	p240–254 (15-mer)	A^b	Mouse CD4	(33,34)
	p262–279 (18-mer)	A^b	Mouse CD4	(20)
	p143–152 (10-mer)	A*0201	Human CD8	(35)
	p199–207 (9-mer)	A*0201	Human CD8	(35)
	p10–27 (18-mer)	DR3, 52, 53	Human CD4	(36)
	p19–36 (18-mer)	Promiscuous	Human CD4	(36)
	p91–108 (18-mer)	Promiscuous	Human CD4	(36)
MPT51	p24–32 (9-mer)	D^d	Mouse CD8	(21)
	p171–190 (20-mer)	A^b	Mouse CD4	(21)
	p53–62 (10-mer)	A*0201	Human CD8	(22)
	p191–202 (12-mer)	Promiscuous	Human CD4	(23)
Hsp65	p489–503 (15-mer)	A^d	Mouse CD4	(37)
	p369–377 (19-mer)	A*0201	Human CD8	(38)
	p3–13 (11-mer)	DR3	Human CD4	(39)
ESAT6	p1–20 (20-mer)	$H2^{b,d}$	Mouse CD4	(40)
	p51–70 (20-mer)	$H2^{a,k}$	Mouse CD4	(40)
	p72–95 (24-mer)	DR52, DQ2	Human CD4	(41)
CFP10	p32–39 (8-mer)	K^k	Mouse CD8	(42)
	p11–25 (15-mer)	A^k	Mouse CD4	(42)
MDP1	p23–31 (9-mer)	D^b	Mouse CD8	(24)
	p41–60 (20-mer)	A^d, E^k	Mouse CD4	(24)
	p111–130 (20-mer)	E^k	Mouse CD4	(24)
	p141–160 (20-mer)	E^k	Mouse CD4	(24)

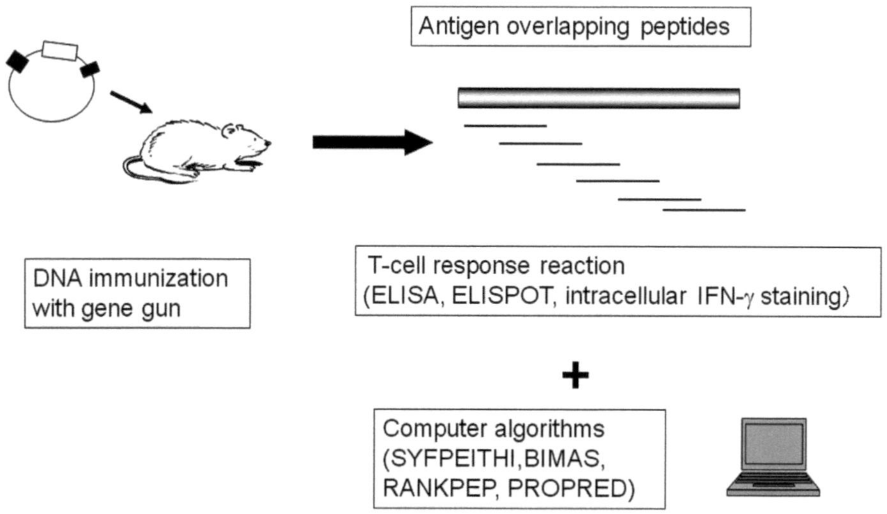

Fig. 1. Schematic diagram for identification of T cell epitopes with DNA immunization.

2. Materials

2.1. Preparation of Plasmid DNA

1. Mammalian expression plasmid such as pCI (Promega, Madison, WI, USA) (see Note 2).
2. Appropriate restriction enzymes.
3. Qiagen Plasmid Midi or Maxi Kit (Qiagen Sciences, MD, USA).

2.2. Preparation of DNA/Gold Cartridge

1. 1.5 mL Microfuge tube.
2. 1.0 μm Gold microcarrier (Bio-Rad Laboratories, Herculus, CA, USA).
3. Plasmid DNA solution in TE buffer (10 mM Tris, pH 8.0, 1 mM EDTA). DNA concentration should be >1 μg/μL.
4. 0.05 M Spermidine (Sigma, St. Louis, MO, USA) in distilled water.
5. 1 M $CaCl_2$ in distilled water.
6. Small variable speed vortex mixer.
7. Microfuge.
8. Ultrasonic cleaner (e.g., Bransonic 1,210 J; Branson Ultrasonics, Danbury, CT, USA).
9. Fresh 100% (v/v) ethanol.
10. Polyvinylpyrrolidone (PVP) (Sigma): Prepare a stock solution of 20 mg/mL PVP in 100% (v/v) ethanol. Dilute this solution with 100% (v/v) ethanol to prepare PVP solution at 0.05 mg/mL (see Note 3). Prepare 3.5 mL of the dilute solution for

each 30-in. length of gold-coat tubing (25 in. to be coated) in the tubing prep station. Keep these solutions tightly capped, when not in use. Prepare the solution freshly.

11. 1.5 mL Disposable polypropylene tube with a screw cap.
12. Nitrogen pressure regulator.

2.3. Loading the DNA/Gold Suspension into Tubing Using the Tubing Prep Station

1. Tefzel tubing (Bio-Rad Laboratories).
2. Tubing prep station (Bio-Rad Laboratories).
3. Nitrogen tank with compressed nitrogen gas.
4. Flowmeter.
5. 10 mL Syringe.
6. Tubing cutter (Bio-Rad Laboratories).

2.4. In Vivo Delivery of DNA-Coated Particle to Epidermis with Gene Gun

1. Inbred mice such as BALB/c or C57BL/6 mice. Mice between 2 and 4 months of age were used for immunization.
2. Razor.
3. Commercial depilatory.
4. Helios gene gun (Bio-Rad Laboratories).
5. Cartridge holder (Bio-Rad Laboratories).
6. Compressed helium gas.
7. Helium pressure regulator.
8. 70% (v/v) Ethanol in a spray bottle.

2.5. Preparation of Immune Spleen Cells and the Immunological Assays

1. Diethyl ether.
2. Scissors and two tweezers.
3. Peptides: Synthesize peptides spanning the entire amino acid sequences of Mtb antigen proteins as approximately 20-mer peptides overlapping by ten residues. All peptides were dissolved in phosphate-buffered saline (PBS) at a concentration of 1 mM and stored at −80°C until use.
4. Sterile Petri dishes (60 × 15 mm). Use bacteriological type, not tissue culture type for cells not to adhere to the dish bottom.
5. 5 mL Syringe.
6. Low-speed centrifuge for sedimenting cells.
7. Stainless metal mesh (wire size ca. 300 μm, pore size ca. 500 μm).
8. ACK lysis solution: 0.15 M NH_4Cl, 1 mM $KHCO_3$, and 0.1 mM EDTA, pH 7.2.
9. Sterile round-bottom 96-well plate for ELISA.
10. 75 mL Flask for MHC stabilization assay.
11. RPMI 1640 medium (Sigma) supplemented with 10% heat-inactivated fetal calf serum (RPMI/10% FCS).

12. CO_2 incubator.
13. 96-Well half-size microwell plate (e.g., EIA RIA Plate A/2; Costar, Cambridge, MA, USA) for ELISA.
14. 96-Well nitrocellulose-backed microwell plate (e.g., MultiScreen 96-well plates; Millipore, Billerica, MA, USA) for ELISPOT.
15. Microplate washer for ELISA (e.g., ImmunoWash 1575; Bio-Rad Laboratories).
16. Microplate reader for ELISA (e.g., IWAKI EZS-ABS Microplate Reader; IWAKI Asahi Techno Glass, Tokyo, Japan).
17. Dissecting microscope or an ELISPOT plate reader for ELISPOT.
18. Coating solution: 0.1 M sodium carbonate, pH 9.6, for ELISA and ELISPOT.
19. Washing buffer: PBS containing 0.05% Tween 20, for ELISA and ELISPOT.
20. FACS buffer: PBS supplemented with 1% FCS, for intracellular cytokine staining.
21. Blocking solution: 10% fetal calf serum or 1% bovine serum albumin in PBS. The blocking solution should be filtered to remove particulates before use. Commercially available reagents such as Blocking One (Nacalai Tesque, Kyoto, Japan) is also usable.
22. Blocking solution/Tween: Blocking solution containing 0.05% Tween 20.
23. Monoclonal antibodies (mAb): anti-murine IFN-γ antibody R4-6A2 as capture antibody and biotin-labeled anti-murine IFN-γ antibody XMG1.2 as detection antibody for ELISA; phycoerythrin (PE)-conjugated anti-IFN-γ antibody XMG1.2 as well as fluorescein isothiocyanate (FITC)-conjugated anti-CD8 and PerCP-Cy5.5-conjugated anti-CD4 antibodies for intracellular IFN-γ staining assay (all from BD Biosciences, Franklin Lakes, NJ, USA).
24. Streptavidin-conjugated horseradish peroxidase (SAv-HRP) (e.g., eBioscience, San Diego, CA, USA).
25. 3, 3′, 5, 5′-Tetramethylbenzidine (e.g., TMB No Hydrogen Peroxide One Component Substrate; BioFX Laboratories, Owings Mills, MD, USA, This solution is supplied as a ready to use solution.).
26. AEC (3-amino-9-ethyl-carbazole) Substrate Set (BD Biosciences).
27. BD Cytofix/Cytoperm Plus Fixation/Permeabilization Kit (BD Biosciences).

28. RMA-S cells or cells transfected with MHC class I gene (e.g., RMAS-K^d, EMAS-D^d, or RMAS-L^d cells for BALB/c mouse).
29. Special equipment: flow cytometry apparatus for intracellular cytokine staining and MHC stabilization assay.

3. Methods

3.1. Preparation of Plasmid DNA

1. Mammalian expression plasmid pCI was digested with appropriate restriction enzymes (see Note 2).
2. Ligate an Mtb antigen-containing DNA fragment to the digested pCI plasmid (see Note 4).
3. Prepare plasmid DNA with Qiagen Plasmid Midi or Maxi Kit according to QIAGEN plasmid purification handbook.

3.2. Preparation of DNA/Gold Cartridge

Prepare DNA/gold cartridge according to Helios gene gun system instruction manual. The method is described briefly below.

1. In a 1.5 mL microfuge tube, weigh out 25 mg gold particle (see Note 5).
2. To the measured gold, add 100 μL of 0.05 M spermidine.
3. Vortex the gold/spermidine mixture for a few seconds, then sonicate for 3–5 s using an ultrasonic cleaner to break up gold clumps.
4. To the gold/spermidine mixture, add the required volume of plasmid to achieve the desired DNA loading rate (DLR) (see Note 6).
5. Mix DNA, spermidine, and gold by vortexing approximately 5 s.
6. While vortexing the mixture at moderate rate on a variable speed vortexer, add 100 μL of 1 M $CaCl_2$ dropwise to the mixture slowly. The volume of 1 M $CaCl_2$ solution added should be equal to that of the spermidine in Step 3.
7. Allow the mixture to precipitate at room temperature for 10 min.
8. Most of the gold will now be in the pellet, but some may be on the sides of the tube. The supernatant should be relatively clear. Spin the DNA/gold solution in a microfuge approximately 15 s to pellet the gold particle. Remove the supernatant and discard.
9. Resuspend the pellet in the remaining supernatant by vortexing briefly. Wash the pellet three times with 1 mL of fresh 100% (v/v) ethanol each time. Spin approximately 5 s in a microfuge between each wash. Discard the supernatants.

10. After the final ethanol wash, resuspend the pellet in 200 μL of the ethanol solution containing 0.05 mg/mL PVP prepared freshly. Transfer this suspension to a 15 mL disposable polypropylene tube with a screw cap. Add 2.8 mL of the ethanol/PVP solution to the centrifuge tube for a 25-in. length of tubing.
11. The suspension is now ready for tube preparation. Alternatively, the DNA/gold particle suspensions can be stored for up to 2 months at −20°C. Prior to freezing, tighten the cap securely.

3.3. Loading the DNA/Gold Suspension into Tubing Using the Tubing Prep Station

Prepare DNA/gold-coating tubing according to Helios gene gun system instruction manual. The method is described briefly below.

1. Set up the Tubing prep station and connect to a nitrogen tank.
2. Prior to preparing cartridges, ensure that the tubing is completely dry by purging with nitrogen. Insert an uncut piece of tubing into the opening on the Tubing prep station.
3. Using the knob on the flowmeter, turn on the nitrogen and adjust the flow to 0.3–0.4 L/min (LPM). Allow nitrogen to flow through the tubing for at least 15 min immediately prior to using it.
4. Turn off the flow of nitrogen to the Tubing prep station using the knob on the flowmeter.
5. From the dried tubing cut about 30 in. (about 75 cm) length of tubing for each 3 mL of DNA/gold suspension.
6. Vortex the DNA/gold suspension in a 15 mL tube and, if necessary, sonicate briefly with an ultrasonic cleaner to achieve an even suspension of gold. Invert the tube several times to resuspend the gold. Immediately remove the cap and quickly draw the DNA/gold suspension into the tubing with 10 mL syringe.
7. Immediately bring the tubing to a horizontal position and slide the loaded tube, with 10 mL syringe attached, into the tubing prep station.
8. Allow the DNA/gold to settle for 3–5 min. Then, remove ethanol at the rate of 0.5–1.0 in. per sec using a syringe (this should require 30–45 s).
9. Detach the syringe from the tubing. Immediately turn the tubing 180° while in the groove.
10. Turn on the switch on the Tubing prep station to start rotating the Tubing prep station.
11. Allow the gold to smear in the tube for 20–30 s. Then, open the valve on the flowmeter to allow 0.35–0.4 LPM of nitrogen to dry the tubing, while it continues to rotate.
12. Continue drying the tubing while rotating for 3–5 min.

13. Turn off the motor on the Tubing prep station. Turn off the nitrogen by closing the valve on the flowmeter. Remove the tubing from the Tubing support cylinder.
14. Examine the coated tubing to verify that the DNA/gold is evenly distributed over the length of the tubing.
15. Using scissors, cut off and discard the ends of the tubing.
16. Use the Tubing cutter to cut the remaining tubing into 0.5-in. pieces.

3.4. In Vivo Delivery of DNA-Coated Particle to Epidermis with Gene Gun

The gold particle coated with plasmid DNA encoding Mtb antigens are injected into mice with the gene gun. The frequency of injections and the interval time between injections depends on the experiments and investigators (see Note 7).

1. Clip animal fur as closely as possible over the desired target area using a razor and brush fur off (see Notes 8 and 9). Spraying the clipping site with 70% (v/v) ethanol makes clipping easy.
2. Load cartridges into the cartridge holder and place it in gene gun.
3. Administer DNA-coated gold particles to the dermis with gene gun using helium discharge pressure of 350–400 psi (pounds per square inch; 350–400 psi is about 24.5–28 kg/cm^2).
4. Remove cartridge holder from gene gun.
5. Remove cartridge from cartridge holder.
6. Turn off the helium pressure to the system.
7. Turn the regulator value counterclockwise to de-pressurize the system.
8. Disconnect the helium gas and gene gun.

3.5. Preparation of Immune Spleen Cells and the In Vitro Stimulation with Mtb Antigen Peptides

Prepare spleen cells aseptically from DNA-immunized mice more than 2 weeks after the last immunization.

1. Sacrifice mice by cervical dislocation or inhalation of diethyl ether.
2. Make an incision in the back skin, grasp either side of the incision with tweezers and pull back the skin until spleen is observed through back membrane.
3. Make an incision over the spleen and transfer the spleen with tweezers to a 60 mm Petri dish containing about 5 mL of RPMI 1640 medium.
4. Put the spleen on sterilized metal mesh and crush it by the head of a 5 mL syringe to prepare a single cell suspension.
5. Transfer the cell suspension to a 15-mL tube.
6. Centrifuge the cell suspension at $400 \times g$ for 5 min at 4°C.

7. Resuspend the cells with 3 mL of ACK lysis solution per spleen to lyse red blood cells.
8. Immediately add 10 mL of RPMI 1640 medium and centrifuge the cells at 400×*g* for 5 min at 4°C. Discard the medium by decantation. Repeat the washing step for three times. Check the cell pellet is whitish after this washing step. In addition, remove the cell debris during this step.
9. Transfer the immune spleen cells to 96-well plates at $1–2 \times 10^6$ cells/well and culture in RPMI/10% FCS in the presence of 7.5 μM of each antigen peptide at 37°C in a humidified 5% CO_2 incubator. Harvest the cell culture supernatants 24–72 h later and store them at −20°C until they are assayed by ELISA.

3.6. Quantification of IFN-γ by ELISA

IFN-γ amounts secreted by immune spleen cells are measured by ELISA. The commercially available ELISA kits such as Quantikine Immunoassay (R&D Systems, Minneapolis, MN, USA) are also usable.

1. Coat half-size 96-well plate with 30 μL/well of 2 μg/mL of capture antibody (anti-murine IFN-γantibody R4-6A2) in coating solution at 4°C overnight.
2. Wash with 100 μL/well of washing buffer three times manually or automatically with a microplate washer.
3. Block with 50 μL/well of blocking solution at 37°C for 2 h.
4. After washing with 100 μL/well of washing buffer three times, add the culture supernatants to the plate and incubate at 4°C overnight.
5. After washing with 100 μL/well of washing buffer three times, add 50 μL/well of 0.5 μg/mL of biotin-labeled anti-murine IFN-γ antibody XMG1.2 in blocking solution/tween to the plate and incubate for 1 h at room temperature.
6. After washing with 100 μL/well of washing buffer five times, add 50 μL/well of 0.1 μg/mL of SAv-HRP in blocking solution/tween to each well.
7. After washing, add 50 μL/well of 3, 3′, 5, 5′-tetramethylbenzidine and incubate for 30 min at room temperature.
8. Determine the absorbency at 450 nm with a microplate reader.

3.7. Detection of IFN-γ-Producing Cells by Enzyme-Linked ImmunoSpot Assay

Single cell suspensions are tested for antigen-specific IFN-γ secretion with ELISPOT. The commercially available ELISPOT kits such as BD ELISPOT kit (BD Biosciences) are also usable.

1. Coat nitrocellulose-backed 96-well plate with 50 μL/well of 2 μg/mL of capture antibody (anti-murine IFN-γ monoclonal antibody R4-6A2) in coating solution at 4°C overnight.

2. Wash with 100 μL/well of washing buffer three times manually or automatically with a microplate washer.
3. Block with 50 μL/well of RPMI 1640 medium with 10% FCS at 37°C for 2 h.
4. Stimulate 200 μL/well of the immune spleen cells at different densities (e.g., 1–2 × 10^6 cells/well) in RPMI/10%FCS medium with 7.5 μM of each peptide in each well.
5. Incubate the plates for 24 h at 37°C in a humidified 5% CO_2 incubator.
6. Aspirate cell suspension and wash wells two times with deionized water. Allow wells to soak for 3–5 min at each wash step.
7. Wash wells three times with 200 μL/well of washing buffer. Discard washing buffer.
8. Add 100 μL/well of 0.5 μg/mL of biotin-labeled detection antibody (anti-murine IFN-γ antibody XMG1.2) and incubate the plates for 2 h at room temperature.
9. Discard detection antibody solution. Wash wells three times with 200 μL/well of washing buffer.
10. After wash, add 100 μL/well of 0.1 μg/mL of SAv-HRP and incubate the plates for 1 h at room temperature.
11. Discard the SAv-HRP solution. Wash wells four times with 200 μL/well of washing buffer.
12. Add 100 μL/well of AEC substrate solution to detect bound SAv-HRP. Monitor spot development for 5–60 min. Do not let color overdevelop as this will lead to high background.
13. Stop substrate reaction by washing wells with deionized water.
14. Air-dry plates at room temperature for 2 h to overnight until it is completely dry.
15. Enumerate spots developed on the nitrocellulose filters manually under a dissecting microscope, or automatically using an ELISPOT plate reader.

3.8. Intracellular IFN-γ Staining Assay

1. Stimulate 200 μL/well of the immune spleen cells (1 × 10^7 cells/mL) in RPMI/10%FCS medium with 7.5 μM of each peptide in nitrocellulose-backed 96-well microwell plates.
2. Incubate the plates for 24–48 h at 37°C in a humidified 5% CO_2 incubator.
3. For the last 6–12 h of incubation, add GolgiStop (containing monensin) or GolgiPlug (containing brefeldin A) solution. GolgiStop solution: the final concentration should be 4 μL of GolgiStop for every 6 mL of cell culture. Prepare first 5 × GolgiStop solution in RPMI 1640 medium (4 μL GolgiStop solution in 1.2 mL RPMI 1640 medium) and add 50 μL of

5 × GolgiStop solution to each well (200 μL). GolgiPlug solution: the final concentration should be 1 μL of GolgiPlug for every 1 mL of cell culture. Prepare first 5 × GolgiPlug solution in RPMI 1640 medium (1 μL GolgiPlug solution in 0.2 mL RPMI 1640 medium) and add 50 μL of 5 × GolgiPlug solution to each well (200 μL).

4. Transfer the cells to 1.5 mL microfuge tubes and wash with 500 μL of FACS buffer.
5. Add 100 μL of FITC-conjugated anti-CD8 and PerCP-Cy5.5-conjugated anti-CD4 mAbs to the 1.5 mL microfuge tubes and incubate on ice for 30 min.
6. Wash twice with 500 μL of FACS buffer (see Note 10).
7. Perform intracellular IFN-γ staining with phycoerythrin (PE)-conjugated anti-IFN-γ mAb XMG1.2 using BD Cytofix/Cytoperm Plus Fixation/Permeabilization Kit according to the manufacturer's instruction.
8. Analyze the cells with a flow cytometer.

3.9. MHC Stabilization Assay

MHC stabilization assay was originally described in Ljunggren et al (26) (see Note 11).

1. Culture RMA-S cells or cells transfected with H2 class I gene (e.g., RMAS-K^d, EMAS-D^d, or RMAS-L^d cells for BALB/c mouse) in RPMI/10%FCS at 26°C overnight using 75-mL flask (see Note 12).
2. Transfer the cells to 96-well round-bottom microwell plates at 2×10^6 cells/well.
3. Incubate the plates for 1 h in the presence or absence of 5–50 μM of respective peptide at 26°C.
4. Incubate the plates at 37°C for 2 h in a humidified 5% CO_2 incubator.
5. Transfer the cells to 1.5 mL microfuge tubes and wash the cells with 500 μL of FACS buffer.
6. Stain cell-surface H2 class I molecules with 100 μL of FITC-conjugated mouse mAbs specific for each H2 class I molecule at appropriate concentrations in FACS buffer and incubate for 30 min on ice.
7. Wash twice the cells with 500 μL of FACS buffer and resuspend in 500 μL of FACS buffer.
8. Analyze the cells with a flow cytometer.
9. To allow comparison between multiple experiments and to reduce inter-experimental variations, the mean fluorescence intensity (MFI) values, which are direct measures of peptide binding, should be converted to percent maximal stabilization

values. The values are calculated using the following formula: (experimental MFI—control MFI)/(maximal MFI—control MFI) × 100. Control MFI is obtained from cells incubated without peptide at 37°C, while the MFI of cells at 26°C is taken as maximal MFI.

3.10. Prediction of T Cell Epitopes by MHC Binding Peptide Prediction Algorithms

For the prediction of potential minimal murine T cell epitopes, which could bind to MHC class I molecules, the following MHC binding peptide prediction algorithms are used through their websites. These are: National Institutes of Health BioInformatics and Molecular Analysis Section (BIMAS) (http://bimas.dcrt.nig.gov/cgi-bin/molbio/ken_parker_comboform) (27), SYFPEITHI (http://www.syfpeithi.de/) (28), RANKPEP program (http://bio.dfci.harvard.edu/Tools/rankpep.html) (29), and Propred (http://www.imtech.res.in/raghava/propred/) (30).

4. Notes

1. Codon usage database of various species is available in internet (http://www.kazusa.or.jp/codon/).
2. pCI mammalian expression plasmid contains a human cytomegalovirus promoter/enhancer element that allows expression of the cloned Mtb gene in mammalian cells and high copy number in the cells. A variety of similar mammalian expression vectors are available and usable for TB DNA vaccine backbone.
3. It is better to optimize PVP concentration first: Prepare tubes with 0, 0.05, 0.1 mg/mL PVP. For most systems, 0.05 mg/mL PVP is a good starting point. PVP is a binding agent of DNA to the tube.
4. Addition of DNA encoding the signal sequence from mammalian genes such as that of interleukin-2 and tissue plasminogen activator to the 5′ of Mtb antigen gene induces secretion of Mtb antigens and may improve expression level of Mtb antigen.
5. Microcarrier loading quantity (MLQ) is the amount of microcarrier (gold) in each cartridge. For most systems, the MLQ of 0.5 mg gold per cartridge is a good starting point. A 25-in. length of tubing gives approximately 50 cartridges. So, the tubing will require 25 mg of gold resuspended in a volume of 3 mL of 100% (v/v) ethanol to give MLQ of 0.5 mg gold/cartridge.
6. DLR is the amount of DNA per mg gold. At MLQ of 0.5 mg gold/cartridge, DLR of 2 μg of plasmid DNA/mg gold results

in loading 1 μg of DNA/cartridge. For the MLQ, preparation of a 25-in. length of gold-coat tubing (25 mg of gold) requires 50 μg of plasmid DNA. For most systems, the DLR usually ranges between 1 and 2.5 μg of plasmid DNA/mg gold.

7. In general, the injection of 1–2 μg of plasmid DNA for 3–6 times with intervals of 1–3 weeks induce sufficient immune responses. We usually inject 1 μg of plasmid DNA four times with 1-week intervals.
8. Razor should be attached to the skin and moved along fur's strike. Otherwise, razor will damage the skin and give bleeding.
9. After clipping, a commercial depilatory can be used to completely remove the animal's fur. This treatment removed the stratum corneum from the skin, completely exposing the epidermis. Carefully rinse the skin with warm water following depilatory treatment. If the target site is wet or dirty, clean and dry with 70% (v/v) ethanol.
10. Before this step, treatment with antibodies specific for mouse FcγII/III receptors that block Fc receptors may be useful for reducing nonspecific staining by fluorescein-conjugated antibodies at the subsequent step.
11. Generally, RMA-S cells (26) and the cells transfected with H2 class I gene or T2 cells (31) and the cells transfected with human HLA class I gene are used for mouse or human system, respectively.
12. Instead of using CO_2 incubator at 26°C, the flask being tightly capped can be incubated at room temperature without CO_2 incubator.

References

1. World Health Organization (2010) WHO report 2010 global tuberculosis control. Geneva, [Online]. http://www.who.int/tb/publications/global_report/2010/pdf/full_report.pdf.
2. Andersen P, Doherty TM (2005) The success and failure of BCG—implications for a novel tuberculosis vaccine. Nat Rev Microbiol 3:656–662
3. Bloom BR, Fine PEM (1994) The BCG experience: implications for future vaccines against tuberculosis. In: Bloom BR (ed) Tuberculosis: pathogenesis, protection, and control. ASM Press, Washington, DC, pp 531–557
4. Fine PEM (1995) Variation in protection by BCG: implications of and for heterologous immunity. Lancet 346:1339–1345
5. Sterne JAC, Rodrigues LC, Guedes IN (1998) Does the efficacy of BCG decline with time since vaccination? Int J Tuberc Lung Dis 2:200–207
6. Kaufmann SHE (2010) Future vaccination strategies against tuberculosis: thinking outside the box. Immunity 33:567–577
7. Stenger SR, Modlin L (1999) T cell mediated immunity to *Mycobacterium tuberculosis*. Curr Opin Microbiol 2:89–93
8. Cooper MA (2009) Cell-mediated immune responses in tuberculosis. Annu Rev Immunol 27:393–422
9. Smith SM, Duckrell HM (2000) Role of $CD8^+$ T cells in mycobacterial infections. Immunol Cell Biol 78:325–333
10. Kaufmann SHE, Flynn JL (2005) CD8 T cells in tuberculosis. In: Cole ST, Eisenach KD,

McMurray DN, Jacobs WR Jr (eds) Tuberculosis and the tubercle bacillus. ASM Press, Washington, DC, pp 465–474
11. Nagata T et al (1999) Codon optimization effect on translational efficiency of DNA vaccine in mammalian cells: analysis of plasmid DNA encoding a CTL epitope derived from microorganisms. Biochem Biophys Res Commun 261:445–451
12. Uchijima M et al (1998) Optimization of codon usage of plasmid DNA vaccine is required for the effective MHC class I-restricted T cell responses against an intracellular bacterium. J Immunol 161:5594–5599
13. Sharp PM, Li WH (1987) The codon adaptation index—a measure of directional synonymous codon usage bias, and its potential applications. Nucleic Acids Res 15:1281–1295
14. Lowrie DB, Silva CL, Tascon RE (1998) Genetic vaccination against tuberculosis. In: Raz E (ed) Gene vaccination: theory and practice. Springer, Berlin, pp 59–71
15. Huygen K (2003) On the use of DNA vaccines for the prophylaxis of mycobacterial diseases. Infect Immun 71:1613–1621
16. Sugawara I et al (2003) Vaccination of guinea pigs with DNA encoding Ag85A by gene gun bombardment. Tuberculosis (Edinb) 83:331–337
17. Tanghe A et al (2000) Tuberculosis DNA vaccine encoding Ag85A is immunogenic and protective when administered by intramuscular needle injection but not by epidermal gene gun bombardment. Infect Immun 68:3854–3860
18. Yoshida A et al (2000) Advantage of gene gun-mediated over intramuscular inoculation of plasmid DNA vaccine in reproducible induction of specific immune responses. Vaccine 18:1725–1729
19. Denis O et al (1998) Vaccination with plasmid DNA encoding mycobacterial antigen 85A stimulates a CD4+ and CD8+ T-cell epitopic repertoire broader than that stimulated by *Mycobacterium tuberculosis* H37Rv infection. Infect Immun 66:1527–1533
20. D'Souza S et al (2003) Mapping of murine Th1 helper T-cell epitopes of mycolyl transferases Ag85A, Ag85B, and Ag85C from *Mycobacterium tuberculosis.* Infect Immun 71:483–1493
21. Suzuki M et al (2004) Identification of murine H2-D^d- and H2-A^b-restricted T-cell epitopes on a novel protective antigen, MPT51, of *Mycobacterium tuberculosis.* Infect Immun 72:3829–3837
22. Aoshi T et al (2008) Identification of an HLA-A*0201-restricted T-cell epitope on MPT51 protein, a major secreted protein derived from *Mycobacterium tuberculosis* by MPT51 overlapping peptide screening. Infect Immun 76:1565–1571
23. Wang LX et al (2009) Identification of HLA-DR4-restricted T-cell epitope on MPT51 protein, a major secreted protein derived from *Mycobacterium tuberculosis* using MPT51 overlapping peptides screening. Vaccine 28:2026–2031
24. Suzuki D et al (2010) Characterization of murine T-cell epitopes on mycobacterial DNA-binding protein 1 (MDP1) using DNA vaccination. Vaccine 28:2020–2025
25. Eweda G et al (2010) Identification of murine T-cell epitopes on low-molecular-mass secretory proteins (CFP11, CFP17, and TB18.5) of *Mycobacterium tuberculosis.* Vaccine 28:4616–4625
26. Ljunggren HG et al (1990) Empty MHC class I molecules come out in the cold. Nature 346:476–480
27. Parker KC, Bednarek MA, Coligan JE (1994) Scheme for ranking potential HLA-A2 binding peptides based on independent binding of individual peptide side-chains. J Immunol 152:163–175
28. Rammensee HG et al (1999) SYFPEITHI: database for MHC ligands and peptide motifs. Immunogenetics 50:213–219
29. Reche PA et al (2004) Enhancement to the RANKPEP resource for the prediction of peptide binding to MHC molecules using profiles. Immunogenetics 56:405–419
30. Singh H, Raghava GPS (2001) ProPred: prediction of HLA-DR binding sites. Bioinformatics 17:1236–1237
31. Baas EJ et al (1992) Peptide-induced stabilization and intracellular localization of empty HLA class I complexes. J Exp Med 176:147–156
32. Huygen K et al (1994) Mapping of TH1 helper T-cell epitopes on major secreted mycobacterial antigen 85A in mice infected with live *Mycobacterium bovis* BCG. Infect Immun 62:363–370
33. Feng CG et al (2001) Priming by DNA immunization augments protective efficacy of *Mycobacterium bovis* Bacille Calmette-Guerin against tuberculosis. Infect Immun 69:4174–4176
34. Kariyone A et al (2003) Immunogenicity of peptide-25 of Ag85B in Th1 development: role of IFN-γ. Int Immunol 15:1183–1194

35. Geluk A et al (2000) Identification of major epitopes of *Mycobacterium tuberculosis* AG85B that are recognized by HLA-A*0201-restricted CD8+ T cells in HLA-transgenic mice and humans. J Immunol 165:6463–6471
36. Mustafa AS et al (2000) Identification of HLA restriction of naturally derived Th1-cell epitopes from the secreted *Mycobacterium tuberculosis* antigen 85B recognized by antigen-specific human CD4+ T-cell lines. Infect Immun 68:3933–3940
37. Nagabhushanam V et al (2002) Identification of an I-A^d restricted peptide on the 65-kilodalton heat shock protein of *Mycobacterium avium*. Immunol Cell Biol 80:574–583
38. Charo J et al (2001) The identification of a common pathogen-specific HLA class I A*0201-restricted cytotoxic T cell epitope encoded within the heat shock protein 65. Eur J Immunol 31:3602–3611
39. Geluk A et al (2000) Binding of a major T cell epitope of mycobacteria to a specific pocket within HLA-DRw17 (DR3) molecules. Eur J Immunol 22:107–113
40. Brandt L et al (1996) Key epitopes on the ESAT-6 antigen recognized in mice during the recall of protective immunity to *Mycobacterium tuberculosis*. J Immunol 157:3527–3533
41. Mustafa AS et al (2000) Multiple epitopes from the *Mycobacterium tuberculosis* ESAT-6 antigen are recognized by antigen-specific human T cell lines. Clin Infect Dis 30(Suppl 3):S201–205
42. Kamath AB et al (2004) Cytolytic CD8+ T cells recognizing CFP10 are recruited to the lung after *Mycobacterium tuberculosis* infection. J Exp Med 200:1479–1489

Chapter 23

Biolistic DNA Vaccination Against Trypanosoma Infection

Marianne Bryan, Siobhan Guyach, and Karen A. Norris

Abstract

Immunization to protect against *Trypanosoma cruzi* infection has the potential to greatly decrease the burden of Chagas' disease in the Americas. Several target antigens have been explored by multiple investigators and show promise, but given that this parasite has multiple stages within the mammalian host, with both intracellular and extracellular forms, a multivalent vaccine will probably be necessary to provide complete immunity and prevent disease. Therefore, DNA immunization is an attractive method for efficient and effective delivery of multiple target antigens. In addition, the target population for a *T. cruzi* vaccine lives predominately in poorer rural areas in South America, making the stable DNA-gold precipitate, which does not require a cold-chain, used in biolistic immunization an attractive method for vaccination. Here we describe a biolistic immunization protocol that is capable of generating high titer antibody responses to recombinant *T. cruzi* vaccine targets and the in vitro preparation of *T. cruzi* for use in experimental vaccine challenge studies.

Key words: *Trypanosoma cruzi*, DNA vaccination, Chagas' disease, Parasite, Infection

1. Introduction

Trypanosoma cruzi, the etiologic agent of Chagas' disease, is a haemoflagellate protozoan parasite. Chagas' disease has uncertain disease prevalence in the United States and is a major health concern in many parts of Latin America, where it remains endemic (1). Despite some successful efforts towards vector control in endemic areas, over 90 million people are considered at risk for infection (2). The World Health Organization estimates that approximately 12 million people are infected with *T. cruzi* resulting in 3.0–3.3 million symptomatic cases and an annual incidence of 200,000 cases in 15 countries (2). Epidemiologic studies estimate that 30% of those infected will develop chronic disease (CD) which is

Stephan Sudowe and Angelika B. Reske-Kunz (eds.), *Biolistic DNA Delivery: Methods and Protocols*, Methods in Molecular Biology, vol. 940, DOI 10.1007/978-1-62703-110-3_23, © Springer Science+Business Media, LLC 2013

characterized by parasite persistence and chronic inflammation of nerve and muscle fibers that lead to disturbances in cardiac conductance and progressive congestive heart failure and/or enlargement of digestive organs such as the esophagus and colon. As most individuals have subclinical symptoms during the early acute stage of disease, when drug treatment would be most effective, diagnoses are not generally made until after clinical symptoms of the chronic disease are apparent and irreversible damage is done.

Historically, vaccine development against *T. cruzi* infections was not pursued due to the possibility of an autoimmune pathology (3). Convincing evidence of an autoimmune etiology remains largely lacking (4), although cross-reactive T cells have been reported (5). Recent studies by Tarleton and coworkers convincingly demonstrate that persistence of parasites at the site of disease correlates with inflammation, suggesting that parasite persistence leads to chronic disease manifestations, such as heart and/or gut enlargement and dysfunction. Thus, immune-mediated clearance of the parasites at early stages of infection would ameliorate or prevent chronic disease, a part of which may be due to cross-reactive cells that develop during chronic infection with *T. cruzi* (6,7), leading to a solid rationale for vaccine development.

Several different types of vaccines have been tested in murine models for protection from *T. cruzi* infection. Vaccines tested in mice include live nonvirulent strains, whole protein preparation, and recombinant subunit vaccines using protein, DNA, or viral vector delivery through several different routes of inoculation (8–11). In this protocol, we describe biolistic DNA vaccination as a means of generating high titer antibody responses to recombinant parasite-derived proteins, which may be tested via intraperitoneal challenge with tissue culture-derived parasites (12,13).

2. Materials

2.1. Bead Preparation Components

1. DNA containing the gene of interest in a eukaryotic expression vector, purified using a Qiagen Mega Endotoxin-free kit. Resuspend the prepared DNA to a final concentration of greater than 1 mg/mL in dH_2O (TE Buffer (10 mM Tris-CL, pH 7.5, 1 mM EDTA) may be used for more concentrated solutions).
2. One micron gold particles (Biorad).
3. Teflon tubing (Biorad).
4. Tubing prep station (Biorad).
5. Nitrogen tank and regulator with attachments for tube prep station.

6. Tube cutter (Biorad).
7. Sonicator water bath.
8. Microfuge rack.
9. A short (5–6″) piece of silicone adaptor tubing on a Luer lock fitting attached to a 10 mL syringe.
10. 0.05 M spermidinein molecular grade water.
11. 10%w/v $CaCl_2$ solution in molecular grade water.
12. New bottle of 200 proof (100%) ethanol (EtOH).
13. Polyvinyl pyrrolidone (PVP) made up in 200 proof EtOH.
14. Microfuge and tubes.
15. Molecular grade water.
16. Molecular grade Agarose.
17. Tris-acetate-EDTA (TAE) Buffer: 40 mM Tris, 20 mM acetic acid, 1 mM EDTA, pH 8.4.
18. Restriction enzyme(s) that linearize the vector or cut out the gene of interest from the expression vector.
19. Agarose gel electrophoresis casting tray, combs, electrophoresis chamber, loading buffer (TAE: 40 mM Tris-acetate, 1 mM EDTA), DNA staining dye, and illuminator to analyze restriction digest of eluted DNA.

2.2. Biolistic Immunization Components

1. Cartridges: Teflon tubing loaded with DNA-coated gold particles and cut to size.
2. Helios Gene Gun (Biorad).
3. Helium tank with helium tank regulator for Gene gun (Biorad).
4. Mice: Balb/c and C57Bl/6.
5. Clippers with fine gauge blade for trimming mouse fur.

2.3. Parasite Culture Components

1. Culture Media: complete Dulbecco's Modified Eagle Medium (cDMEM) (plus Glutamax) with 10% FBS (fetal bovine serum), 10 mM HEPES buffer, 0.2 mM Sodium Pyruvate, 50 μg/mL gentamicin.
2. Phosphate Buffered Saline: 3.2 mM Na_2HPO_4, 0.5 mM KH_2PO_4, 1.3 mM KCl, 135 mM NaCl, pH 7.4.
3. Freezing Media: 90% FBS, 10% dimethyl sulfoxide (DMSO).
4. NIH 3T3 fibroblast cells.
5. Infection vehicle: 1 × PBS + 1% glucose.
6. T150 Tissue culture flasks with vented caps.
7. Parafilm.

8. Aerated Incubator with water bath and carbon dioxide (5%) set to 37°C.
9. Anaerobic Incubator set to 34°C.
10. Neubauer chamber.
11. 70% EtOH for cleaning the Neubauer chamber.
12. Microscope with 40× objective.
13. Table top centrifuge equipped to handle 50 and 14 mL conical tubes at 1,000 ×*g* (with covers to prevent contamination in the case of a leak or tube breakage).

3. Methods

3.1. Preparation of DNA-Gold Loaded Cartridges

1. Prior to preparing the DNA-gold precipitate, load Teflon tubing into the tube prep station and dry with nitrogen gas for at least 15 min.
2. Determine the amount of DNA (μg) per amount of gold (mg) (DNA loading rate) and gold (mg) per cartridge (bead-loading rate). Calculate the number of Teflon tubing lengths (tubes) you will need, estimating that each tube will generate 20 cartridges (see Note 1). Prepare 3 mL of PVP solution per tube. Calculate the amount of gold (Au) needed by estimating 17.54 cartridges/mL of PVP-Au suspension. Weigh out the gold directly into a 1.5 mL microfuge tube(s) (see Note 2).
3. To the weighed out gold in a microtube add the appropriate volume of 0.05 M spermidine (see Note) and break up the clumps of gold in the spermidine solution by placing the microfuge tube containing the solution in a sonicator water bath for 3–5 s. Then, add DNA to the gold/spermidine solution (see Note 3). Cap and invert the tube several times. Vortex the tube for about 1 s.
4. Add 1 M $CaCl_2$ to precipitate the DNA onto the gold by adjusting the vortexer speed down and while gently vortexing, add drop wise a volume of 1 M $CaCl_2$ equal to the volume of the spermidine added to the dry gold. Once the entire amount of $CaCl_2$ has been added, vortex the solution at high speed for 5 s (see Note 4). Allow the solution to precipitate at room temperature for at least 10 min. After the 10 min precipitation, most gold will be in a pellet and some on the sides of the tube. The supernatant will be relatively clear.
5. While the gold solution is precipitating, prepare the calculated volume of PVP solution (into 15 mL conical tubes) at a concentration appropriate for the animal model to be immunized. Use a fresh bottle of 100% EtOH for each bead preparation. PVP concentrations vary based on animal model:

(a) 0.01 mg PVP/mL EtOH for cell culture.

(b) 0.03 mg PVP/mL EtOH for mice and monkeys.

(c) 0.05 mg PVP/mL EtOH for humans and pigs.

6. Wash the DNA-gold precipitate by first centrifuging the microfuge tube for 10–15 s at high speed to pellet all the gold and aspirate the supernatant. Loosen the pellet by raking the tube across a microfuge rack. Add 800 μL 100% EtOH and invert the tube several times to wash the DNA-coated gold. Centrifuge and aspirate the supernatant. Repeat twice.
7. To transfer the DNA-gold precipitate into the PVP solution, add 1 mL of the PVP solution to the DNA-coated gold and transfer the solution to a new 15 mL conical tube. Repeat twice more with the remaining 2 mL of PVP solution.
8. Load the PVP-DNA-Au solution into the Teflon tubing. First, prepare a piece of dry Teflon tubing (see step 1) that is several inches longer than the tube holder on the prep station. Attach a short (5–6″) piece of silicone adaptor tubing to the Tetzel tubing that will be inserted into the prep station on one end and a 10 mL syringe on the other. Sonicate DNA/Gold/PVP solution for 3–5 s just prior to drawing the solution into the tubing. Draw the precipitate into the tubing. Detach the syringe from the tubing and load the gold loaded tubing into the prep station. Allow the solution to settle in the tubing for 10 s then rotate the tubing 45°. Wait 30 s then rotate 45° again. Repeat six times. Reattach the tubing with the syringe to the tubing prep station and slowly (0.5–1 in./s) draw off the liquid (see Note 5). Rotate the tubing in the tubing prep station for 1–3 min. Dry the tubing with N_2 (0.35–0.4 LPM air flow) for 5 min.
9. Remove the DNA-Au loaded tube from the prep station and cut the tube into 1.2 cm pieces (cartridges) using the tube cutter kit. Store in a 50 mL conical tube with a desiccation tablet.
10. To verify the loading of the DNA onto the beads, place a cartridge in a 0.6 mL microtube and add 10 μL of dH_2O to the top of the tub (see Note 6). Spin at top speed in a microfuge to wash the tube and elute the DNA. Use restriction enzyme digest(s) to lineralize or cut out segments of DNA from the vector (confirm size, insert in vector), then run the digest on a 1% TAE agarose gel to visualize the DNA (Fig. 1).

3.2. Biolistic DNA Immunization

1. Load the gold cartridges into the cartridge holder by placing the cartridge holder on a flat surface with the numbered side facing up (see Note 7).
2. Load the cartridge holder into the gun (see Note 8), attach the gene gun to the regulator and set the pressure to 400 psi.

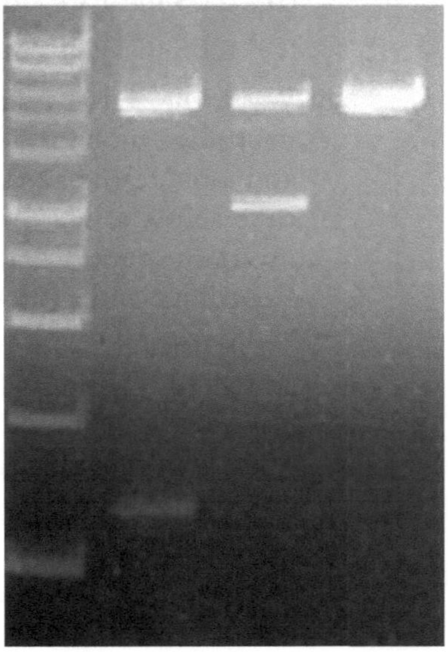

Fig. 1. Analysis of bead preparation by agarose gel electrophoresis. After preparation of cartridges, DNA was eluted and digested. Lanes 1 and 2 contain DNA from two different bead preps, with differing insert size, resolved on a 1% TAE agarose gel. Lane 3 contains the digested vector control, which is the same size as the vector fragment from the two constructs containing the genes of interest.

3. To deliver DNA to the target tissue, shave the site of inoculation to remove fur. Touch the targeted area with the spacer so that the spacer is flush and the Gene Gun is perpendicular to the target surface. Activate the safety interlock switch and press the trigger button to deliver the DNA-coated gold to the target. Ratchet to the next cartridge by pulling in and releasing the cylinder advance lever (see Note 9). After approximately 5 s, the Gene Gun is ready to deliver the next cartridge.
4. This method allows for immunization with multiple targets and generates highly reproducible high titer antibody responses in the mouse model of infection (Fig. 2).

3.3. Preparation of Tissue Culture-Derived T. Cruzi Parasites for Challenge of DNA Immunized Animals

1. *T. cruzi* tissue culture trypomastigotes (TCT) (see Note 10) are grown in NIH 3T3 fibroblasts. To prepare the fibroblasts for infection, seed cells at a density of 3×10^6 cells in 32 mL of cDMEM media in a vented T150 tissue culture flask. Incubate overnight in a tissue culture incubator at 37°C, 5% CO_2. After 16–24 h, check NIH 3T3 fibroblasts and confirm that they are 40–50% confluent.
2. Prepare cryopreserved TCT for infection of fibroblast culture by removing the TCT containing cryovial from liquid nitrogen and thawing quickly in a 37°C water bath. Transfer the TCT from cryovial into 10 mL of PBS in a 15 mL conical tube. Pellet TCT

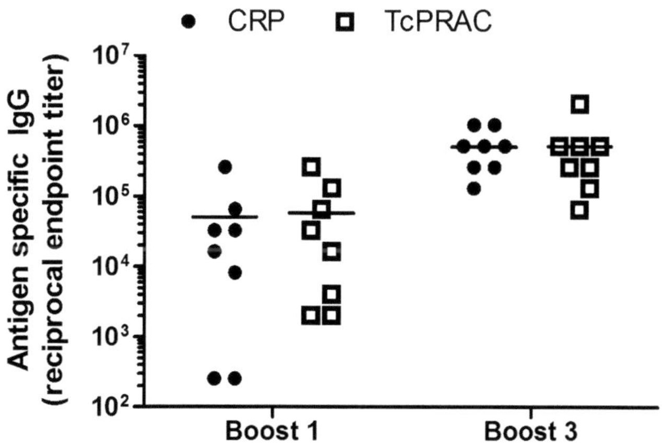

Fig. 2. Mice receiving two different antigens generate robust responses to both antigens. Sera from mice immunized with a combination of two antigens, *T. cruzi* complement regulatory protein (CRP) and *T. cruzi* proline racemase (TcPRAC) were analyzed for specific response to the recombinant proteins via ELISA after the first and third boost. Reciprocal endpoint titer was defined as the first dilution of sera that was not two times above background (sera from naïve/vector control animals).

by centrifugation (1,000 × *g*) for 5 min. at 4°C. Remove supernatant by aspirating, then suspend TCT pellet in cDMEM.

3. Infect NIH 3T3 fibroblasts with TCT by adding 1×10^7 TCT to 40–50% confluent fibroblast culture in T150 flask (see Note 11). Incubate TCT with NIH 3T3 fibroblasts in a tissue culture incubator at 37°C, 5% CO_2 for 36–48 h.
4. To culture infected NIH 3T3 fibroblasts for TCT harvest remove the spent media and replace with fresh cDMEM. Transfer the flask to a 34°C incubator and maintain under anaerobic conditions (cover filter cap on flask with Parafilm). Change media every 2 days, checking for emergence of trypomastigote parasites (see Note 12).
5. Harvest the TCT culture supernatant from fibroblast culture into a 50 mL conical tube. Remove the larger amastigote stage parasites by low speed centrifugation, 180 × *g* for 5 min. Remove supernatant, containing the TCT, to a new 50 mL conical tube and repeat centrifugation. Remove the supernatant to a new 50 mL conical tube and count trypomastigotes (see Note 13).
6. Prepare TCT for infection by making an appropriate aliquot of TCT into a 15 mL conical tube and centrifuge for 10–15 min at 1,000 × *g* to pellet TCT (see Note 14). Resuspend TCT in PBS + 1% glucose at the appropriate concentration for infection (see Notes 15 and 16).
7. Response to challenge of immunized animals may be monitored by assessing disease condition, parasite load in the blood (parasitemia), animal mortality, and secondary immune responses, and tissue (heart, liver) at the experimental endpoint (Fig. 3).

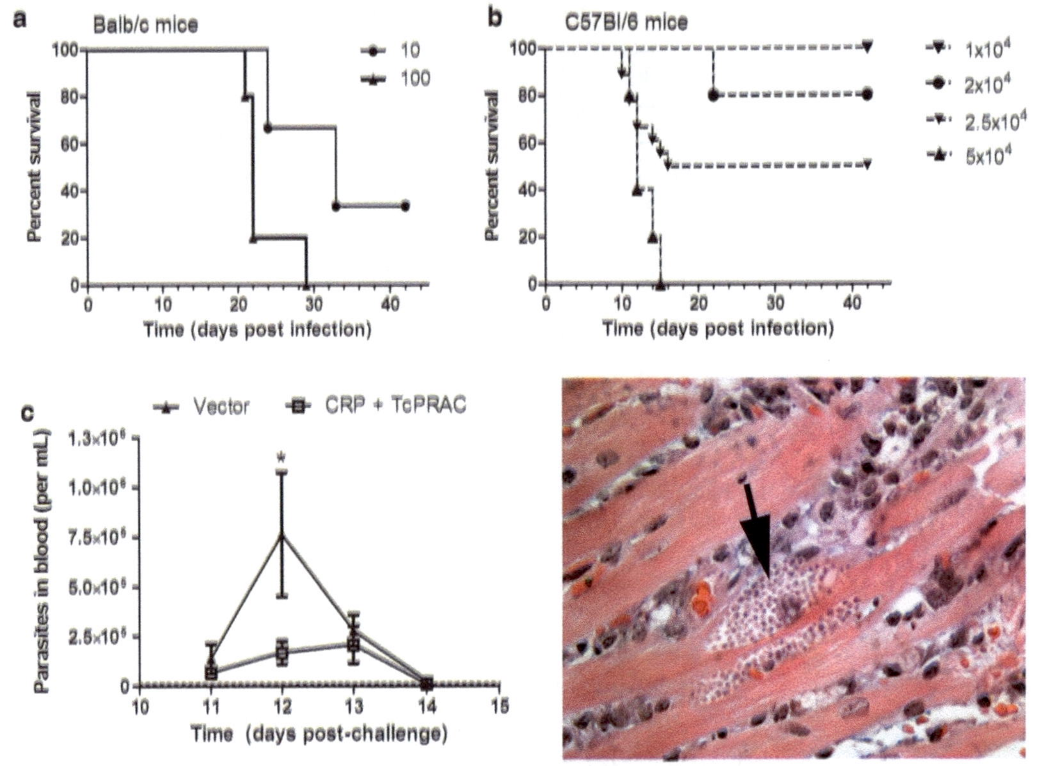

Fig. 3. *T. cruzi* infectious challenge of immunized mice. (**a**) Balb/c mice are more susceptible than C57Bl/6 mice to challenge with TCT (i.p.). Plots show the survival of mice inoculated with the indicated dose of parasites. (**b**) Parasitemia after *T. cruzi* challenge of biolistic DNA immunized (CRP and TcPRAC) vs. vector control mice. (**c**) Amastigote parasite nest (*arrow*) in heart tissue of *T. cruzi* challenged mouse (H&E stain, 800× magnification).

4. Notes

1. In our experiments, we used a loading rate of 8 μg DNA per mg gold, bead-loading rate of 0.25 mg gold per cartridge. Sample calculation of gold amount needed for two tubes: 6 mL PVP × 0.25 mg Au/cartridge × 17.54 cartridges/mL PVP-Au suspension = 26.31 mg Au. Sample calculation of stock DNA needed: 8 μg DNA/mg Au (DNA loading rate) × 26.31 mg Au/3.9 (stock concentration DNA (mg/mL)) = 53.7 μL of stock DNA.
2. Add 400 μL of 0.05 M spermidine if the total volume of DNA stock solution to be added is less than or equal to 50 μL. Add 500 μL of 0.05 M spermidine if the volume of DNA stock solution to be added is greater than 50 μL.
3. The volume of DNA added should not exceed the volume of spermidine added.
4. In preparations using at least 1 μg DNA/mg of gold, precipitation should be evident by clumping of gold beads and rapid falling of gold into a pellet immediately after the addition of $CaCl_2$.

5. Slowly drawing off the DNA-Au-PVP solution from the tubing leads to a more uniform coating.
6. DNA will elute from the gold particles, which will settle in the bottom of the tube.
7. Starting at position 1, load up to 12 cartridges into the cartridge holder. Invert the cartridge holder and push the cartridges against a flat surface so that they are flush with numbered side of the cylinder.
8. Insert the cartridge by moving the cylinder lock on the gene gun so it is latched in the forward position and the barrel pin does not protrude behind the barrel. Unlatch the push bar by pulling it outward. Pull back and hold the cylinder advance lever to retract the inner barrel sleeve into the gun barrel. Place the loaded cartridge holder into the Gene Gun with the position 12 label facing up and the numbered side of the cartridge holder facing the exit nozzle of the barrel. When the cartridge holder is in its correct position, the knob on the backside of the cartridge holder will slip into the notch on the barrel plate and the cartridge holder will be flush with the barrel plate. Release the cylinder advance lever; the O-ring on the inner barrel sleeve should hold the cartridge holder in position. Unlatch the cylinder lock; it should snap into position and the barrel pin should be inserted into the center hole in the cartridge holder. Push the push bar in; it should snap back into position and engage the cartridge holder in one of the deep crevices. Push in and release the cylinder advance lever to ratchet the cartridge holder one position. The number 12 should be visible at the top point of the cartridge holder. The Gene Gun is now ready for pressurizing with helium.
9. We usually deliver two shots to each mouse. We have successfully delivered two antigens by giving two shots with the first antigen on the left side of the mouse, and two shots with the second antigen on the right side of the mouse. To prevent potential hearing damage in the 110 dB range at which the gene gun is fired, persons handling or in the same room with the Gene Gun during discharge should wear earmuffs or ear plugs. Eye protection should always be worn when working with high-pressure gases. All other personal protection garb required in the animal facility should be worn.
10. *T. cruzi* is a biosafety level 2 (BSL2) pathogen. This work is conducted in accordance with the guidelines recommended in "Biosafety in Microbiological and Biomedical Laboratories," Fifth edition, U.S. Department of Health and Human Services Centers for Disease Control and Prevention and National Institutes of Health, 2009.

11. Ratio of 4 tryps/cell average, can go up to 10 tryps/cell (in 2 days they'll all be infected), if want slower infection, 2 tryps/cell. After adding TCT to NIH 3T3 fibroblasts, check that the parasites are mobile. During this step, the parasites infect the cells and begin intracellular replication (amastigote stage) and should mostly be cleared from the culture media.
12. These conditions allow the fibroblasts to senesce, while the parasites replicate within the cells. As infection progresses, amastigotes will fill and then rupture the cells. Trypomastigotes will typically begin to emerge from culture by 6–8 days after infection.
13. To count TCT, dilute the culture into cDMEM and count at 400× on a hemocytometer. Culture supernatant diluted 1:5 into DMEM (5 mL total volume, 15 mL conical tube) usually is appropriate. Count parasites in all four corner grids of the Neubauer chamber, focusing up and down in each square in order to see all the, highly motile, parasites. Clean the hemocytometer with 70% EtOH (do not use a strong detergent, as residual detergent on the chamber will kill the parasites and prevent accurate counting).
14. It is important to use the smaller conical tube at this point to avoid loss of parasites, which will compromise the final dilution used for infection. Spinning for greater than 10 min is highly advised to pellet all these small motile parasites.
15. To maintain TCT by infection of fresh 3T3 fibroblasts, prepare 3T3 fibroblasts as in step 1. Infect 3T3 fibroblast, culture, and harvest TCT as in steps 3–5. Cryopreserve TCT in liquid nitrogen by making 1 mL aliquots of 1×10^7 TCT in labeled cryovials, freeze at −80°C for 24–48 h, then transfer to liquid nitrogen.
16. For intraperitoneal parasite inoculation into a mouse, inject TCT in 100–200 μL of PBS + 1%glucose.

References

1. Hotez PJ (2008) Neglected infections of poverty in the United States of America. PLoS Negl Trop Dis 2:e256
2. Morel CM, Lazdins J (2003) Chagas disease. Nat Rev Microbiol 1:14–15
3. Kierszenbaum F (2003) Views on the autoimmunity hypothesis for Chagas disease pathogenesis. FEMS Immunol Med Microbiol 37:1–11
4. Buckner FS, Van Voorhis WC (2000) Immune response to *Trypanosoma cruzi*: control of infection and pathogenesis of Chagas disease. In: Cunningham MW, Fujinami RS (eds) Effects of microbes on the immune system. Lippincott Williams & Wilkins, Philadelphia, pp 569–592
5. Girones N et al (2007) Role of Trypanosoma cruzi autoreactive T cells in the generation of cardiac pathology. Ann N Y Acad Sci 1107:434–444
6. Tarleton RL (2001) Parasite persistence in the aetiology of Chagas disease. Int J Parasitol 31:550–554
7. Zhang L, Tarleton RL (1999) Parasite persistence correlates with disease severity and localization in chronic Chagas' disease. J Infect Dis 180:480–486

8. Miyahira Y et al (1999) Induction of CD8+ T cell-mediated protective immunity against Trypanosoma cruzi. Int Immunol 11:133–141
9. Quanquin NM et al (1999) Immunization of mice with a TolA-like surface protein of Trypanosoma cruzi generates $CD4^+$ T-cell-dependent parasiticidal activity. Infect Immun 67:4603–4612
10. Schnapp AR et al (2002) Cruzipain induces both mucosal and systemic protection against Trypanosoma cruzi in mice. Infect Immun 70:5065–5074
11. Garg N, Tarleton RL (2002) Genetic immunization elicits antigen-specific protective immune responses and decreases disease severity in Trypanosoma cruzi infection. Infect Immun 70:5547–5555
12. Bryan MA, Norris KA (2010) Genetic immunization converts the Trypanosoma cruzi B-Cell mitogen proline racemase to an effective immunogen. Infect Immun 78:810–822
13. Bryan MA, Guyach SE, Norris KA (2010) Specific humoral immunity versus polyclonal B cell activation in Trypanosoma cruzi infection of susceptible and resistant mice. PLoS Negl Trop Dis 4:e733

Chapter 24

Biolistic DNA Vaccination Against Melanoma

Julia Steitz and Thomas Tüting

Abstract

We describe here the use of particle-mediated gene transfer for the induction of immune responses against melanoma antigens in murine tumor models using the melanocyte differentiation antigen tyrosinase-related protein 2 (TRP2) as an antigen in a murine B16 melanoma model. We have utilized marker genes such as β-galactosidase (βgal) and EGFP, which can be readily detected, as control antigens to establish the gene delivery and to detect antigen-specific humoral and cellular immune responses. After biolistic DNA vaccination with plasmids encoding the TRP2 gene we observed the induction of TRP2-specific T-cells and antibodies associated with vitiligo-like fur depigmentation and tumor immunity against B16 melanoma cells. Here we describe the preparation of cartridges with DNA-coated gold beads and the in vivo gene transfer into skin using the Helios Gene Gun system. We also describe protocols for the measurement of humoral and cellular immune responses against the melanocyte differentiation antigen TRP2. These protocols can subsequently be adapted to other antigens.

Key words: DNA vaccine, TRP2, CD8+ T-cells, Vitiligo, Melanoma

1. Introduction

The successful biolistic DNA vaccination of mice with plasmids expressing foreign proteins was first described in 1992 by Johnston and colleagues where the induction of potent cellular and humoral immune responses against a human growth hormone was demonstrated (1). It has been well established that DNA immunization by particle-mediated gene transfer promotes broad-based and long-lasting antigen-specific immune responses capable of protecting against challenges with infectious agents and tumor cells in rodents (reviewed in refs. (2–4)). After first investigations of DNA immunization in mice using model tumor antigens such as chicken ovalbumin or β-galactosidase (βgal) (5–7), clinically relevant melanoma antigens were evaluated for the induction of potent antigen-specific cellular and humoral immune responses associated with

Stephan Sudowe and Angelika B. Reske-Kunz (eds.), *Biolistic DNA Delivery: Methods and Protocols*, Methods in Molecular Biology, vol. 940, DOI 10.1007/978-1-62703-110-3_24, © Springer Science+Business Media, LLC 2013

tumor rejection activity in vivo. With the molecular identification of tumor antigens (8, 9), protocols and vaccination strategies to induce and to measure tumor antigen-specific immune responses have been developed and murine models employing clinically relevant melanocyte differentiation antigens such as gp100, TRP1, and TRP2 (10, 11) were established. Initial investigations of our group, using cDNA encoding autologous TRP2 as a model antigen, have demonstrated that biolistic immunization of C57Bl/6 mice was unable to induce significant protective immunity against B16 melanoma cells naturally expressing TRP2 (6). Subsequently, we and others found that genetic immunization of mice with cDNA encoding xenogeneic homologous human TRP2 was able to circumvent peripheral tolerance and mediate marked protective immunity against metastatic growth of B16 melanoma cells (12, 13). Importantly, immunization with xenogeneic human, but not with autologous murine TRP2 cDNA was able to stimulate CD8+ T-cells in vivo recognizing the H2-K^b-binding peptide SVYDFFVWL which derives from an evolutionary conserved region of both murine and human TRP2 corresponding to amino acids 180–188. Additionally, we found that mice immunized with human, but not with murine TRP2 cDNA, produced antibodies reactive with human TRP2 and cross-reactive with murine TRP2. Furthermore, induction of immunity to TRP2 was associated with vitiligo-like fur depigmentation as a sign of autoimmune-mediated destruction of melanocytes. Effective immune responses were also induced by immunizing animals with a fusion protein between enhanced green fluorescent protein (EGFP) and autologous murine TRP2 (EGFP.mTRP2), providing a strong CD4 helper sequence and circumventing tolerance to enhance the tumor immune response (14).

We describe here the use of biolistic DNA immunization for the induction of immune response against melanoma antigens in a murine B16 melanoma model. We recommend to use marker genes such as β-galactosidase and EGFP, which can be readily detected, in parallel as control for effective gene transfer and induction of immune responses. Here we describe the preparation of cartridges with DNA-coated gold particles and the in vivo gene transfer into skin using the Helios Gene Gun system. We also describe protocols for the measurement of humoral and cellular immune responses as well as for the analysis of vaccine efficiency in lung metastases or subcutaneous B16 tumor models.

2. Materials

Prepare all solutions using ultrapure water and analytical grade reagents. Prepare and store all reagents at room temperature (unless indicated otherwise). Diligently follow all waste disposal regulations when disposing waste materials. For animal experiments follow regulations for housing and handling of animals according to the Institutional Animal Care and Use Committee guidelines and do not start before all animal work was approved by the appropriate committee. Use sterile equipments if possible. If not indicated otherwise, materials and reagents are from common commercial laboratory providers.

2.1. Gene Gun Bullets Preparation

1. Helios Gene Gun System (Bio-Rad, Hercules, CA, USA), 100/120 V, handheld biolistic system, includes helium hose assembly with regulator, tubing prep station, syringe kit, Tefzel tubing, tubing cutter, optimization kit (Kit for preparing Helios gene gun cartridges, includes 0.5 g PVP, 5 desiccant pellets, 5 cartridge collection/storage vials, 15 m of Tefzel tubing (see Note 1)).
2. Microcentrifuge.
3. Vortex.
4. 1.6 μm Gold microcarriers (Bio-Rad).
5. 15 mL conical polypropylene tubes.
6. 10 mL syringes.
7. 1.5 mL microcentrifugation tubes.
8. Parafilm.
9. Compressed nitrogen—purity grade >4.8 (99.998%); maximum pressure, 2,600 psi—with single-stage, low-pressure nitrogen tank regulator, maximum pressure reading of 30–50 psi.
10. 1 μL purified expression plasmid DNA in dH_2O (50 μg): Expression plasmid encoding the murine melanoma antigen TRP2 (pCI-mTRP2) was constructed by inserting the fragment encoding the melanoma differentiation protein TRP2 into the expression vector pCI (Promega, Madison, WI, USA). Expression plasmid encoding the human TRP2 protein (pCMV-hTRP2), a kind gift from Dr. S. Shibahara (Miyagi, Japan). We grow plasmids in *Escherichia coli* strain DH5α, using Qiagen Endofree Plasmid Maxi Kits (Qiagen, Chatsworth, CA, USA) for purification, resuspend in dH_2O, and store plasmids at −20°C (see Note 2).
11. 1 M $CaCl_2$ in dH_2O.
12. 100% Ethanol (see Note 3).
13. 0.1 M Spermidine: Store aliquots of pure Spermidine liquid at −20°C. Prepare 0.05 M Spermidine solution freshly before use by diluting 75 μL of 0.1 M Spermidine solution in 75 μL of dH_2O.
14. Dissolve PVP at 1 mg/mL in 100% Ethanol and seal with parafilm. Stock solution can be stored frozen at −20°C. Before use dilute PVP (1 mg/mL) solution to 0.075 mg/mL (75 μL/mL in 100% Ethanol).

2.2. Gene Gun Immunization of Mice

1. Female C57BL/6 (H-2^b), 6–10 weeks old.
2. Electrical Clipper or razor blades.
3. Hearing protection.
4. Compressed helium—purity grade >4.5 (99.995%); maximum pressure, 2,600 psi—with high-pressure helium tank gas regulator.
5. Helios Gene Gun System.
6. Cartridges containing DNA-coated gold beads.

2.3. Detection of TRP2-Specific Antibodies by Western Blot

1. Microvette CB300 with lithium-heparin for blood sampling (Sarstedt, Nümbrecht-Rommelsdorf, Germany) (see Note 4).
2. Needle 18G.
3. B16 melanoma cells.
4. Microcentrifuge.
5. Vortex.
6. Block heater (for 1.5 mL tubes, 100°C).
7. Vertical electrophoresis unit.
8. Power supply.
9. Semi-dry blotter and Sample Twister.
10. X-ray film cassette and radiofilm development system.
11. Crushed ice.
12. SDS-gels.
13. Immobilon P Membrane PVDF.
14. Blotting paper 1.5 mm.
15. Hyperfilm MP 265 × 3,750 or Hyperfilm ECL if using ECL substrate system.
16. PEP8: anti-TRP2 antibody. A rabbit antiserum reactive with the C-terminus of murine TRP2 (αPep8) generously provided by Dr. Hearing (NIH, Bethesda, MD, USA) (15) (see Note 5).
17. Chemiluminescent development solution (e.g., Super Signal West Pico Substrate, Thermo Fischer Pierce, Rockford, IL, USA or ECLplus, GE Lifescience, Munich, Germany).
18. HRP-conjugated goat anti-rabbit IgG antibody and HRP-conjugated goat anti-mouse IgG antibody.
19. Complete protease inhibitor (Roche Applied Sciences, Mannheim, Germany). Dissolve according to manufacturer's instructions and freeze aliquots at −20°C.
20. TNEN-Lysis buffer: Mix 1 mL of 1 M Tris–HCl pH 7.5 solution, 3 mL of 5 M NaCl solution, 1 mL of 1% NP 40, 1 mL of 0.5 M ethylenediaminetetraacetic acid (EDTA) solution (see Note 6). Prior to use, add complete protease inhibitor at a ratio of 1:25.

21. Laemmli-sample buffer (Bio-Rad). Prior to use, add 2-Mercaptoethanol to an aliquot of Laemmli-sample buffer at 2% (v/v).
22. 10× Gel running buffer: Dissolve 121 g Tris–HCl (MW = 121), 238 g Hepes (free acid MW = 238), 10 g SDS in 1 L dH_2O. Before use, dilute tenfold with dH_2O.
23. Transfer buffer: Dissolve 2.42 g Tris–HCl, 11.38 g Glycine, 200 mL Methanol in 1 L dH_2O.
24. TPBS: 1× PBS, 0.1% Tween 20.
25. Blotto: 5% nonfat powdered skim milk (see Note 7) in TPBS (25 g in 500 mL of TPBS). 50 mL aliquots can be stored at −20°C.

2.4. Detection of TRP2-Specific T-Cell Responses

1. Surgical scissors and anatomical forceps (use sterile instruments!).
2. Pipet aid.
3. Sterile 5 mL and 10 mL serological pipettes.
4. Incubator (37°C, 5% CO_2, high humidity).
5. Laminar flow hood (class II).
6. Inverse microscope.
7. Neubauer cell counting chamber.
8. Cell strainer 70 μm.
9. 50 mL centrifugation tubes.
10. Centrifuge (with adapter for 50 mL tubes, 528 × *g*).
11. Erythrocyte-Lysis-Buffer: 155 mM NH_4Cl, 10 mM $KHCO_3$, 0.1 mM EDTA pH 7.3 (see Note 6). Dissolve in 750 mL dH_2O 8.29 g NH_4Cl, 1.0 g $KHCO_3$, 0.0372 g EDTA, adjust pH to 7.3. Fill up to 1 L with dH_2O.
12. Peptides: H2-K^b-binding TRP2-encoded peptide SVYDFFVWL ($TRP2_{aa180-188}$), H2-K^b-binding βgal-encoded peptide DAPIYTNV ($βgal_{96-103}$). Peptides were dissolved at 10 μg/mL in sterile PBS containing 10% DMSO and were stored at −20°C (see Note 8).
13. Splenocytes from naïve syngenic animals.

2.4.1. IFNγ Elispot Assay

1. Millipore MAHA ELISPOT plates (MAHA S4510, Millipore, Billerica, MA, USA).
2. 10 mL syringe with rubber seal.
3. Dissecting microscope for manual counting of spots or Elispot reader.
4. Multichannel pipette (12-channel, 50–300 μL).
5. Sterile 50 mL reagent reservoir.
6. Sterile 90 mm Petri dishes.

7. Purified rat anti-mouse IFN-γ antibody (clone R4-6A2, BD Biosciences PharMingen, Heidelberg, Germany) for capture.
8. Biotinylated rat anti-mouse IFN-γ antibody (clone XMG1.2, BD Biosciences PharMingen) for detection.
9. Streptavidin-peroxidase conjugate (Roche Applied Sciences, Mannheim, Germany).
10. Peroxidase substrate kit DAB (Vector Laboratories, Burlingame, CA, USA).
11. Recombinant human IL-2 (e.g., clinical product Proleukin from Novartis). The cytokine should be stored at high concentration, for example 10^6 U/mL diluted in lymphocyte-medium (containing serum!) (see item 14 below) at −80°C. Once thawed, cytokines can be stored at 4°C and should be used within 4 weeks.
12. Wash Buffer: 1× PBS, 0.1% Tween 20. (Add 500 μL of Tween 20 into 500 mL of PBS).
13. Lymphocyte-medium: 500 mL RPMI 1640, 50 mL heat inactivated fetal bovine serum (FBS) (10%) (see Note 9), 5 mL of 200 mM L-glutamine solution, 5 mL of 10^4 IU/mL Penicillin/10 mg/mL Streptomycin solution, 5 mL 100× nonessential amino acids solution, 5 mL 100 mM sodium-pyruvate solution.
14. Assay-buffer: 0.5% BSA, 0.01% Thimerosal, 0.05% Tween 20 in 1× PBS. (Add 5 g BSA, 0.1 g Thimerosal, and 0.5 mL Tween 20 to 1 L of 1× PBS).

2.4.2. In Vivo Kill Assay

1. Flow cytometer capable to measure *Ex/Em*: 492/517 nm (FL1 channel).
2. 12×75 mm polystyrene tubes for flow cytometry.
3. CFSE (Carboxyfluorescein Succinimidyl Ester) (Invitrogen Molecular Probes, Darmstadt, Germany). Dissolve lyophilized CFSE in DMSO to prepare a 10 mM stock solution. Small aliquots should be stored at −20°C. CFSE solution is stable for up to 6 months.
4. FACS Buffer: 500 mL 1× PBS, 2 mL of 0.5 M EDTA solution (final concentration 2 mM) (see Note 6), 10 mL heat inactivated FBS (2%) (see Note 9). Store at 4°C.

2.5. Assessment of Antitumor Immunity In Vivo

1. 75 cm^2 tissue culture flask.
2. 50 mL tubes.
3. 1 mL syringe.
4. 27G needle.
5. Surgical scissors and anatomical forceps (use sterile instruments!).

6. 5 mL and 10 mL serological pipettes.
7. Pipet aid.
8. Dissection microscope (if available).
9. Incubator (37°C, 5% CO_2, high humidity).
10. Laminar flow hood (class II).
11. Inverse microscope.
12. Neubauer cell counting chamber.
13. 1× Trypsin-EDTA solution.
14. DMEM culture medium: 500 mL DMEM, 50 mL heat inactivated FBS (10%) (see Note 9), 5 mL of 200 mM L-glutamine solution, 5 mL of 10^4 IU/mL penicillin/10 mg/mL streptomycin solution.

3. Methods

Carry out all procedures at room temperature unless otherwise specified.

3.1. Preparation of Gene Gun Bullets

We routinely use 0.5 inch cartridges containing 1 μg DNA coated onto 0.5 mg gold beads.

Each preparation will generate 50 cartridges which requires 25 mg gold and 50 μg purified DNA.

1. Weigh 25 mg gold into a 1.5 mL microcentrifugation tube.
2. Add 100 μL of 0.05 M Spermidine (freshly prepared) and vortex to disrupt clumps (if available sonicate briefly in an ultrasound cleaner bath).
3. Mix 50 μg of 1 mg/mL plasmid DNA with the Spermidine-gold solution and vortex to disrupt clumps (see Note 10).
4. While vortexing, slowly add 100 μL of 1 M $CaCl_2$ dropwise to precipitate DNA onto gold beads. Allow gold beads to settle for 10 min.
5. After precipitation, centrifuge the DNA-gold mix for 1 min at 13,000 rpm.
6. Discard supernatant.
7. Wash three times with 1 mL of 100% Ethanol to remove dH_2O (vortexing for 15–20 s, followed by centrifugation for 1 min at 13,000 rpm).
8. Resuspend gold beads in 200 μL of 100% Ethanol containing 0.075 mg/mL of PVP and transfer to a 15 mL polypropylene tube. In order to remove all the gold particles in the microcentrifuge tube rinse the 1.5 mL microcentrifuge tube several

times with a total of 2.8 mL of PVP/Ethanol solution. Keep vortexing in between.

9. For loading of tefzel tubing connect the nitrogen hose to the tubing prep station, measure and cut the tefzel tubing using a razor blade and insert into the machine (see Note 11).
10. Pre-dry tefzel tubing by blowing nitrogen at 0.3–0.4 L/min for 15 min (see Note 12).
11. Take the tubing out and attach a syringe with a piece of soft tubing to one end.
12. Vortex and/or sonicate the gold-DNA-Ethanol solution and immediately draw suspension into the tubing using the attached syringe, leaving several cm on the ends. NO BUBBLES!!! Immediately turn the tubing horizontal and reintroduce it into the tubing prep station, with the syringe still attached.
13. Allow beads to settle for 10 min.
14. Remove Ethanol at 1.5–3 cm/s (30–45 s to empty the tubing) from the Tefzel tubing, quickly turn the tubing by hand, and start to rotate the tubing to distribute the gold beads evenly. Briefly press the switch to II, to rotate tubing by 180°. Remove the syringe, wait for 4 s, switch to I and let the tubing rotate.
15. After 30 s, turn on the nitrogen flow (0.4 L/min) and dry gold beads inside the tubing for about 10–15 min (see Note 13).
16. Stop rotation and gas flow and remove the tubing. Trim the ends using the tubing cutter and remove any large obviously empty areas. Cut tubing in cartridges of 0.5 inches. Store cartridges at 4°C in sealed containers with a dessicant pellet. Bullets are stable for several months (see Note 14).

3.2. Gene Gun Immunization of Mice

1. Prepare the gene gun and test its function with an empty cartridge holder, setting the discharge pressure to 400 psi (see Note 15).
2. Shave abdominal skin of mice with an electrical clipper (see Note 16).
3. Now load the gene gun bullets carrying the DNA-gold particle into the cartridge and fix full cartridge holder into the gene gun system.
4. Put on the hearing protection.
5. Position the spacer directly onto the skin and pull the trigger to immunize mice. Turn cartridge holder to the next bullet and shoot again. We usually shoot twice, resulting in the delivery of 2 μg of plasmid DNA at each immunization.
6. We routinely immunize by bombarding the abdomen. In this way, one can conveniently hold the mouse and shoot without anesthesia.

7. We usually immunize animals on a weekly basis up to five immunizations (2 shots per immunization) in order to induce a strong immunity against pigment cell antigens like TRP2 (see Note 17).

3.3. Detection of TRP2-Specific Antibody Responses by Western Blot Analysis

Collect blood at various time points after gene gun immunization from the tail vein or sephaneous vein using the Lithium-Heparin-coated Microvette tubes. Obtain plasma samples by centrifugation of collected blood by 13,000 rpm for 10 min. Plasma can be stored at −20°C for subsequent analysis.

For the detection of TRP2 specific antibody responses Western Blot assays using B16 cell lysates as antigen source can be used.

1. Harvest confluent B16 cells, wash three times with PBS and count cells.
2. Transfer cells into 1.5 mL centrifugation tube and centrifuge for 10 min, at 13,000 rpm.
3. Resuspend cells in TNEN-lysis buffer (+ protease inhibitor) to final concentration of 10^7 cells/mL.
4. Vortex and incubate lysates for 1 h on ice followed by centrifugation for 10 min at 13,000 rpm.
5. Collect supernatant and determine protein concentration using a standard protein assay kit.
6. Set block heater to 100°C.
7. Set up vertical electrophoresis unit and place SDS-page gel into the chamber, remove calm and fill up with running buffer until top edge of gel is covered (see also unit manual).
8. Remove air bubbles in sample slots by pipetting running buffer into slots.
9. Mix 25 μL of sample (equivalent to 75–150 μg protein) with 15 μL of Laemmli buffer (+2% 2-Mercaptoethanol [ME]) (see Note 18).
10. Heat up sample for 3 min at 100°C for denaturation of proteins.
11. Spin down samples shortly and put back on ice.
12. Start loading SDS-Page gel. Use 15 μL of full range rainbow marker for size control.

    Load one sample (40 μL) into one slot (see Note 19). Close electrophoresis unit and connect cables to power supply. Run gel at a constant voltage of 80 V for 1 h and 15 min, or until blue sample line is visible at the lower edge of the gel (see Note 20).
13. Prepare PVDF membrane and filter paper for protein transfer by cutting rectangles in size of the gel. Moisturize PVDF membrane first in 70% Ethanol for about 15 s followed by dH_2O

storage. Soak six rectangles of filter paper in transfer buffer until protein transfer.

14. When SDS-page run is finished set up semi-dry blotter for protein transfer. Place three filter papers on transfer field. Put the PVDF membrane on top (mark one edge and remember!). Remove gel carefully from plates and place on top of PVDF membrane. Cover with three filter papers and make sure that no air bubbles are present. Use a round instrument (e.g., a cut 5 mL serologic pipette) to roll over the membrane and filter papers to remove air bubbles. Close lid of semi-dry blotter and connect cable to the power supply. Transfer proteins for 30 min at a constant voltage of 15 V (see Note 21).
15. At the end remove PVDF membrane with transferred proteins immediately and proceed with blocking (see Note 22).
16. Block unspecific binding sites at PVDF membrane for 1 h (or overnight) by bathing in 10 mL Blotto solution using a sample twister.
17. Wash three times with TPBS for 5 min.
18. Incubate membrane for 1 h (or overnight) either with 10 mL of 1:400 dilution of plasma samples or 1:1,000 dilution of PEP8 antibody (positive control) in PBS containing 2% skim milk and 0.1% Tween 20 (mix 6 mL of TPBS solution with 4 mL Blotto and add 10 μL of antibody). Use sample twister (see Note 23).
19. Wash three times with TPBS for 5 min.
20. Incubate membrane for 1 h with either 10 mL of 1:10,000 dilution of HRP-conjugated goat anti-mouse IgG (in case of mouse plasma) or HRP-conjugated goat anti-rabbit IgG (in case of αPep8) antibody in PBS containing 2% skim milk and 0.1% Tween 20 (mix 6 mL of TPBS solution with 4 mL Blotto and add 1 μL of antibody). Use sample twister.
21. Wash three times with TPBS for 5 min.
22. Add chemiluminescence substrate (mix 1.5 mL solution A with 1.5 mL of solution B). Move vessel with the membrane carefully to cover whole membrane for 5 min (sample twister can be used).
23. Place membrane into transparent plastic bag and fix into X-ray film cassette.
24. Move to the dark room with cassette to develop the X-ray film with the radiofilm developer. Mark X-ray film sheet by cutting one corner in order to remember orientation of the sheet and subsequently the position of the samples. Apply sheet directly onto the covered membrane and close cassette. After different time intervals (10 s, 30 s, 1 min, 2 min) remove sheet and develop in radiofilm developer (see Note 24) (Fig. 1).

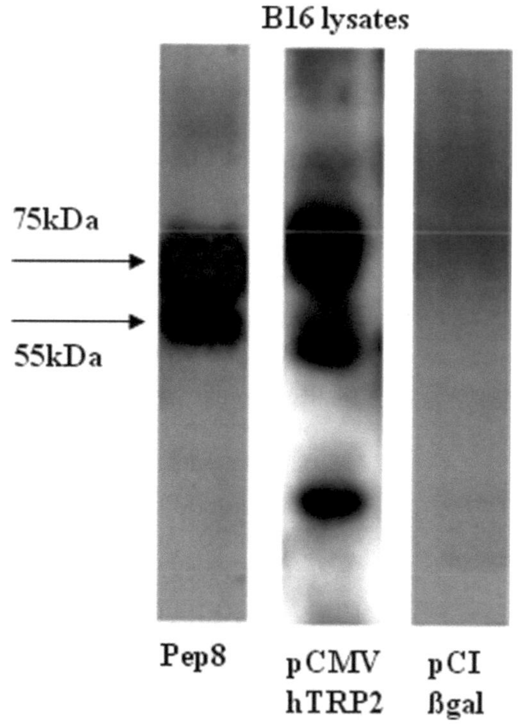

Fig. 1. Detection of TRP2-specific antibody responses after gene gun immunization with expression plasmids encoding for human tyrosinase-related protein 2 (hTRP2) or β-galactosidase (pCI-βgal). B16 melanoma cell lysates were used as TRP2 antigen source in western blot analysis and membrane-bound TRP2 protein was detected by incubation with the polyclonal rabbit antiserum against TRP2 (Pep8) (positive control). TRP2-specific antibody responses in mice could be demonstrated in plasma after immunization with pCMV-hTRP binding to the 55 kDa and the 75 kDa band of TRP2 in B16 lysates, while no reactivity was seen in plasma of pCI-βgal immunized animals.

TRP2-specific antibodies can be detected also in Elisa assays using B16 lysates. However, optimal dilutions of lysates used as immobilized antigen must be evaluated prior titration analysis. Alternatively, TRP2-specific antibodies in plasma from immunized animals can be demonstrated in flow cytometry using B16 cells (see Note 25).

3.4. Detection of TRP2-Specific T-Cell Responses

3.4.1. IFNγ Release (ELISPOT) Assay

1. Prepare ELSIPOT plates by coating sterile Millipore HA plates overnight at 4°C with 10 μg/mL purified anti-IFN-γ coating antibody in sterile PBS (50 μL/well).
2. Wash plates with 200 μL/well three times with sterile PBS and block with 200 μL/well of complete medium for at least 2 h at 37°C.
3. Harvest and pool splenocytes from two immunized mice in each group at various time points after immunization. Disrupt spleens with scissors, with a stainless steel mesh or directly on a 70 μm cell strainer using the rubber end of a 10 mL syringe in

a Petri dish with a small amount of medium. Filter again through a 70 μm cell strainer and centrifuge at 528 ×*g* for 5 min.

4. Deplete red blood cells by adding 1 mL of erythrocyte-lysis buffer and incubate for about 2 min. Stop reaction by adding 9 mL of PBS followed by centrifugation for 5 min at 528 ×*g*.
5. Wash splenocytes twice with 50 mL of PBS (see Note 26).
6. Count cells and adjust splenocytes to 10^7 cells/mL complete medium containing 50 IU/mL of IL-2 (see Note 27).
7. Prepare triplicates of each group and condition by pipetting 100 μL/well of splenocyte preparation.
8. Restimulate splenocytes for about 22 h by adding 100 μL/well of medium containing 1 μg/mL $\beta gal_{876-884}$ or $TRP2_{aa180-188}$ peptide. Include always control peptides and medium alone.
9. Wash ELISPOT plates three times in PBS containing 0.1% Tween 20. Add 2.5 μg/mL biotinylated anti-IFN-γ antibody (50 μL/well) and incubate for 2 h at 37°C.
10. Wash ELISPOT plates three times in PBS containing 0.1% Tween 20. Add HRP-conjugated streptavidin diluted 1:1,500 in assay-buffer (100uL/well) for 30 min at RT.
11. Wash ELISPOT plates three times in PBS containing 0.1% Tween 20.
12. Mix DAB substrate according to the manufacturer's instructions and develop color with 50 μL/well. Color development takes place between 5 and 20 min.
13. Rinse plates thoroughly under tap water and let dry overnight. You can detach filter membranes by mounting them on an adhesive membrane normally used to seal ELISA plates. *Only needed if using a dissecting microscope for analysis!*
14. Count spots at filter membrane under the dissecting microscope or use a professional Elispot reader and follow manufacturer's instructions (see Note 28).

3.4.2. Detection of TRP2-Specific Cytotoxic T-Cells In Vivo

Antigen-specific cytotoxicity of T-cells in animals after gene gun immunization can be detected with different methods. A classical test represents the chromium release assay using the radioisotope ^{51}Cr as described by Tüting et al. (3). Alternatively we used in vivo *kill* assays to detect antigen-specific cytotoxic T-cells in vivo with the advantage to work without radioactive materials (16–18).

Before starting, calculate the amount of lymphocytes you need to inject. A minimum of 2 animals per group should be tested in your experiment at one time point. For the calculation consider the following: 1 donor spleen contains about 50–100 million of lymphocytes. CFSE labeling results in 40% loss of lymphocytes. Calculate a total of 10^7 cells per recipient mouse to inject.

1. Isolate spleens from syngenic naïve mice as described above (see Subheading 3.4.1, steps 3 and 4).
2. Count cells, centrifuge, and resuspend cells in 30 mL PBS. Divide cells in three groups (10 mL each) and add peptides.
   (a) + Relevant peptide ($TRP2_{aa180-188}$) (1 μg/mL)
   (b) + Control peptide (1 μg/mL)
   (c) Without peptide
3. Incubate for 15 min at 37°C.
4. Prepare 10 mL of CFSE high and low solutions (avoid light exposure)

   CFSE high: 1:6,666 dilution of stock CFSE (10 mM) in PBS = 1.5 μM

   CFSE low: 1:10 dilution of CFSE high (1.5 μM) in PBS = 0.15 μM
5. Add 10 mL of CFSE solution to 10 mL cell suspension.
   (a) + CFSE high
   (b) + CFSE low
6. Incubate for 15 min at 37°C.
7. Wash cells three times with 50 mL of *cold* PBS (see Note 29).
8. Count cells and adjust cell suspensions to equal concentrations (e.g., 10^7/mL).
9. Check labeling of cells in flow cytometry before injection by mixing equal amounts of labeled and unlabeled cells. Analyze using a flow cytometer with 488 nm excitation and emission filters appropriate for fluorescein (see Note 30).
10. Mix CFSE high and low cells 1:1 and centrifuge at 528 × *g* for 5 min.
11. Resuspend in PBS and adjust concentration to 10^8/mL.
12. Inject 10^7 total cells intravenously into the tail vein of prior immunized mice (see Note 31).
13. 4 h later, isolate splenocytes from single mice as described above (see subheading 3.4.1, step 3 and 4).
14. Resuspend cells in FACS buffer to detect the presence of CFSE-labeled cells using a flow cytometer with 488 nm excitation and emission filters appropriate for fluorescein.
15. To calculate specific lysis, following formula should be used:

$$\text{Ratio} = (\text{percentage CFSE}^{\text{high}} / \text{CFSE}^{\text{low}})$$

$$\text{Percentage specific lysis} = 100 - [(\text{ratio relevant immunized} / \text{ratio irrelevant immunized}) \times 100]$$

3.5. Evaluation of the Induced Immune Response in a B16 Melanoma Model

For the expansion of B16 melanoma cells, culture cells in several flasks in complete DMEM medium in a cell culture incubator at 37°C and 5% CO_2 and high humidity. Examine cells daily, observing the morphology, the color of the medium and the density of the cells. Split cells 1:10 when confluent (every 3–4 days) by Trypsin-EDTA treatment (1 mL) until cells detach (see Note 32).

In order to evaluate the induction of protective immune responses against the growth of melanoma by gene gun immunization, challenge animals by injection of B16 melanoma cells 2–4 weeks after final immunization. Two models can be used: (a) lung metastasis model and (b) subcutaneous model.

1. Harvest B16 cells when confluent by Trypsin-EDTA treatment.
2. Wash cells three times with 10 mL PBS.
3. Count cells and adjust cell concentration to 10^6/mL in PBS.
4. Vortex and fill 1 mL syringe with cell suspension just prior to injection.

3.5.1. Lung Metastasis Model

1. For better visibility of the tail veins, warm up mouse tail with an infrared light or water bath just prior to injection.
2. Using a 27G needle, inject 400 μL (=4×10^5 cells) intravenously into tail vein (see Note 33).
3. After 2 weeks, sacrifice animals and isolate complete lungs from each animal and transfer into a 50 mL Falcon tube filled with 70% Ethanol (10–25 mL).
4. Count pigmented (black) metastasis by eye or with the help of a dissecting microscope.

   Lungs with and without B16 melanoma metastasis are shown in Fig. 2a. In Fig. 2b, numbers of B16 lung metastases in mice after gene gun immunization with pCMV-hTRP2 or pCI-βgal (control) are demonstrated.

3.5.2. Subcutaneous Melanoma Model

1. For subcutaneous injection lift the skin of the hind flank and prepare a "tent."
2. Insert needle into the subcutaneous tissue. Aspirate prior to making the injection. Proper placement should yield no aspirate.
3. Inject with a 27G needle 100 μL (=10^5 cells) subcutaneously into the hind flank of the animal.
4. Observe the animals and measure tumor size with a caliper twice a week (see Note 34).

3.6. Induction of Vitiligo-Like fur Depigmentation

Vitiligo is a common pigmentation disorder where melanocytes are focally destroyed in the epidermis leading to patchy hypopigmentation of the skin in affected patients. Vitiligo-like depigmentation

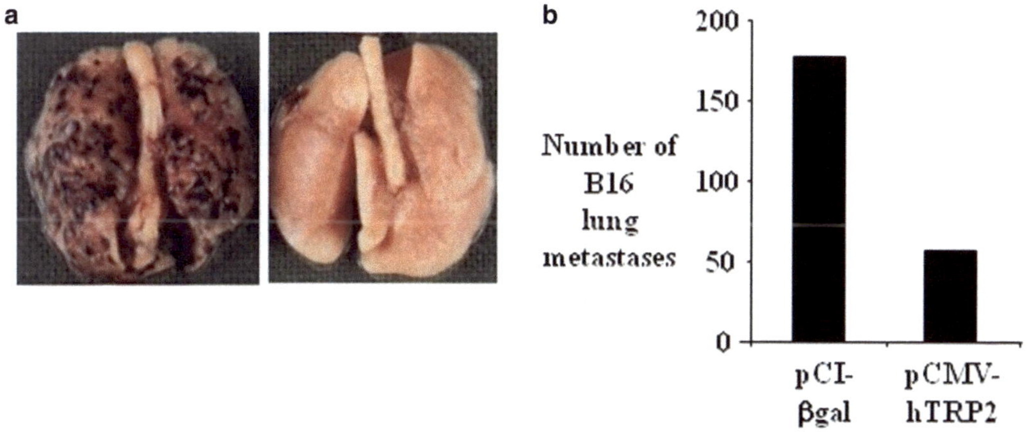

Fig. 2 B16 lung metastasis in C57Bl/6 mice. B16 melanoma cells were cultured in DMEM culture medium and 4×10^5 cells were injected intravenously. Two weeks after tumor inoculation animals were sacrificed and lungs isolated. (**a**) Shown are lungs of animals injected with B16 cells (*left*) or with PBS (*right*). (**b**) Shown are numbers of B16 lung metastases in mice after gene gun immunization with pCI-βgal or pCMV-hTRP2.

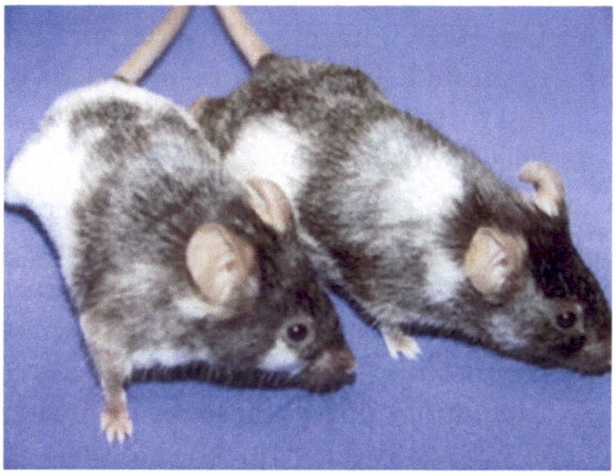

Fig. 3. Vitiligo-like fur depigmentation after immunization with expression plasmids encoding for human TRP2. Female C57Bl/6 mice were immunized with 2 mg of plasmid DNA on a weekly basis for a total of 5 weeks. Animals were shaved several times and regrown hair appeared white leading to a "salt and pepper" like fur color.

also appears not infrequently during immunotherapy of melanoma and indicates a favorable clinical course (19–21). Therefore, we observed the induction of vitiligo-like fur depigmentation in immunized mice as indication for effective induction of anti-melanoma immune responses. Animals showed first white spots at the side of immunization as early as after the second immunization. Regrown hair appeared white leading to a "salt and pepper" like fur color. If animals were housed for long period after immunization, white spots spread over the whole mouse as shown in Fig. 3 (see Note 35).

4. Notes

1. Gold particles can be obtained of different sizes and shapes. Each lot should be microscopically examined. One must bear in mind that the shape and size of the gold beads will determine their impulse and, consequently, their penetration into tissues and cells. For in vivo transfection, a mixture of different sizes may be beneficial since gold will penetrate the skin to various depths. Therefore, the Helios Gene Gun optimization kit contains gold particles of different sizes (0.6 μm, 1 μm, 1.6 μm) to investigate the optimal set up for your experiment.
2. A clean preparation of plasmid DNA resuspended in dH_2O at 1 mg/mL should be employed. In our hand, better quality of DNA and higher transfection efficiency could be obtained by using the Qiagen Endofree-Maxi-Kit instead of Midi- and Mini-Kits. Be sure that the DNA is Ethanol-free. If DNA concentration is less than 1 mg/mL, perform standard phenol–chloroform precipitation and dilute in lower volume in order to get a concentration of 1 mg/mL.
3. Use 100% water-free Ethanol (pure Ethanol incubated 3× over dessicated beads, kept in sealed bottle in a dessicator). Alternatively Ethanol can be stored at −20°C in 50 mL tubes sealed with parafilm.
4. For blood sampling through the sephaneous vein fix mouse in a 50 mL tube. Remove one leg and shave the outer part to reveal the sephaneous vein. Put some grease onto the shaved skin before puncturing vein with an 18G needle. Immediately collect blood with the lithium-heparin coated CB300 Microvette tube until sufficient amount is obtained. This method has the advantage of good quality and quantity of blood/plasma samples, easy use and reduced stress for the animals.
5. Alternatively to the Pep8 antiserum, commercial antibodies are now available from Abcam (#ab74073).
6. To prepare 0.5 M EDTA solution (MW 292.24), dissolve 73.06 g EDTA in 50 mL dH_2O. Dissolving of EDTA requires pH > 8. Therefore, adjust pH to >8 until EDTA is completely dissolved. Afterwards pH can be changed to final level. For long-term storage and for the use in cell culture sterile filter EDTA solution and store at 4°C.
7. Nonfat powdered skim milk can be purchased in regular food stores and is therefore cheaper than ready to use Blotto (5% nonfat skim milk in TBS) solution from biotechnology companies.
8. Depending on the hydrophobicity of the respective peptide, you should try to dissolve the peptide in a smaller amount of

either PBS or DMSO. If you have problems getting the peptide into solution, you can then adjust accordingly. A good strategy is to dissolve the peptide first in a small amount of pure DMSO followed by subsequent dilution in PBS until final concentration. The DMSO content should be as low as possible (usually 10–50%), since it is toxic to cells.

9. Heat inactivate FBS for 45 min at 56°C in a water bath. Heat inactivation is the method of choice to destroy complement, and to ensure that the cells will not be lysed by antibody binding. We usually inactivate 500 mL at once and freeze 50 mL aliquots at -20°C, adequate for 500 mL culture medium.
10. 50 μg per coating (may be less—down to 10 μg—but not much more because of clumps). One tefzel tube will produce 50 cartridges, 50 mg DNA = 1 mg/cartridge.
11. Before starting the gas flow, check that the system is closed. Then open nitrogen bottle valve and set pressure at regulator between 3 and 6 and finally tune flow on tube loader between 2 and 3. Close black valve only when transiently stopping. Always close bottle valve first, then the rest.
12. Especially during humid weather conditions, it is important to dry the tubing first. The major problem preventing an even loading of the tubing with gold beads appears to be residual water in the Ethanol. For this reason you seal PVP and beads in 100% Ethanol with parafilm and wait until they acquire room temperature before opening. Try to use fresh 100% Ethanol whenever possible.
13. Loading rarely turns out perfect. The quality of the DNA appears to influence the fine distribution of the beads in the cartridge. Surprisingly, even rather unevenly loaded cartridges work well for immunization. You can test the DNA precipitation onto the gold beads. Place three bullets in a 1.5 mL microcentrifuge tube with 600 μL dH_2O. Vortex, sonicate, rinse, and centrifuge. Measure $OD_{260\ nm}$. It should be close to 0.05 (=1.5 μg/mL).
14. You can also test the cartridges by discharge onto parafilm or into agar plates (you can use plates to grow bacteria or prepare some with 3% water agar). Cut sections and observe the distribution of gold beads under the microscope.
15. We have had best results at 400 psi in mice. You should start the experiments with β-gal and EGFP as marker genes and simultaneous test antigens. Expression in the skin can be conveniently detected and immune responses analyzed.
16. Skin should be always free of hairs before the bombardment with DNA. Electrical clipper, razor blades or depilatory cream can be used. It might be advisable to perform immunization of mice under anesthesia.

17. For the induction of strong immune responses against the growth of transplanted B16 melanoma cells we immunized always five times (once per week with a total of 2 μg of DNA). However, extensive dose escalation studies were not performed in our group.
18. Optimal separation and intensity results are depending on the optimal loading amount of protein. Therefore, it is recommended to test different concentrations of protein in initial tests to establish the final protocol.
19. Avoid overflow of samples into neighbor slots. Use specific gel loading tips. It might be beneficial to have an empty slot between samples. Protein bands can be overlapped by high intensities from neighbor samples.
20. 80 V for 1 h 15 min was established for the SDS-page gel run with our equipment. Conditions might be different with other machines and manufacturer's instructions should be considered. Voltage should be low enough that the gels are not getting too hot with the consequence of melting and loss of electricity flow. Conditions might change during gel run and should be observed. It is possible to reuse gel running buffer after SDS-page gel run. However, new buffer must be used if running conditions change.
21. 15 V for 30 min was established for the protein transfer with our equipment. Conditions might be different with other machines and manufacturer's instructions should be considered. The successful transfer of proteins can be verified with the rainbow marker, which should be visible at the PVDF membrane after transfer.
22. It is important to remove the PVDF membrane immediately from the semi-dry blotter when transfer is finished. Absence of electricity in the presence of transfer buffer and filter papers will result in randomly diffusion of proteins and fuzzy bands at the end.
23. Incubation with antibodies can be done in size matched buckets, 10 or 50 mL tubes or membrane can be wrapped into transparent plastic bags together with the antibody solution. Importantly, membrane-bound protein has to be accessible for the antibody and antibody has to be provided in excess. Antibody solution can be reused once when stored at 4°C and if storage does not exceed a few days.
24. For the development of X-ray film sheets different radiofilm developer can be used. Depending on the sensitivity of the machine and the used substrate (ECLplus, West pico, or others) it might be necessary to test different incubation times in order to obtain optimal results.
25. Using whole cell lysates as antigen resource to detect specific antibodies in polyclonal sera in Elisa assays requires good

blocking of unspecific binding sites and a detailed analysis of optimal concentrations of antigen and secondary antibodies. For the demonstration of TRP2-specific antibodies in flow cytometry, B16 cells have to be permeabilized and fixed as described in standard intracellular staining protocols.

26. If analysis of IFNγ secretion by specific cell subtypes (CD4+ or CD8+ T-cells) is required, a separation step, using anti-CD4 and anti-CD8 antibodies (hybridoma supernatants or purified antibody) followed by separation with magnetic beads or complement-mediated depletion, can be performed to remove (or enrich) effectors.
27. Consider and calculate that 1 spleen contains about 50–100 million cells. Splenocytes can be frozen for subsequent analysis at −80°C in freezing medium (RPMI complete containing 20% FBS and 10% DMSO).
28. If dissecting microscope is used, it is recommended that the same person is analyzing Elispot results of all experimental groups and experiments in one project since there is a high variability in identifying real spots over background staining between personnel.
29. It is important to use ice-cold PBS because it stops CFSE enzyme reaction.
30. The intensity levels of CFSE high and low cells should be 1 log apart otherwise peaks are overlapping in flow cytometry and calculation of specific cytotoxicity is not possible.

    The intensity level of CFSE high cells should be around 10^4 (FL1 channel) before injection. If the intensity level is not high enough at the beginning, it is difficult to detect divided CFSE-labeled cells in whole lymphocyte population after in vivo incubation.
31. For the analysis of T-cells in in vivo cytotoxicity assays we injected cells 1 week after last gene gun immunization.
32. Before start of the experiment calculate amount of cells necessary to inject all animals at once. 3–5 million cells/flask can be roughly obtained when cells are confluent. B16 cells are able to grow in different media like RPMI and DMEM. However, we observed that only in DMEM cultured B16 cells developed in vivo pigmented (black) metastasis. The growth kinetic of B16 cells might vary depending on clones (B16F10) and passage numbers of the cell line. Therefore, it is recommended to test different cell amounts for transplantation to establish the tumor model prior final experiment. Metastasis should be defined and number of metastasis should be countable. Be also sure that cells are free of mycoplasm which can influence the tumor growth in vivo.

33. B16 cell suspension should be at RT and injected slowly. Too fast injection can cause death of animals. Mice can fall easily into hypothermia when cold and big amounts of fluids are injected. Keep animals warm after injection (infrared light).
34. Palpable tumors can be detected as early as 1 week after injection, animals with tumors bigger than 1 cm in diameter should be sacrificed for animal welfare reasons. Animals with a weight loss more than 15% should also be sacrificed for animal welfare reasons.
35. The induction of vitiligo-like fur depigmentation is not only dependent on the effective induction of immune responses against melanocytic antigens, but also requires an inflammatory signal in the skin (18, 22). The bombardment with DNA and shaving of skin results in a stimulation of effector cells which migrate to the skin to destroy melanocytes in the hair follicle. The observation of vitiligo-like fur depigmentation can be enhanced by additional shaving of non-DNA bombarded skin.

Acknowledgment

This work was supported by grants from the Deutsche Forschungsgemeinschaft (DFG) to T. Tüting and by a Lise-Meitner stipend of the NRW government to J. Steitz.

References

1. Tang DC, DeVit M, Johnston SA (1992) Genetic immunization is a simple method for eliciting an immune response. Nature 356:152–154
2. Donnelly JJ et al (1997) DNA vaccines. Annu Rev Immunol 15:617–648
3. Tüting T et al (1998) Autologous human monocyte-derived dendritic cells genetically modified to express melanoma antigens elicit primary cytotoxic T cell responses in vitro: enhancement by cotransfection of genes encoding the Th1-biasing cytokines IL-12 and IFN-alpha. J Immunol 160:1139–1147
4. Tüting T, Albers A (2000) Particle-mediated gene transfer into dendritic cells: a novel strategy for the induction of immune responses against tumor antigens. Methods Mol Med 35:27–47
5. Condon C et al (1996) DNA-based immunization by in vivo transfection of dendritic cells. Nat Med 2:1122–1128
6. Tüting T et al (1999) Dendritic cell-based genetic immunization in mice with a recombinant adenovirus encoding murine TRP2 induces effective anti-melanoma immunity. J Gene Med 1:400–406
7. Corr M et al (1996) Gene vaccination with naked plasmid DNA: mechanism of CTL priming. J Exp Med 184:1555–1560
8. Boon T, van der Bruggen P (1996) Human tumor antigens recognized by T lymphocytes. J Exp Med 183:725–729
9. Rosenberg SA (1997) Cancer vaccines based on the identification of genes encoding cancer regression antigens. Immunol Today 18:175–182
10. Schreurs MW et al (1998) Genetic vaccination against the melanocyte lineage-specific antigen gp100 induces cytotoxic T lymphocyte-mediated tumor protection. Cancer Res 58:2509–2514

11. Weber LW et al (1998) Tumor immunity and autoimmunity induced by immunization with homologous DNA. J Clin Invest 102:1258–1264
12. Steitz J et al (2000) Genetic immunization of mice with human tyrosinase-related protein 2: implications for the immunotherapy of melanoma. Int J Cancer 86:89–94
13. Bowne WB et al (1999) Coupling and uncoupling of tumor immunity and autoimmunity. J Exp Med 190:1717–1722
14. Steitz J et al (2002) Genetic immunization with a melanocytic self-antigen linked to foreign helper sequences breaks tolerance and induces autoimmunity and tumor immunity. Gene Ther 9:208–213
15. Kameyama K et al (1993) Pigment production in murine melanoma cells is regulated by tyrosinase, tyrosinase-related protein 1 (TRP1), DOPAchrome tautomerase (TRP2), and a melanogenic inhibitor. J Invest Dermatol 100:126–131
16. Steitz J et al (2006) Comparison of recombinant adenovirus and synthetic peptide for DC-based melanoma vaccination. Cancer Gene Ther 13:318–325
17. Gaffal E et al (2007) Comparative evaluation of CD8 + CTL responses following gene gun immunization targeting the skin with intracutaneous injection of antigen-transduced dendritic cells. Eur J Cell Biol 86:817–826
18. Lane C et al (2004) Vaccination-induced autoimmune vitiligo is a consequence of secondary trauma to the skin. Cancer Res 64:1509–1514
19. Scheibenbogen C, Hunstein W, Keilholz U (1994) Vitiligo-like lesions following immunotherapy with IFN alpha and IL-2 in melanoma patients. Eur J Cancer 30A:1209–1211
20. Irvine KR et al (1996) Cytokine enhancement of DNA immunization leads to effective treatment of established pulmonary metastases. J Immunol 156:238–245
21. Okamoto T et al (1998) Anti-tyrosinase-related protein-2 immune response in vitiligo patients and melanoma patients receiving active-specific immunotherapy. J Invest Dermatol 111:1034–1039
22. Steitz J et al (2004) Initiation and regulation of CD8 + T cells recognizing melanocytic antigens in the epidermis: implications for the pathophysiology of vitiligo. Eur J Cell Biol 83:797–803

Chapter 25

Biolistic DNA Vaccination Against Cervical Cancer

Michal Šmahel

Abstract

The development of cervical cancer is associated with infection by oncogenic human papillomaviruses (HPVs), of which type 16 (HPV16) is the most prevalent in HPV-induced malignant diseases. The viral oncoproteins E6 and E7 are convenient targets for anti-tumor immunization. To adapt the corresponding genes for DNA vaccination, their oncogenicity needs to be reduced and immunogenicity enhanced. The main modifications for achieving these aims include mutagenesis, rearrangement of gene parts, and fusion with supportive cellular or viral/bacterial genes or their functional parts.

As HPVs are strictly human specific, an animal model of HPV infection does not exist. Therefore, immunization against HPV-induced tumors is most frequently tested in mouse models utilizing transplantable syngeneic tumor cells producing the HPV16 E6/E7 oncoproteins. In this chapter, one such cell line designated TC-1 is characterized and the effect of immunization with the modified E7 fusion gene against TC-1-induced subcutaneous tumors is described. As down-regulation of MHC class I molecules is one of the most important escape mechanisms of cervical carcinoma cells, the TC-1/A9 clone with reversibly reduced MHC class I expression has been developed and, herein, its response to DNA vaccination is also shown and compared with that of the TC-1 cells.

Key words: Human papillomavirus, E6/E7 oncogenes, MHC class I, TC-1 cells, Immune escape, Oncogenicity, Fusion gene, Cluster immunization

1. Introduction

Cervical carcinoma (CC) is the second most common malignancy in women worldwide, with approximately 500,000 new cases diagnosed each year (1). The mortality rate is about 60%. The development of CC is etiologically linked to the infection with human papillomaviruses (HPVs). These viruses comprise over 120 genotypes. Some of them, so-called high-risk HPVs, are oncogenic. The type most frequently associated with human cancers is HPV16 that has been proved to be present in 50–60% of CC cases (2, 3).

Stephan Sudowe and Angelika B. Reske-Kunz (eds.), *Biolistic DNA Delivery: Methods and Protocols*,
Methods in Molecular Biology, vol. 940, DOI 10.1007/978-1-62703-110-3_25, © Springer Science+Business Media, LLC 2013

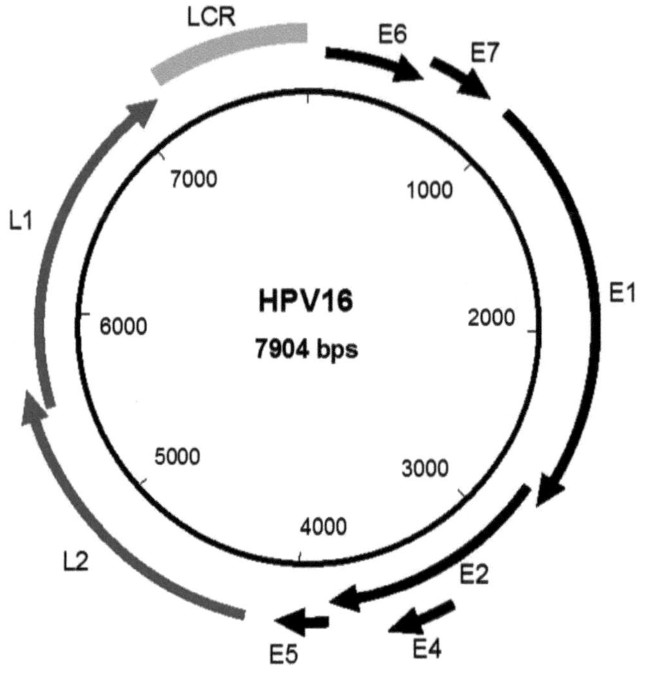

Fig. 1. The genome of HPV16.

The genome of HPV16 (Fig. 1) that is formed by circular dsDNA of approximately 8 kbp codes for six early (E) proteins with regulatory functions and two late (L) proteins constituting the viral capsid. Transcription of all HPV genes is regulated by the long control region (LCR) located upstream of the E6 gene. Three proteins have oncogenic abilities. While E6 and E7 are the primary transforming proteins, E5 may support their activity and thus contribute to tumor progression (4). The E6 and E7 oncoproteins are responsible not only for oncogenic transformation of infected cells but also for the maintenance of the transformed phenotype and they are the only viral proteins constitutively expressed in CC cells as well as in cell lines derived from these tumors. Therefore, they are suitable targets for anti-tumor immunization.

HPVs are strictly human specific and epitheliotropic. They infect basal keratinocytes either of the skin or of the anogenital and oropharyngeal mucosa. Their replication depends on the differentiation state of infected cells. Progression from pre-neoplastic lesions to invasive CC is often associated with integration of the viral genome that results in enhanced E6 and E7 expression (5).

Because of host specificity, no animal model of HPV infection exists. Tumorigenesis caused by HPV16 E6 and/or E7 oncogenes can be studied by using transgenic mice carrying viral oncogenes under the control of the keratinocyte 14 promoter (6, 7). Chronic treatment of these mice with estrogen resulted in squamous cell carcinoma development in the cervix and vagina (8, 9). Transgenic mouse models producing the HPV16 E6 and/or E7 oncoproteins as self-antigens have been also used for the analysis of the induction of specific cell-mediated immune reactions, especially with respect to overcoming tolerance (7, 10–13). However, the efficacy of therapeutic vaccines against HPV-induced tumors is most frequently tested in mice by using transplantable syngeneic tumor cells producing the HPV16 E6/E7 oncoproteins. These cell lines, reviewed by Bubeník (14), were prepared either by transduction of tumor cell lines with the E6/E7 oncogenes or by co-transformation of non-oncogenic cells with the viral oncogenes and another oncogene (ras or myc). In the majority of laboratories, TC-1 cells are currently utilized. These cells were derived from primary lung cells obtained from a C57BL/6 mouse (H-2^b haplotype) and transduced with the E6/E7 oncogenes and H-ras activated by the G12V mutation (15). Molecular characteristics of TC-1 cells are summarized in Table 1. Probably due to their high production of MHC class I, CD80, and CD54 molecules, the cells are highly sensitive to immunization against E6/E7 (and other potential tumor antigens). Moreover, TC-1 cells are easily subcultured and propagated in vitro and they reliably form subcutaneous tumors after administration of $3–5 \times 10^4$ cells. They do not metastasize spontaneously, but after intravenous injection, experimental lung metastases develop readily (16).

As tumor cells can be recognized and removed by the host immune system (17), transformed cells are under the pressure of an immune attack during tumor development which leads, thanks to genetic instability of tumor cells, to the selection of malignant cells with adaptations that provide them with protection and a survival advantage (18). Evasion of tumor cells from the host immune responses is probably markedly implicated in the low efficacy of antitumor vaccines in clinical trials. The main escape mechanisms include down-regulation or loss of expression of MHC class I molecules on the cell surface, down-regulation or loss of tumor-antigen production, up-regulation of immunosuppressive factors and/or reduction of immunostimulatory factors secreted by tumor cells or cells infiltrating the tumor, and resistance to apoptotic stimuli (19, 20).

The escape mechanisms used by HPVs and found in CC have been reviewed by Kanodia et al. (21). HPV oncogenes affect production and/or function of numerous cellular proteins involved in modulation of immune reactions, including MHC class I molecules, components of the antigen processing machinery (APM), cytokines, chemokines, toll-like receptor 9, and adherence molecules. Reduced levels of surface MHC class I molecules have been recorded in up to 70–90% of CC patients (22, 23) and metastatic cells have been shown to have lower MHC class I expression in comparison with cells from primary tumors (24). Down-regulated MHC class I production has been also found associated with a worsened prognosis in patients at early stages of tumor development (25).

Table 1
Molecular characterization of TC-1 cell line (– = no expression, + = expression)

Phenotype	Remarks	References
HPV16 E7 +		(15)
HPV16 E6 +[a]	Protein undetectable	(50)
Cytokeratin +	Epithelial marker	(15)
MHC class I +	Both loci (K and D) expressed	(15, 30)
MHC class II –		(15)
CD80 (B7-1) +		(51)
CD86 (B7-2) –		(52)
CD54 (ICAM-1) +		(52)
CD274 (B7-H1) +[a]	Protein not detected	(52)
CD275 (B7-H2) +[a]	Protein not detected	(52)
H2-T23 (Qa-1) –	Nonclassical MHC class I	(52)
Ccl2 (MCP-1) +		(35)[b]
VCAM-1 +	Up-regulated in TC-1 P3 cells	(33)
VLA-4 –		(33)[b]
Cxcl12 (SDF-1) +		(33)

MHC major histocompatibility complex, *CD* cluster of differentiation, *ICAM-1* intercellular adhesion molecule-1, *MCP-1* monocyte chemoattractant protein-1, *VCAM-1* vascular cell adhesion molecule-1, *VLA-4* very late antigen-4, *SDF-1* stromal cell-derived factor-1

[a]mRNA found by RT-PCR

[b]Results of cDNA array analysis of TC-1 cells are included in the study

Several mechanisms can be responsible for MHC class I down-regulation (26, 27), with defects in APM being one of the most common (28). In CC, the loss of the transporter associated with antigen processing-1 (TAP-1) has been demonstrated to correlate to the loss of MHC class I molecules (29).

To study escape mechanisms in the TC-1 model and to adapt this model to clinical conditions, cell lines with heterogenous MHC class I expression were derived from tumors induced by TC-1 cells in mice immunized against the E7 antigen. From these cell lines, clones TC-1/A9, TC-1/D11 (30), and TC-1 P3 (A15) (31) with markedly reduced MHC class I expression were isolated in vitro. The clones were characterized by increased resistance to immunization. For the TC-1/A9 and TC-1/D11 clones, the level of MHC class I reduction correlated with the level of immunoresistance and also with the reduction of TAP-1 expression. MHC class I down-regulation is reversible in both systems. Surface expression of MHC class I molecules can be up-regulated by IFN-γ (30, 31). Sustained incubation with IFN-γ is needed for the preservation of increased MHC class I level. Similarly, tumor cells isolated from TC-1/A9-induced tumors carried surface MHC class I molecules that gradually decreased in vitro. Examination of epigenetic regulatory mechanisms in TC-1 and TC-1/A9 cells revealed that MHC class I reduction in TC-1/A9 cells was associated with epigenetic down-regulation of APM components (TAP-1, TAP-2, LMP-2, and LMP-7) (32).

Depletion analysis of immune reactions induced by DNA vaccines against E7 showed that while $CD8^+$ T lymphocytes were crucial for the elimination of the TC-1 cells, NK cells considerably supported $CD8^+$ cells in protection against the TC-1 P3 (A15) cells (31) and were even most important for the anti-tumor effect against the TC-1/A9 cells (Smahel et al., unpublished data).

Further analysis of the TC-1 P3 cell line (derived directly from a tumor) and its A15 clone revealed that the up-regulation of vascular cell adhesion molecule-1 (VCAM-1) (33) and down-regulation of Fas expression (34) contribute to immunoresistance in this system. Our study of TC-1-tumor-derived cell lines and clones showed altered expression of immunomodulatory cytokines/chemokines Ccl2 (MCP-1), osteopontin, and midkine, but we did not find correlation with immunoresistance of the cells. However, sequencing of the E7 oncogene in TC-1 clones that were quite resistant to anti-E7 immunization detected the N53S mutation in the immunodominant $H\text{-}2^b$ epitope (35). Functional analysis confirmed the responsibility of this mutation for immunoresistance (36).

The HPV16 E6 and E7 genes have two features disadvantageous for the development of DNA vaccines: oncogenicity and low immunogenicity. Numerous modifications of the E6 and E7 genes enhancing their immunization potency have been reported and the majority of them have been recently reviewed (37). The principal mechanisms contributing to the enhanced immunogenicity include: (1) codon optimization that does not change protein sequence, but substantially increases production of antigens, (2) alterations of cellular localization that improve antigen processing and presentation by MHC class I or class II molecules, (3) alterations of antigen stability that either support antigen degradation and subsequent presentation in antigen-presenting cells (APC) or, conversely, improve antigen stability and thus increase the amount of an antigen available for cross-presentation, (4) targeting antigens to dendritic cells, and (5) activation of $CD4^+$ T cells by the addition of helper epitopes. For most modifications, chimeric genes containing parts of cellular genes (e.g., signal sequences) or foreign genes (viral or bacterial, e.g., toxins or chaperones) are constructed.

Oncogenicity is inconsistent with the use of DNA vaccines in clinical practice. Two approaches were applied to reduce the transformation potential of the HPV16 E6 and E7 oncogenes. The first one is based on the knowledge of binding to tumor suppressor proteins p53 and pRb, respectively, which results in the inhibition of their functions and thus contributes to malignant transformation of cells. The binding sites of the E6 and E7 oncoproteins have been eliminated by mutagenesis of amino acid residues playing the decisive role in binding (38–41). In the second approach, the genes were divided into several parts that were rearranged ("shuffled") and supplemented with junction sequences (42–44).

To determine the effect of DNA vaccination against HPV16-induced tumors, the C57BL/6 mouse model of subcutaneous tumors is used as presented in this chapter. For tumor induction, either highly responsive TC-1 cells producing the viral E6 and E7 oncoproteins or their TC-1/A9 clone with reduced MHC class I expression are administered. The candidate DNA vaccines should be based on modified E6 and/or E7 oncogenes characterized by improved immunogenicity and inhibited oncogenicity.

In our laboratory, we use the Bio-Rad's Helios Gene Gun System with the Tubing Prep Station and Tubing Cutter for the preparation of cartridges coated with DNA/gold particles and immunization of mice. In the producer's manual, all procedures are thoroughly described including the explanation of the methodological principles, equipment handling, optimization of the procedure, and its possible modifications. Furthermore, as other chapters of this book focus on the individual aspects of gene gun immunization, I give here a concise protocol with some modifications and details as performed by our laboratory.

Based on the expected efficacy of immunization, we use either TC-1 or TC-1/A9 cells that differ in the expression of MHC class I molecules (Fig. 2) and perform either preventive or therapeutic vaccination.

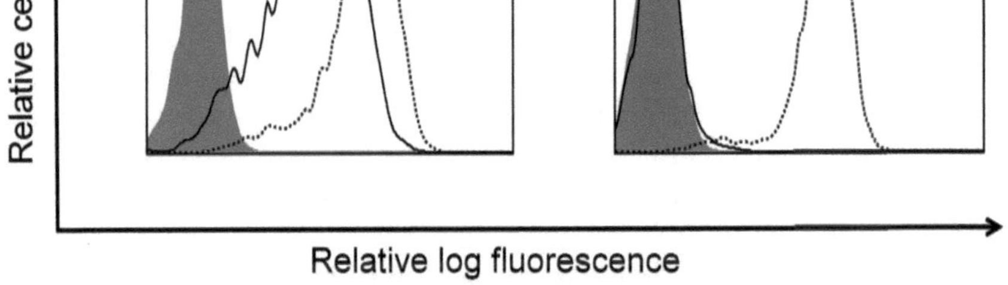

Fig. 2. Expression of MHC class I molecules on TC-1 and TC-1/A9 cells. MHC class I molecules were determined by flow cytometry after staining with anti-H-2K^b/H-2D^b antibody (*solid lines*) or with isotype control antibody (*filled histograms*). The effect of IFN-γ treatment (200 U/mL for 40 h) is also shown (*doted lines*).

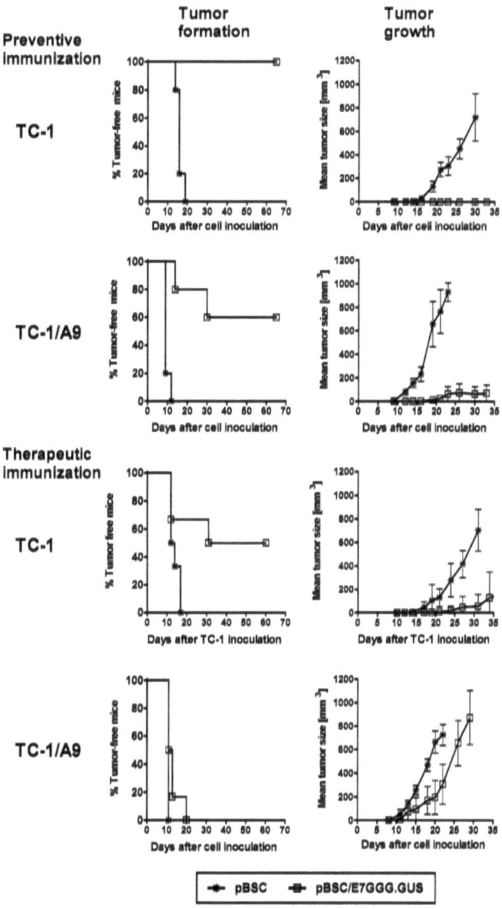

Fig. 3. Formation and growth of TC-1- or TC-1/A9-induced tumors after preventive and therapeutic immunization with the E7GGG.GUS fusion gene. In preventive immunization, mice ($n=5$) were immunized twice with 2 μg of plasmid DNA at a 1-week interval and challenged s.c. with 3×10^4 TC-1 or TC-1/A9 cells 1 week after the second immunization dose. In therapeutic immunization, mice ($n=6$) were s.c. inoculated with 3×10^4 TC-1 or TC-1/A9 cells and immunized with 2 μg of plasmid DNA at days 3, 7, and 10 after cell administration. The plasmid pBSC was used as a negative control.

It is much more difficult to elicit an anti-tumor effect against TC-1/A9-induced tumors with reduced surface expression of MHC class I molecules than against TC-1-induced tumors. As noted above, different effector lymphocyte subpopulations should also be considered. Figure 3 shows representative anti-tumor effects after immunization with the E7GGG.GUS gene. This DNA vaccine was constructed by the fusion of the modified E7GGG gene containing three substitutions in the pRb-binding site of E7 (39) and the GUS gene encoding *E. coli* β-glucuronidase (45). When compared with fusions of E7GGG with lysosome-associated membrane protein-1 (LAMP-1) or mouse heat shock protein 70 (Hsp70), the E7GGG.GUS gene showed superior anti-tumor potency (45, 46).

The number of immunizations and their timing are other important parameters of DNA vaccination. As the E7 gene is more immunogenic in C57BL/6 mice than the E6 gene, two doses of anti-E7 DNA vaccines administered 1 week apart usually result in a significant anti-tumor effect while three doses of anti-E6 vaccines are needed for significant prevention of TC-1-induced tumors (41). In therapeutic setting, short-interval (cluster) immunization (47, 48) with three doses is recommended to reduce the relatively rapid growth of tumors (Fig. 3).

2. Materials

2.1. Prerequisites

1. Cell culture laboratory.
2. Facility for mouse experiments.

2.2. Preparation of Cartridges for Gene Gun Immunization

1. Helios Gene Gun System (Bio-Rad).
2. Nitrogen tank (grade 4.8 or higher) with regulator.
3. Tefzel tubing (Bio-Rad).
4. Ultrasonic cleaner.
5. Vortexer with a holder for 1.5 mL Eppendorf tubes.
6. Analytical balance.
7. Microcentrifuge.
8. Screw-cap scintillation vials (for cartridge storage).
9. Desiccant pellets.
10. 1.5 mL Eppendorf tubes.
11. 50 and 15 mL polypropylene conical tubes.
12. 5 mL pipettes and pipette-aid.
13. 20, 200, and 1,000 μL micropipettors and tips.
14. Minute timer.
15. Scissors.
16. Parafilm.
17. 1-μm gold particles.
18. Polyvinylpyrollidone (PVP), MW = 360,000.
19. 100% ethanol (see Note 1).
20. 1 M spermidine (see Note 2).
21. 1 M calcium chloride.
22. 1–1.1 μg/mL plasmid DNA (see Note 3).

2.3. Delivery of DNA Vaccines by the Helios Gene Gun

1. Mice, 7–8-week-old C57BL/6 females.
2. Cartridges coated with DNA/gold particles.
3. Helios gene gun and accessories.
4. Helium tank (grade 4.5 or higher) with regulator.
5. Ear protection.
6. Shaving cream.
7. Razor blades.
8. Parafilm.
9. Tinfoil.

2.4. Administration of Tumor Cells and Observation of Tumor Growth

1. C57BL/6 female mice, 7–8 weeks old or after immunization.
2. TC-1 and/or TC-1/A9 cell lines.
3. Cell culture medium: Dulbecco's modified Eagle's medium (DMEM) with sodium pyruvate and 4.5 g/L glucose, 10% fetal bovine serum, 2 mM L-glutamine, 100 U/mL penicillin, 100 μg/mL streptomycin. Store at 4°C.
4. Phosphate-buffered saline (PBS): 1.7 mM KH_2PO_4, 5 mM Na_2HPO_4, 150 mM NaCl, pH 7.4.
5. Trypsin/EDTA: 0.05% w/v trypsin, 0.02% w/v EDTA in PBS. Store at 4°C.
6. 0.4% trypan blue in PBS.
7. 25, 75, and 150 cm^2 culture flasks and 15 cm culture dishes.
8. 15 and 50 mL polypropylene conical tubes.
9. 1.5 mL Eppendorf tubes.
10. 2 mg/mL etomidate (Hypnomidate; Janssen Pharmaceutica, Beerse, Belgium).
11. 1 mL syringes and 0.50 × 16 mm needles.
12. Calipers for tumor measurements.

3. Methods

3.1. Preparation of Cartridges for Gene Gun Immunization

We prepare cartridges loaded with 0.5 mg of 1 μm gold particles coated with 1 μg of plasmid DNA (DNA loading rate 2 μg DNA/mg gold). To support the attachment of DNA/gold suspension to the Tefzel tubing, PVP is added to ethanol at a final concentration of 0.05 mg/mL.

1. In 1.5 mL microfuge tubes, weigh out 4 mg of PVP and 25 mg of 1 μm gold microcarriers.
2. Connect the Tubing Prep Station to the nitrogen tank using a nitrogen gas regulator.

3. Remove caps from the Tefzel tubing, insert the uncut tubing into the Tubing Prep Station, and allow the nitrogen gas to flow through the tubing at a rate of 0.3–0.4 LPM for at least 15 min.
4. Turn off the nitrogen using the knob on the Tubing Prep Station, remove the tubing from the Tubing Prep Station, and close its ends with the caps.
5. Add 200 μL of ethanol to PVP (20 mg/mL) and seal the cap of the tube with Parafilm. Mix thoroughly and repeatedly by vortexing. Dilute this solution to a final concentration of 0.05 mg/mL with ethanol—add 9 μL of 20 mg/mL PVP to 3.6 mL of ethanol per tubing to be coated.
6. Dilute 1 M spermidine to a concentration of 0.05 M with deionized water. Add 100 μL of 0.05 M spermidine to the measured gold particles. Vortex gold particles in spermidine for ~5 s and sonicate in an ultrasonic cleaner for ~10 s.
7. Add 50 μg of 1–1.1 μg/μL plasmid DNA to gold with spermidine and vortex for ~5 s.
8. Put a holder for 1.5-mL Eppendorf tubes on a vortexer and set up speed to 1,000–1,200 rpm.
9. While vortexing, add 100 μL of 1 M $CaCl_2$ dropwise to the center of the tube containing the mixture of gold, DNA, and spermidine.
10. Allow the mixture to precipitate by incubating for 10 min at room temperature. Shake the tube after the initial 5 min.
11. Spin the tube at 2,000×g in a microcentrifuge for ~15 s at room temperature and discard the supernatant.
12. Add 1 mL of fresh 100% ethanol, resuspend the pellet by vortexing briefly, spin for ~15 s at room temperature and discard the supernatant. Repeat twice step 12.
13. Add 180 μL of ethanol with PVP (from step 5), resuspend the pellet by repeated aspiration into a 200 μL tip with a cut off end, and transfer the suspension to a 15 mL conical tube. Repeat twice step 13.
14. Add 2.5 mL of ethanol with PVP to the 15-mL polypropylene conical tube with the DNA/gold suspension.
15. From the tubing dried with nitrogen, cut a ~70-cm piece and insert its one end into the silicone adaptor tubing fitted to a 10 mL syringe.
16. Vortex the DNA/gold suspension briefly and sonicate the tip of the 15 mL tube containing gold clumps for ~3 s. Then briefly vortex the tube again, remove its cap immediately, and quickly aspirate the suspension (without bubbles) into the Tefzel tubing leaving the ~3–4 cm ends unfilled.

17. Bring the loaded tubing to a horizontal position immediately and insert it into the Tubing Prep Station. Do not remove the syringe.
18. Allow the DNA/gold suspension to settle for 3 min.
19. Evenly draw off the ethanol into the syringe at a rate of ~2 cm/s (this should take about 30 s).
20. Turn the tubing 180°, let it stand for ~3 s, and start rotating the Tubing Prep Station. After 30–40 s, allow the nitrogen to flow through the rotating tubing at a rate of 0.35–0.4 LPM for 3 min.
21. Turn off tubing rotation and nitrogen flow. Remove the tubing from the Tubing Prep Station and verify even spreading of the gold (mark unevenly coated sections).
22. Cut off one uncoated or sparsely coated end of the tubing using scissors.
23. Place the labeled scintillation vial with a desiccant pellet into the Tubing Cutter and cut the evenly coated tubing. Seal the cap of the vial with Parafilm.
24. Store the vials containing cartridges coated with DNA/gold particles at 4°C.

3.2. Delivery of DNA Vaccines by the Helios Gene Gun

1. Mix shaving cream with water on a piece of tinfoil.
2. Coat the abdomen of the mouse with adulterated shaving cream using a soft toothbrush.
3. Carefully shave the abdomen with razor blades. Wipe the cut fur from the razor blades with cotton and wash the razor blades in a beaker filled with water.
4. Wipe the abdomen with cotton to remove the rest of the cut fur and shaving cream.
5. Connect the helium tank to the regulator and the gene gun.
6. Set discharge pressure to 400 psi. Use hearing protection.
7. Warm vials with cartridges coated with DNA/gold particles to room temperature. Fill a cartridge holder with cartridges and place it into the gene gun.
8. Restrain the mouse by the scruff of the neck and stretch the shaved target skin.
9. Touch the spacer perpendicularly to the target site and trigger the gene gun (see Note 4).

3.3. Administration of Tumor Cells and Observation of Tumor Growth

1. Aspirate the medium from the 15 cm tissue culture dish containing TC-1 or TC-1/A9 cells (Fig. 4, see Note 5).
2. Rinse cells with 5 mL of PBS.
3. Add 3 mL of trypsin/EDTA, cover the cells, and incubate for 3–4 min.

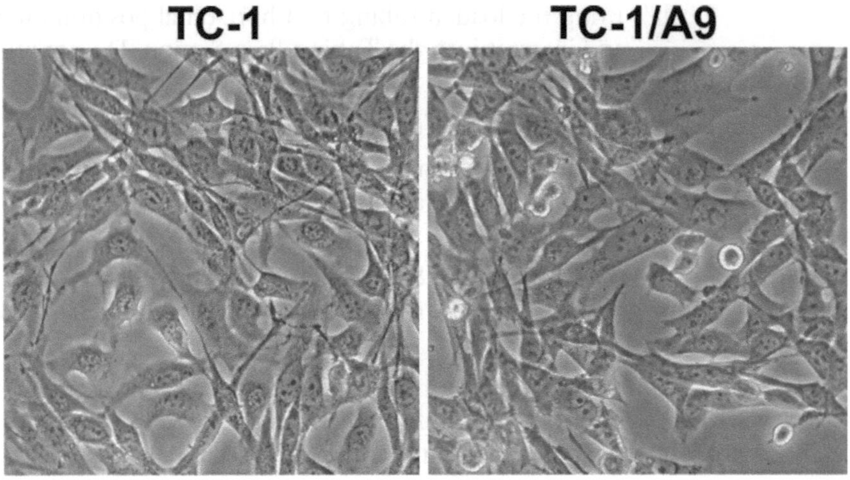

Fig. 4. Morphology of TC-1 and TC-1/A9 cells. Phase contrast; 10× objective.

4. Add 12 mL of cold complete medium, resuspend the cells, and transfer the suspension into a 50 mL polypropylene conical tube.
5. Spin the tube at $200 \times g$ for 10 min at 4°C and discard the supernatant.
6. Wash the cells twice with ~45 mL of cold PBS.
7. Resuspend the cells with 10 mL of PBS on ice and adjust the concentration of the live cells to 2×10^5/mL in a 15 mL polypropylene conical tube. Keep cells on ice.
8. Anesthetize mice by intraperitoneal injection of 250 μL of Hypnomidate.
9. Mark the anesthetized mice on the ears.
10. Place the animals on a cage lid and spray the interscapular area of each animal with disinfection solution.
11. Mix the cells by inverting the tube three to five times and fill a syringe—aspirate the cells into the syringe without a needle attached.
12. Lift the skin in the interscapular area by the thumb and forefinger and slowly inject 150 μL of the cell suspension subcutaneously.
13. Inspect tumor formation by palpation two to three times a week (see Note 6).
14. Measure tumor growth using calipers every 3–4 days (see Note 7).
15. Sacrifice mice when the largest diameter of the tumor exceeds 15 mm or when moribund.

4. Notes

1. Use only fresh 100% ethanol and do not prepare cartridges at high humidity. Close bottles and tubes with 100% ethanol immediately after use. To minimize absorption of moisture from the air, two persons should perform the procedure from step 12.
2. Spermidine is air-sensitive and is delivered under argon. Incubate spermidine at 37°C to melt it. Then, mix 1 mL of spermidine with 5.3 mL of water to prepare 1 M solution. Load 0.5-mL Eppendorf tubes with this solution, seal them with Parafilm, and store as single-use aliquots at −20°C for no more than 6 month.
3. Prepare plasmid DNA of high purity using a commercial Plasmid Prep Kit, adjust the concentration to 1–1.1 μg/μL with 10 mM Tris (pH 8.0), 1 mM EDTA (TE), aliquot the adjusted DNA (55 or 105 μg for coating of one or two tubings, respectively), and store at −20°C for repeated cartridge preparations.

   Although any mammalian expression vector can be used for immunization of mice, it is advisable to utilize a specialized vector achieving high levels of antigen production. The design of plasmid DNA vaccine vectors with respect to plasmid production, antigen expression, and vaccine efficacy and safety has been thoroughly reviewed (49).
4. After the shot, there should be a brown spot of about 5–7 mm in diameter in the target site (Fig. 5). No disruption of the skin should be macroscopically apparent, but microscopic changes in the epidermis can be observed 1 day after immunization and a scab develops in some mice (Fig. 6).

   The typical dose for gene gun immunization of mice is 1–2 μg of plasmid DNA, i.e., 1 or 2 shots in our conditions. When performing 2 shots, the target sites should not overlap and should be situated on opposite sides of the abdomen (Fig. 5). In repeated immunization(s), the target sites should not overlap with those from the previous immunization(s), as the skin regenerates after shots and residual scabs can hinder the penetration of DNA/gold particles into the skin, especially in cluster immunization.
5. The cells should be in logarithmic growth phase—seed 2×10^6 cells per a 15 cm tissue culture dish 2 days before preparation of cell suspension (we use the third passage after thawing the cells).

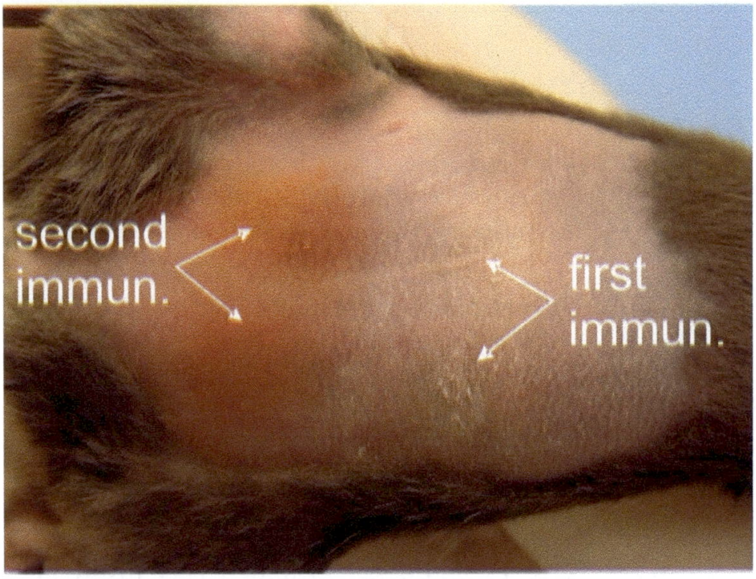

Fig. 5. Mouse abdomen after immunization by the gene gun. The mouse was twice immunized with two shots. The interval between immunizations was 1 week. The picture was taken 30 min after the second immunization.

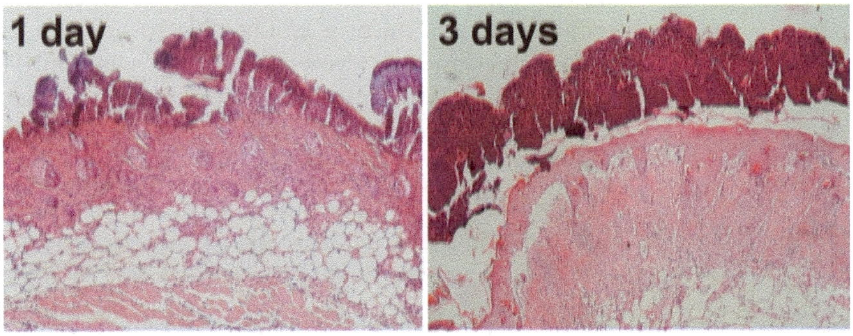

Fig. 6. Microscopic examination of the skin after gene gun shots. The skin was stained with hematoxylin and eosin 1 and 3 days after immunization. While not apparent 1 day after immunization, the epidermis was thickened 3 days after the shot. A scab was formed in both samples examined.

6. TC-1 and TC-1/A9 cells form palpable tumors usually in 12–18 and 9–12 days, respectively. We consider the tumor formed, when a nodule of 1–2 mm in diameter can be found and it is increased in the subsequent inspection. Tumors exceeding 10 mm in length often become necrotic in the center.

7. Measure three perpendicular diameters (a, b, c) and calculate the tumor volume using the formula $\pi abc/6$. Tumors developed from TC-1/A9 cells grow substantially faster than TC-1-derived tumors (Fig. 3).

Acknowledgement

Our immunization by the gene gun was supported by grants no. NT11541-4/2010 and NT13862-4/2012 from the Ministry of Health of the Czech Republic.

References

1. Parkin DM et al (2005) Global cancer statistics, 2002. CA Cancer J Clin 55:74–108
2. Bosch FX et al (1995) Prevalence of human papillomavirus in cervical cancer: a worldwide perspective. International biological study on cervical cancer (IBSCC) Study Group. J Natl Cancer Inst 87:796–802
3. Walboomers JM et al (1999) Human papillomavirus is a necessary cause of invasive cervical cancer worldwide. J Pathol 189:12–19
4. Moody CA, Laimins LA (2010) Human papillomavirus oncoproteins: pathways to transformation. Nat Rev Cancer 10:550–560
5. Chow LT, Broker TR, Steinberg BM (2010) The natural history of human papillomavirus infections of the mucosal epithelia. APMIS 118:422–449
6. Arbeit JM et al (1994) Progressive squamous epithelial neoplasia in K14-human papillomavirus type 16 transgenic mice. J Virol 68:4358–4368
7. Melero I et al (1997) Immunological ignorance of an E7-encoded cytolytic T-lymphocyte epitope in transgenic mice expressing the E7 and E6 oncogenes of human papillomavirus type 16. J Virol 71:3998–4004
8. Arbeit JM, Howley PM, Hanahan D (1996) Chronic estrogen-induced cervical and vaginal squamous carcinogenesis in human papillomavirus type 16 transgenic mice. Proc Natl Acad Sci USA 93:2930–2935
9. Elson DA et al (2000) Sensitivity of the cervical transformation zone to estrogen-induced squamous carcinogenesis. Cancer Res 60:1267–1275
10. Borchers A et al (1999) E7-specific cytotoxic T cell tolerance in HPV-transgenic mice. Arch Virol 144:1539–1556
11. Riezebos-Brilman A et al (2005) Induction of human papilloma virus E6/E7-specific cytotoxic T-lymphocyte activity in immune-tolerant, E6/E7-transgenic mice. Gene Ther 12:1410–1414
12. Souders NC (2007) Listeria-based vaccines can overcome tolerance by expanding low avidity CD8+ T cells capable of eradicating a solid tumor in a transgenic mouse model of cancer. Cancer Immun 7:2
13. Sewell DA, Pan ZK, Paterson Y (2008) Listeria-based HPV-16 E7 vaccines limit autochthonous tumor growth in a transgenic mouse model for HPV-16 transformed tumors. Vaccine 26:5315–5320
14. Bubenik J (2002) Animal models for development of therapeutic HPV16 vaccines. Int J Oncol 20:207–212
15. Lin KY et al (1996) Treatment of established tumors with a novel vaccine that enhances major histocompatibility class II presentation of tumor antigen. Cancer Res 56:21–26
16. Ji H et al (1998) Antigen-specific immunotherapy for murine lung metastatic tumors expressing human papillomavirus type 16 E7 oncoprotein. Int J Cancer 78:41–45
17. Swann JB, Smyth MJ (2007) Immune surveillance of tumors. J Clin Invest 117: 1137–1146
18. Pettit SJ et al (2000) Immune selection in neoplasia: towards a microevolutionary model of cancer development. Br J Cancer 82: 1900–1906
19. Pawelec G et al (2000) Escape mechanisms in tumor immunity: a year 2000 update. Crit Rev Oncog 11:97–133
20. Pawelec G (2004) Tumour escape: antitumour effectors too much of a good thing? Cancer Immunol Immunother 53:262–274
21. Kanodia S, Fahey LM, Kast WM (2007) Mechanisms used by human papillomaviruses to escape the host immune response. Curr Cancer Drug Targets 7:79–89
22. Keating PJ et al (1995) Frequency of down-regulation of individual HLA-A and -B alleles in cervical carcinomas in relation to TAP-1 expression. Br J Cancer 72:405–411
23. Koopman LA et al (2000) Multiple genetic alterations cause frequent and heterogeneous human histocompatibility leukocyte antigen class I loss in cervical cancer. J Exp Med 191:961–976

24. Cromme FV et al (1994) Differences in MHC and TAP-1 expression in cervical cancer lymph node metastases as compared with the primary tumours. Br J Cancer 69:1176–1181
25. Connor ME et al (1993) Evaluation of multiple biological parameters in cervical carcinoma: high macrophage infiltration in HPV-associated tumours. Int J Gynecol Cancer 3:103–109
26. Hicklin DJ, Marincola FM, Ferrone S (1999) HLA class I antigen downregulation in human cancers: T-cell immunotherapy revives an old story. Mol Med Today 5:178–186
27. Tait BD (2000) HLA class I expression on human cancer cells. Implications for effective immunotherapy. Hum Immunol 61: 158–165
28. Seliger B, Maeurer MJ, Ferrone S (2000) Antigen-processing machinery breakdown and tumor growth. Immunol Today 21:455–464
29. Cromme FV et al (1994) Loss of transporter protein, encoded by the TAP-1 gene, is highly correlated with loss of HLA expression in cervical carcinomas. J Exp Med 179:335–340
30. Smahel M et al (2003) Immunisation with modified HPV16 E7 genes against mouse oncogenic TC-1 cell sublines with downregulated expression of MHC class I molecules. Vaccine 21:1125–1136
31. Cheng WF et al (2003) CD8+ T cells, NK cells and IFN-gamma are important for control of tumor with downregulated MHC class I expression by DNA vaccination. Gene Ther 10:1311–1320
32. Manning J et al (2008) Induction of MHC class I molecule cell surface expression and epigenetic activation of antigen-processing machinery components in a murine model for human papilloma virus 16-associated tumours. Immunology 123:218–227
33. Lin KY et al (2007) Ectopic expression of vascular cell adhesion molecule-1 as a new mechanism for tumor immune evasion. Cancer Res 67:1832–1841
34. Cheng WF et al (2005) Antigen-specific CD8+ T lymphocytes generated from a DNA vaccine control tumors through the Fas-FasL pathway. Mol Ther 12:960–968
35. Smahel M et al (2005) Characterization of cell lines derived from tumors induced by TC-1 cells in mice preimmunized against HPV16 E7 oncoprotein. Int J Oncol 27:731–742
36. Smahel M et al (2008) Mutation in the immunodominant epitope of the HPV16 E7 oncoprotein as a mechanism of tumor escape. Cancer Immunol Immunother 57:823–831
37. Lin K et al (2010) Therapeutic HPV DNA vaccines. Immunol Res 47:86–112
38. Boursnell ME et al (1996) Construction and characterisation of a recombinant vaccinia virus expressing human papillomavirus proteins for immunotherapy of cervical cancer. Vaccine 14:1485–1494
39. Smahel M et al (2001) Modified HPV16 E7 genes as DNA vaccine against E7-containing oncogenic cells. Virology 281:231–238
40. Cassetti MC et al (2004) Antitumor efficacy of Venezuelan equine encephalitis virus replicon particles encoding mutated HPV16 E6 and E7 genes. Vaccine 22:520–527
41. Polakova I et al (2010) DNA vaccine against human papillomavirus type 16: modifications of the E6 oncogene. Vaccine 28:1506–1513
42. Osen W et al (2001) A DNA vaccine based on a shuffled E7 oncogene of the human papillomavirus type 16 (HPV 16) induces E7-specific cytotoxic T cells but lacks transforming activity. Vaccine 19:4276–4286
43. Ohlschlager P et al (2006) An improved rearranged human papillomavirus type 16 E7 DNA vaccine candidate (HPV-16 E7SH) induces an E7 wildtype-specific T cell response. Vaccine 24:2880–2893
44. Oosterhuis K et al (2011) Preclinical development of highly effective and safe DNA vaccines directed against HPV 16 E6 and E7. Int J Cancer 129:397–406
45. Smahel M et al (2004) Enhancement of immunogenicity of HPV16 E7 oncogene by fusion with E. coli beta-glucuronidase. J Gene Med 6:1092–1101
46. Pokorna D et al (2005) Combined immunization with fusion genes of mutated E7 gene of human papillomavirus type 16 did not enhance antitumor effect. J Gene Med 7:696–707
47. Bins AD et al (2005) A rapid and potent DNA vaccination strategy defined by in vivo monitoring of antigen expression. Nat Med 11:899–904
48. Peng S et al (2008) Cluster intradermal DNA vaccination rapidly induces E7-specific CD8+ T-cell immune responses leading to therapeutic antitumor effects. Gene Ther 15:1156–1166
49. Williams JA, Carnes AE, Hodgson CP (2009) Plasmid DNA vaccine vector design: impact on efficacy, safety and upstream production. Biotechnol Adv 27:353–370
50. Samorski R, Gissmann L, Osen W (2006) Codon optimized expression of HPV 16 E6

renders target cells susceptible to E6-specific CTL recognition. Immunol Lett 107:41–49
51. Janouskova O, Sima P, Kunke D (2003) Combined suicide gene and immunostimulatory gene therapy using AAV-mediated gene transfer to HPV-16 transformed mouse cell: decrease of oncogenicity and induction of protection. Int J Oncol 22:569–577
52. Reinis M et al (2007) Immunization with MHC class I-negative but not -positive HPV16-associated tumour cells inhibits growth of MHC class I-negative tumours. Int J Oncol 30:1011–1017

Chapter 26

Efficiency of Biolistic DNA Vaccination in Experimental Type I Allergy

Verena Raker, Joachim Maxeiner, Angelika B. Reske-Kunz, and Stephan Sudowe

Abstract

Gene gun-mediated delivery of allergen-encoding plasmid DNA has been in focus for many years now as being a needle-free alternative to the protein-based desensitization regimen used in specific immunotherapy. Biolistic immunization with the Helios gene gun has proven to be potent in the induction of antigen-specific $CD4^+$ and $CD8^+$ T cells. Here we describe biolistic vaccination in experimental mouse models of IgE-mediated type I allergy as well as allergen-induced airway inflammation.

Key words: Allergy, DNA vaccination, Gene gun, IgE, Allergic airway inflammation

1. Introduction

The crucial event in the pathogenesis of type I allergy (atopic disease) is excessive production of allergen-specific IgE antibodies, promoted by the elicitation of a distinct Th2 immune response. Since Raz et al. (1) reported for the first time that intradermal injection of antigen-encoding DNA prior to sensitization efficiently suppressed specific IgE immune responses, genetic immunization has been in focus as an attractive alternative to specific immunotherapy (SIT) which until now represents the only causative treatment of allergy. Various routes of DNA administration have been tested to modulate allergic sensitization in animal models of type I allergies (reviewed in ref. (2)) including bombardment with DNA-coated gold particles using the helium-powered gene gun. In contrast to intramuscular or intradermal needle injection, biolistic

Stephan Sudowe and Angelika B. Reske-Kunz (eds.), *Biolistic DNA Delivery: Methods and Protocols*,
Methods in Molecular Biology, vol. 940, DOI 10.1007/978-1-62703-110-3_26, © Springer Science+Business Media, LLC 2013

transfection of the skin allows for the direct introduction of plasmids into cutaneous antigen-presenting cells (epidermal Langerhans cells, dermal dendritic cells) and therefore requires 10–100 times lesser amounts of DNA to elicit an immune reaction (3, 4). We and others have shown that gene gun-mediated immunization using antigen-encoding plasmid DNA activates CD4⁺ T cells, which in addition to interleukin-(IL-)4 and IL-5 release high amounts of interferon-(IFN)-γ (1, 5–8). In addition, considerable numbers of antigen-specific IFN-γ–producing CD8⁺ effector T cells were demonstrated in gene gun–immunized mice (5, 6, 9). IFN-γ producing CD4⁺ as well as CD8⁺ T cell populations are known to play important roles in the inhibition of IgE responses. Consequently, IgE production induced by sensitization with β-galactosidase (βGal) protein as model antigen was significantly suppressed by preceding biolistic immunization with βGal-encoding plasmid vectors (Fig. 1) (6, 10, 11). Moreover, eosinophilic infiltration into the lungs after intranasal challenge of βGal-sensitized mice was strongly reduced by prophylactic application of βGal-encoding plasmids in an experimental model of allergic airway inflammation (Fig. 2) (12). Although biolistic transfection improved Th2-associated pathology, antigen-induced airway hyperreactivity (AHR) was not alleviated by biolistic transfection (Fig. 3), probably due to immigration of neutrophils and/or IFN-γ producing T cells into the lung (12). In conclusion, gene gun-mediated biolistic transfection represents an attractive alternative therapeutic approach to genetic immunization by intramuscular or intradermal needle injection.

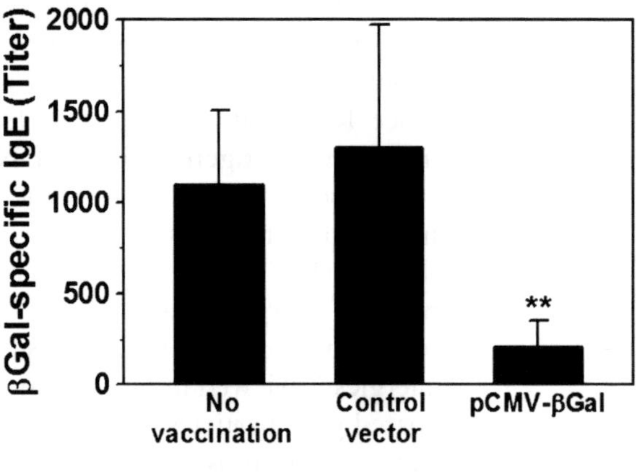

Fig. 1. Inhibition of IgE production by prophylactic gene gun-mediated vaccination. BALB/c mice ($n=4$) were vaccinated every week by a total of five biolistic transfections with pCMV-βGal or a non-encoding control vector or were left unvaccinated. Subsequently mice were sensitized by six intraperitoneal injections of 1 μg βGal adsorbed to aluminum hydroxide as adjuvant at intervals of 2 weeks. Four days after the sixth application of βGal, levels of βGal-specific IgE in sera were determined by ELISA. Data represent mean βGal-specific IgE titer ± SD. $^{**}P<0.01$.

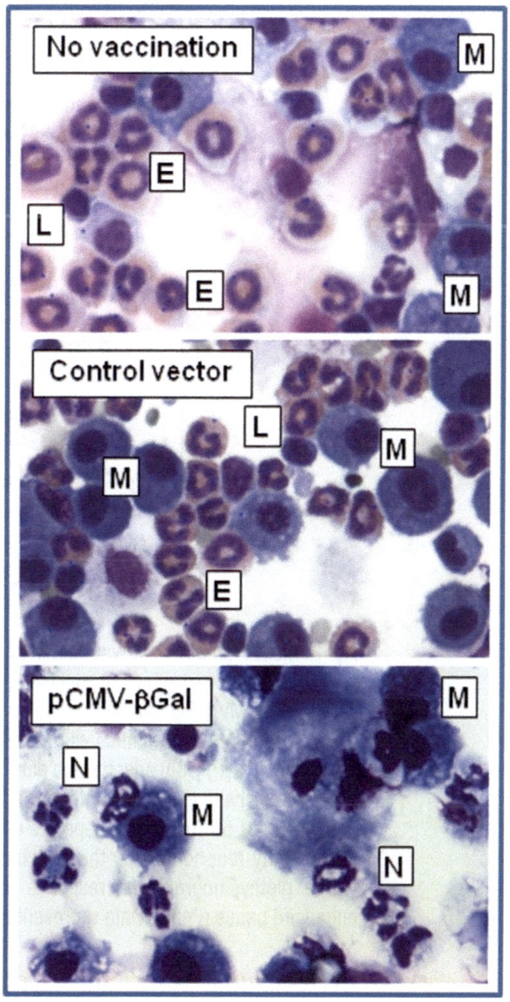

Fig. 2. Reduction of pulmonary eosinophilic infiltration by prophylactic gene gun-mediated vaccination. BALB/c mice ($n=4$) were vaccinated every week by a total of three biolistic transfections with pCMV-βGal or a non-encoding control vector. Subsequently vaccinated mice as well as untreated control mice were sensitized at a 10 day interval by two intraperitoneal injections of 1 μg βGal adsorbed to aluminum hydroxide as adjuvant. One week later all mice were challenged with antigen on three consecutive days by intranasal application of 50 μg βGal. Twenty-four hour after the last challenge BAL was performed and BAL cells were differentiated using Diff-Quick staining. *M* monocyte/macrophage; *L* lymphocyte; *E* eosinophil; *N* neutrophil.

2. Materials

Unless otherwise indicated use ultrapure water for preparation of the solutions (Milli-Q-Integral System, Millipore, Schwalbach, Germany) and reagents of analytical grade.

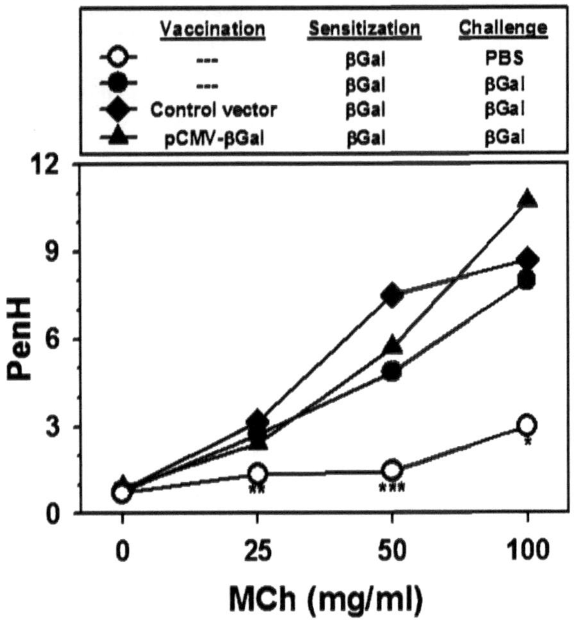

Fig. 3. Measurement of airway reactivity after prophylactic gene gun-mediated vaccination. BALB/c mice ($n=4$) were vaccinated every week by a total of three biolistic transfections with pCMV-βGal or a non-encoding control vector. Subsequently vaccinated mice as well as untreated control mice were sensitized at a 10 day interval by two intraperitoneal injections of 1 μg βGal adsorbed to aluminum hydroxide as adjuvant. One week later (all) mice were challenged with antigen or saline on three consecutive days by intranasal application of 50 μg βGal, a separate control group of sensitized mice was treated with PBS. Twenty-four hours after the last challenge airway reactivity was measured by noninvasive plethysmography in response to increasing doses of methacholine (MCh) as enhanced pause (Penh). Data represent means. *$P<0.05$, **$P<0.01$, ***$P<0.001$.

2.1. Mice

1. Female BALB/c mice (see Note 1).
2. BALB/c mice were vaccinated three to five times in weekly intervals with a plasmid encoding βGal as model antigen under control of the ubiquitously active CMV promoter (pCMV-βGal) (9) (see also Chapter 17 in this volume).

2.2. Mouse Model of IgE-Mediated Type I Allergy

1. Recombinant βGal from *Escherichia coli* (Sigma-Aldrich, Deisenhofen, Germany) as antigen. Prepare a 1 mg/mL βGal stock solution in phosphate buffered saline (PBS).
2. Aluminum hydroxide (e.g., Imject® Alum, Thermo Fisher Scientific, Rockford, IL, USA) as adjuvant.
3. 15 mL polypropylene tubes.
4. 1 mL syringe with 0.7 mm cannula.
5. 96 well ELISA microtiter plates (see Note 2) and ELISA plate seals.
6. Precision microplate reader.
7. Microplate washer.
8. PBS: 137.5 mM NaCl, 10 mM NaH_2PO_4, pH 7.2.

9. Coating Buffer: 0.1 M $NaHCO_3$, pH 8.2.
10. Detection antibody: monoclonal rat anti-mouse-IgE (clone EM-95.3 (13)).
11. Secondary antibody: biotin-conjugated mouse anti-rat-IgG (e.g., Dianova, Hamburg, Germany).
12. Strepatavidin-conjugated horseradish peroxidase (e.g., ExtrAvidin-Peroxidase, Sigma-Aldrich).
13. Dilution buffer: PBS with 1% bovine serum albumin.
14. Washing buffer: PBS with 0.05% Tween 20.
15. Substrate buffer: 0.2 M NaH_2PO_4, 0.1 M sodium citrate, pH 5.0.
16. Ortho-phenylenediamine (OPD) (Sigma-Aldrich).
17. 30% H_2O_2.
18. Stop solution: 1 M H_2SO_4.

2.3. Mouse Model of Allergen-Induced Airway Inflammation

2.3.1. Intranasal Challenge with βGal

1. Recombinant βGal (Sigma-Aldrich). Prepare a 1 mg/mL βGal stock solution in PBS.
2. 1.5 mL Eppendorf tubes.
3. Forene® (Abbott GmbH & Co. KG, Wiesbaden, Germany) as narcotic agent (active substance: isoflurane).
4. Exsiccator.

2.3.2. Bronchoalveolar Lavage

1. 1 mL syringe with 0.7 mm cannula.
2. Intravenous catheter Vasofix Safety® (Braun, Melsungen, Germany).
3. 1.5 mL LoBind tubes (Eppendorf, Hamburg, Germany).
4. PBS: 137.5 mM NaCl, 10 mM NaH_2PO_4, pH 7.2.
5. Bronchoalveolar lavage (BAL) solution: PBS with 10% fetal calf serum.
6. Gey's lysis buffer: 10 mM $KHCO_3$, 155 mM NH_4Cl, 100 μM EDTA, pH 7.4; sterile by filtration (0.2 μm).
7. Microscope glass slides with frosted ends and 16 mm cover glasses.
8. Cytospin centrifuge with cell funnels and filters.
9. 50 mL polypropylene tubes.
10. Diff-Quick staining kit (Siemens Healthcare Diagnostics, Eschborn, Germany).
11. Entellan® rapid embedding agent for microscopy (Merck, Darmstadt, Germany).
12. Hemocytometer.
13. Stereo microscope with digital camera unit.
14. Trypan blue.
15. Foamed polystyrene plate.

16. Canullas 0.7 mm.
17. Tweezers.
18. Anatomical scissors.

2.3.3. Noninvasive Measurement of Lung Function

1. Four-chamber whole-body plethysmography system (Buxco Research Systems, Wilmington, NC, USA).
2. Methacholine (Acetyl-β-methylcholine chloride) (Sigma-Aldrich).
3. PBS: 137.5 mM NaCl, 10 mM NaH_2PO_4, pH 7.2.
4. Nebulizer.
5. BioSystem XA Software (Buxco Research Systems).

2.3.4. Invasive Measurement of Lung Function

1. Four-chamber resistance and compliance (RC) system (Buxco Research Systems).
2. Methacholine (Acetyl-β-methylcholine chloride) (Sigma-Aldrich).
3. PBS: 137.5 mM NaCl, 10 mM NaH_2PO_4, pH7.2.
4. Narcoren® (Merial GmbH, Halbergmos, Germany) as narcotic agent (active substance: pentobarbital). Prepare 1:5 Narcoren® stock solution in PBS.
5. 2 mL Eppendorf tubes.
6. Tweezers.
7. Anatomical scissors.
8. Scalpel.
9. Spatula.
10. Medical yarn.
11. 18 gauche tracheal tube.
12. Esophagus tube.
13. Ventilation pump.
14. BioSystem XA Software (Buxco Research Systems).

3. Methods

3.1. Mouse Model of IgE-Mediated Type I Allergy

3.1.1. Treatment Protocol

1. *Sensitization*: Dilute βGal stock solution to 10 μg/mL in PBS. Thoroughly mix βGal working solution with an equal volume of aluminum hydroxide. Immunize gene gun-vaccinated as well as naïve control mice up to five times in intervals of 2 weeks by intraperitoneal injection of 200 μL of βGal/adjuvant mixture (see Note 3) into the lower right quadrant of the abdomen (10, 11) (see Note 4).

2. Collect blood various times during the vaccination and sensitization period by puncture of the retro-orbital plexus or by tail vain bleeding (see Note 5). Remove clot of blood after 60 min at room temperature (RT) and centrifuge the cell suspension for 5 min at 10,000 × *g*. Store the supernatant as serum in safe lock tubes at −20°C until thawing for determination of βGal-specific IgE titers (see Note 6).

3.1.2. Detection of βGal-Specific IgE in Sera by ELISA

1. Dilute βGal stock solution to 5 μg/mL in coating buffer. Add 100 μL of antigen solution to the wells of a 96-well ELISA microtiter plate. Keep plate at 4°C overnight.
2. Wash plate with washing buffer using the ELISA washer (see Note 7).
3. Add 200 μL of dilution buffer to each well for 1 h at RT to block residual binding sites.
4. Wash plate with washing buffer.
5. Add 100 μL of dilution buffer to each well. Prepare serial dilutions (1:2) of the serum: pipet 100 μL of serum into the first well and after careful mixing transfer 100 μL of solution to the next well. Repeat seven times. To evaluate background reaction by the reagents, add diluent buffer only to separate antigen-coated wells as blank control. After sealing the plate, incubate 2 h at RT or overnight at 4°C.
6. Wash plate twice with washing buffer.
7. Add 100 μL of detection rat anti-mouse IgE antibody (1:1,000 in dilution buffer) to the wells and incubate plate for 45 min at RT.
8. Wash plate twice with washing buffer.
9. Add 100 μL of secondary biotin-conjugated mouse anti-rat-IgG antibody (1 μg/mL in dilution buffer) to each well and incubate plate for 45 min at RT.
10. Wash plate twice with washing buffer.
11. Add 100 μL of strepatavidin-conjugated horseradish peroxidase (1:2,000 in dilution buffer) as enzyme (see Note 8) to each well for 30 min at RT.
12. Wash plate four times with washing buffer.
13. Prepare OPD solution (1 mg/mL) in substrate buffer. Immediately before use add 1 μL 30% H_2O_2/mL OPD solution. Apply 100 μL of the mixture to the wells until color changes into yellow.
14. Stop the enzymatic reaction by adding 100 μL of 1 M H_2SO_4 to each well (see Note 9).
15. Measure optical density (OD) of the samples at a wave length of 490 nm using the Microplate reader.

16. Subtract OD of blank wells from OD of sample wells. Use linear regression analysis to calculate the antibody titer of the serum as the reciprocal serum dilution yielding an absorbance value of OD = 0.2.

3.2. Mouse Model of Allergen-Induced Airway Inflammation

3.2.1. Treatment Protocol

1. *Sensitization*: Immunize gene gun-vaccinated as well as naïve control mice three times in weekly intervals by intraperitoneal injection of βGal as described in Subheading 3.1.1, step 1.
2. *Challenge*: Anesthetize mice 7 days after the last sensitization by sedation with Forene® in an exsiccator. Sprinkle paper towel on the bottom of the exsiccator with the narcotic and place a mouse in the tank until it is quiescent (see Notes 10 and 11). Challenge the mice by intranasal instillation of 50 μL βGal stock solution (see Notes 12 and 13). Control mice receive PBS alone intranasally (see Note 14). Repeat the procedure on the two subsequent days.

3.2.2. Differential Staining of BAL Cells

Perform a BAL to analyze the cells infiltrating the lung after intranasal challenge, and differentiate the cells recovered according to their morphology and staining behavior.

1. Kill the mouse by intraperitoneal injection of a lethal dose of 200 μL of Narcoren® working solution (1:5 in PBS).
2. Place the mouse in a supine position on a foamed polystyrene plate and fix the extremities with cannulas to obtain a slight tension on the thorax.
3. Prepare the thorax by opening the skin of the mouse with an Y-cut, starting ventral of the body and ending at the forelimbs. Fix the skin with cannulas on the plate.
4. Open the thorax by cutting the dermal layer directly below the sternum and further along the costal arch.
5. Carefully retain the liver from the diaphragm. Perforate the diaphragm without injuring the lobes of the lungs by cutting a hole in the diaphragm where the heart is visual through the diaphragm (marked by a darker spot).
6. Carefully prepare the trachea by removing the adjacent tissue and dermal layer (see Note 15).
7. Carefully perforate the trachea in the upper quarter using a 0.7 mm canulla.
8. Place an intravenous catheter (without needle) into the perforated hole and steadily inject 1 mL of BAL solution into the lung.
9. Aspirate the BAL solution in a continuous procedure using the syringe.
10. Determine the volume of the BAL solution recovered from the lung (see Note 16).

11. Centrifuge the BAL solution (10 min, 300×*g*, 4°C). Collect the supernatant as BAL fluid (BALF) in 1.5 mL LoBind tubes and immediately freeze at −20°C (see Note 17).
12. If lysis of erythrocytes is necessary (see Note 18), resuspend the pellet in 0.5 mL Gey's lysis buffer and incubate 45 s. Subsequently, add 10 mL PBS, centrifuge cells as outlined in step 11 and discard the supernatant. Wash the BAL cells two additional times as described.
13. Resuspend the cell pellet in 0.5 mL PBS. Mix an aliquot of the cell suspension 1:10 with trypan blue and determine the total living cell number under a stereo microscope using a hemocytometer by trypan blue exclusion (see Note 19).
14. Add 2 mL of PBS and centrifuge the cell suspension (10 min, 300×*g*, 4°C). Discard the supernatant and adjust the cell number with PBS to 1×10^7 cells/mL.
15. Cytospin (5 min, 500 Upm) 5×10^5 cells in a volume of 50 μL onto a microscope glass slide using the cytospin centrifuge according to the manufacturer's instructions (see Note 20).
16. Fill the three Diff-Quick staining solutions into separate 50 mL polypropylene tubes and slowly dip the microslide into each solution five times as follows:
    (a) Diff-Quick Fixation.
    (b) Diff-Quick I.
    (c) Diff-Quick II.
    (d) Rinse the slide with water.
17. Air dry smear of cells at RT.
18. Add one drop of Entellan® rapid embedding agent onto the smear and apply a suitable cover glass on this position.
19. Differentiate at least 300 BAL cells per slide based on conventional criteria like morphology and staining behavior (see Note 21) using a microscope with digital camera unit.
20. Calculate the relative frequencies of the various cell populations in the BAL (monocytes/macrophages, lymphocytes, eosinophils, neutrophils). To determine the absolute cell number (in cells/ml BAL), multiply the percentage of the respective cell population by the total cell number (extrapolated to 1 mL) as determined before.

3.2.3. Noninvasive Measurement of Lung Function

To quantify the degree of airway reactivity after intranasal challenge with the antigen, noninvasive whole-body plethysmography is a quick and easy-to-handle method to measure the lung function in mice. The normally breathing mice are not anesthetized and the system provides the chance for repetitive and longitudinal measurements like kinetic experiments. No tracheal surgery is necessary. Four mice can be analyzed at the same time.

1. Calibrate the noninvasive whole-body plethysmography system according to the manufacturer's guidelines (to calibrate the system has the outcome to validate all measurements and to reach a proper and consistent dynamic range).
2. Prepare a 50 mg/mL stock solution of methacholine in PBS and serially dilute (1:2) the stock solution (25, 12.5, 6.25 mg/mL).
3. Place individual mice into each of the four chambers of the plethysmography system and start the measurement protocol.
4. Start the nebulizer to deliver an aerosol of PBS to each chamber for 30 s.
5. Record the baseline airway reactivity as enhanced pause (Penh) response for 5 min (see Note 22).
6. Start the nebulizer to deliver an aerosol of the lowest concentration of methacholine (6.25 mg/mL) to each chamber for 30 s.
7. Record the change of lung function by provocation to aerosolized methacholine as enhanced pause (Penh) response for 5 min (see Note 22).
8. Recovery time (3 min).
9. Repeat steps 6–8 with increasing doses of methacholine (12.5–50 mg/mL).
10. End of measurement.
11. Analyze the raw data with the BioSystem XA Software. Calculate the fold increase of Penh measured for each concentration of methacholine in relation to baseline Penh (see Note 23).

3.2.4. Invasive Measurement of Lung Function

The invasive measurement of the lung function is a very sensitive and effective method, characterized by controlled ventilation, bypassing the upper airways and a local delivery of the aerosolized bronchoconstrictor via a tracheal tube (see Note 24). It is a technically demanding method that requires anesthetization and tracheal surgery of the mice. Therefore no repetitive measurements are possible.

1. Calibrate each of the four chambers of the RC system according to the manufacturer's guidelines.
2. Prepare a 50 mg/mL stock solution of methacholine in PBS and serially dilute (1:2) the stock solution (25, 12.5, 6.25, 3.125 mg/mL).
3. Prepare Narcoren® working solution for anesthetization of mice by thoroughly mixing Narcoren® stock solution and PBS (1:5) in a 2 mL Eppendorf tube on a shaker.
4. Depending on the body weight of the mouse inject an appropriate volume of Narcoren® working solution intraperitoneally.

The recommended dose of Narcoren® for anesthetization of mice is 75 mg/kg body weight.

5. Pinch the foot of the mouse with tweezers to check the surgical tolerance of the mouse.
6. Lift the coat in the upper part of the cervix with the tweezers and cut off a piece of coat with the anatomical scissors, until a coat-free circle of skin can be seen.
7. Perform a vertical cut of the skin with the scalpel above the thymus and use two tweezers to carefully pull the tissue apart until the esophagus and the trachea are visible.
8. Remove the skin and muscle tissue covering the trachea using surgical scissors, two tweezers and a scalpel.
9. Carefully pull up the trachea with the tweezers and put the flat part of the spatula underneath the trachea but on top of the esophagus (see Note 25). Insert a medical yarn under the trachea, which will subsequently be used to tie the trachea to the tracheal tube.
10. Perform a tiny horizontal cut between two tracheal rings
11. Place the 18 gauge tracheal tube into the cut and carefully push it in the airways until you feel a resistance. Then pull it a few millimeters back out (see Note 26).
12. Pull up both ends of the medical yarn with the tweezers to bind the tracheal tube tight to the trachea with a surgical knot.
13. Place the mouse into a chamber of the RC system. Depending on the measurement method an esophagus tube, connected to a transducer, is placed well.
14. Connect the implanted tracheal tube to the free adjustable ventilation pump and close the chamber to avoid environmental influences.
15. Start the following study protocol:
    (a) Record the baseline resistance and compliance for 60 s.
    (b) Deliver aerosolized methacholine at the lowest concentration (3.125 mg/mL) for 30 s.
    (c) Record the change of resistance and compliance for 5 min.
    (d) Repeat step 16b, c with increasing doses of methacholine (6.25, 12.5, 25, 50 mg/mL).
16. Analyze the raw data with the BioSystem XA Software. Calculate the fold increase of resistance and compliance measured for each concentration of methacholine in relation to baseline resistance and compliance (see Note 23).

4. Notes

1. BALB/c mice are known to be high responders in terms of IgE production and are therefore recommended for type I allergy models. Female mice tend to produce higher amounts of immunoglobulins after sensitization. Follow the principles of laboratory animal care (NIH publication no. 85–23, revised 1985). Elicit approval of the experiments by the local ethics commission.
2. ELISA microtiter plates with high binding polystyrene surface (e.g., Microlon ELISA plates, Greiner bio-one, Frickenhausen, Germany) are recommended.
3. The quantity of βGal injected per application should account for 1 μg.
4. Alternatively, mice can be sensitized without adjuvant by repeated subcutaneous application of 10 μg βGal dissolved in 200 μL PBS in weekly intervals (14). Subcutaneous injection is performed at the scruff of the neck and is verified by the observation of a fluid bubble forming under the skin during injection.
5. Minimum volume of blood sample should be 0.5 mL. Puncture of the retro-orbital plexus usually yields more blood than tail vain bleeding.
6. Usually substantial titers of specific IgE in unvaccinated control mice can be determined at the earliest after the third application of βGal. IgE production drastically increases in mice with progressing time and with number of immunizations.
7. Remove the residuary washing buffer after each wash step by firmly blotting the plate on paper towel to avoid background substrate metabolism.
8. To avoid unspecific metabolism of the substrate, pipet the enzyme solution without touching the walls of the well.
9. Reaction should be stopped when the substrate solution in the blank control starts to become yellow. After addition of the stop solution color changes into orange/brown.
10. Because isoflurane is volatile and potentially harmful, always perform sedation under the laboratory hood, otherwise the indoor air may be contaminated.
11. Keep an eye on the mouse during the whole sedation procedure, which usually lasts 30–45 s. Be aware that the mouse is vital before intranasal application of antigen.
12. The quantity of βGal applicated by intranasal installation should account for 50 μg.

13. For intranasal treatment, position the mouse supine and fix the head between two fingers. Be aware that the mouse breathes normally. Continuously pipet 50 μL of the antigen solution drop by drop into one nose hole. Stop applying the solution when the mouse is breathing out to prevent the formation of bubbles and thus to avoid incomplete inoculation of the antigen.
14. We recommend the use of PBS-treated control mice to assess antigen-independent side effects of the intranasal application itself.
15. Take care that arteries are not disrupted by preparation of the trachea, because massive bleeding might hamper the differentiation of BAL cells.
16. Usually 0.8–0.9 mL of BAL solution can be recovered from the lung.
17. We routinely determine cytokine levels (IL-5, IL-13, IFN-γ) in the BALF using ELISA (12).
18. Massive bleeding during preparation of the trachea often leads to an excess of erythrocytes in the BAL. If the pellet has a red color, erythrocytes have to be lysed.
19. Living cells, which exclude trypan blue, appear colorless under the microscope, while dead cells, which incorporate trypan blue, appear blue.
20. 5×10^5 cells is the maximum number of BAL cells that should be cytospun onto the glass slide. If the cells are too densely located on the glass slide, try different cell numbers for centrifugation to ensure optimal identification of cells.
21. The various cell populations in the BAL can be differentiated as follows by conventional criteria such as morphology and staining behavior (see Fig. 2): monocytes/macrophages appear as large cells with violet, sometimes indented nucleus and light blue-gray cytoplasm, lymphocytes appear as small cells with a round, dense and dark-violet nucleus and a thin rim of peripheral light-blue cytoplasm, the polymorphonuclear granulocytes (with lobulated purple nucleus) are either eosinophils (with cytoplasm containing granules that are stained brightly pinkish-red) or neutrophils (with unstained cytoplasm).
22. Data are reported as the mean value of 30 values of Penh recorded every 10 s for 5 min.
23. To determine AHR, calculate provocative concentration value PC_{100} for every individual mouse by interpolation of the dose–response curve as the methacholine dose causing 100% increase above baseline reactivity.

24. The invasive measurement of airway resistance and compliance is the recommended standard for the determination of airway reactivity to be published in scientific journals.
25. It is important for the measurement to separate the trachea from the esophagus.
26. It is essential that this step is performed correctly to ensure delivery of the methacholine aerosol into the lower airways.

References

1. Raz E et al (1996) Preferential induction of a Th1 immune response and inhibition of specific IgE antibody formation by plasmid DNA immunization. Proc Natl Acad Sci USA 93:5141–5145
2. Chua KY, Kuo IC, Huang CH (2009) DNA vaccines for the prevention and treatment of allergy. Curr Opin Allergy Clin Immunol 9:50–54
3. Fynan EF et al (1993) DNA vaccines: protective immunizations by parenteral, mucosal, and gene-gun inoculations. Proc Natl Acad Sci USA 90:11478–11482
4. Pertmer TM et al (1995) Gene gun-based nucleic acid immunization: elicitation of humoral and cytotoxic T lymphocyte responses following epidermal delivery of nanogram quantities of DNA. Vaccine 13:1427–1430
5. Sudowe S et al (2003) Transcriptional targeting of dendritic cells in gene gun-mediated DNA immunization favors the induction of type 1 immune responses. Mol Ther 8:567–575
6. Sudowe S et al (2009) Uptake and presentation of exogenous antigen and presentation of endogenously produced antigen by skin dendritic cells represent equivalent pathways for the priming of cellular immune responses following biolistic DNA immunization. Immunology 128:e193–e205
7. Pertmer TM, Roberts TR, Haynes JR (1996) Influenza virus nucleoprotein- specific immunoglobulin G subclass and cytokine responses elicited by DNA vaccination are dependent on the route of vector DNA delivery. J Virol 70:6119–6125
8. Yoshida A et al (2000) Advantage of gene gun-mediated over intramuscular inoculation of plasmid DNA vaccine in reproducible induction of specific immune responses. Vaccine 18:1725–1729
9. Ross R et al (2003) Transcriptional targeting of dendritic cells for gene therapy using the promoter of the cytoskeletal protein fascin. Gene Ther 10:1035–1040
10. Ludwig-Portugall I et al (2004) Prevention of long-term IgE antibody production by gene gun-mediated DNA vaccination. J Allergy Clin Immunol 114:951–957
11. Sudowe S et al (2006) Prophylactic and therapeutic intervention in IgE responses by biolistic DNA vaccination primarily targeting dendritic cells. J Allergy Clin Immunol 117:196–203
12. Zindler E et al (2008) Divergent effects of biolistic gene transfer in a mouse model of allergic airway inflammation. Am J Respir Cell Mol Biol 38:38–46
13. Baniyash M, Eshhar Z (1984) Inhibition of IgE binding to mast cells and basophils by monoclonal antibodies to murine IgE. Eur J Immunol 14:799–807
14. Conrad ML et al (2009) Comparison of adjuvant and adjuvant-free murine experimental asthma models. Clin Exp Allergy 39:1246–1254

Chapter 27

Safety Assessment of Biolistic DNA Vaccination

Barbara Langer, Matthias Renner, Jürgen Scherer, Silke Schüle, and Klaus Cichutek

Abstract

DNA-based vector systems have been widely studied as new modalities for the prevention and treatment of human diseases. As for all other medicinal products, safety is an important aspect in the evaluation of such products. In this chapter, we reflect on the basic safety issues which have been raised with respect to preventive and therapeutic DNA vaccines, including insertional mutagenesis in case of chromosomal integration, possible formation of anti-DNA antibodies, induction of autoimmune responses and/or immunological tolerance. In addition, local reactions at the site of administration and adverse effects resulting from plasmid DNA spread to nontarget tissues are discussed. Most importantly, however, the benefit-risk profile of a medicinal product is crucial for a decision on granting marketing authorization or not. A product has an acceptable benefit-risk profile if the benefits of the product outweigh its risks for the treated patient.

Key words: Biolistic DNA delivery, Biosafety, Mutagenesis, Biodistribution

1. Introduction

DNA-based vectors have been widely studied to develop new, efficient, and safe medicinal products for the prevention and treatment of human diseases (1, 2). The underlying concept of DNA-based vaccination is the administration of a nucleic acid sequence encoding an antigen of interest. The antigen is then expressed by recipient cells and thus entering the antigen processing and presentation pathways required for elicitation of a cellular and humoral immune response.

There are several possible mechanisms by which the gene product is processed and presented to the immune system (for review see ref. (3)). Myocytes and keratinocytes constitute the predominant

Stephan Sudowe and Angelika B. Reske-Kunz (eds.), *Biolistic DNA Delivery: Methods and Protocols*, Methods in Molecular Biology, vol. 940, DOI 10.1007/978-1-62703-110-3_27, © Springer Science+Business Media, LLC 2013

cell type transfected after DNA inoculation into muscle or skin, respectively. In addition, inoculation of plasmid DNA may lead to direct transfection of professional antigen presenting cells (APCs), in particular dendritic cells in muscle and epidermal Langerhans cells or dermal dendritic cells in skin. Professional APCs are manifold more effective than keratinocytes or myocytes in eliciting an immune response (4, 5). As an alternative mechanism priming particularly a cellular immune response, the antigen produced in transfected somatic cells may also be secreted or released due to cell apoptosis and then taken up, processed, and presented by professional APCs. Dendritic cells have been identified as a potent mediator of such cross-presentation of antigens derived from phagocytosed apoptotic bodies from non-APCs to induce an MHC class I-restricted cytotoxic T cell response (3, 6). However, besides the intended antigen-directed immune responses, immunological presentation of self-components derived from apoptotic cells may also lead to induction of autoimmune reactions. Thus, the actual mode of action of DNA vectors may have an impact on the safety profile of such a vaccine, and potential risks have to be addressed in respective safety evaluation studies (see below).

While DNA vaccination has been shown to be effective at eliciting T helper cell, cytotoxic T cell as well as antibody responses in preclinical studies, initial clinical trials in humans failed to demonstrate sufficient immunogenicity (7). The comparatively low magnitude of the immune responses observed has been attributed to inefficient plasmid uptake, i.e., low transfection efficiency, low levels of transgene expression, or insufficient recruitment of professional APCs to the site of administration. Magnitude and type of the immune response resulting from DNA vaccination may be influenced by several factors, including

- Quantity and sequence of the administered DNA,
- Site and method of administration,
- Vaccination regimen,
- Level and duration of antigen expression,
- Localization of antigen synthesis,
- Site of antigen presentation (intracellular, membrane-bound, secreted),
- Co-expressed molecules (e.g., cytokines, chemokines, co-stimulatory molecules),
- Overall quality of the vaccine.

Current data suggest that intramuscular and intradermal needle injection preferentially elicit a type 1 T helper cell (Th1) response. In contrast, a shift towards a T helper cell type 2 (Th2) response pattern and predominant IgG1 production has been observed with particle-mediated epidermal DNA delivery (8).

Among the approaches to improve vaccine delivery, electroporation and gene gun technology have led to promising results (9). Electroporation exposes the target tissue to short bursts of electric current causing a temporary and reversible increase in cell membrane permeability, thereby facilitating plasmid uptake into cells with improved transfection efficiency and increased levels of antigen expression (10). In addition, in vivo electroporation leads to a low level local inflammatory reaction with recruitment of professional APCs to the injection site. Electroporation-mediated DNA vaccine delivery induced an up to 100-fold stronger immune response compared to conventional needle injection. In addition, electroporation does not seem to skew the response pattern. Gene gun technology uses a gas-driven biolistic bombardment device to propel plasmid-coated gold beads directly into the cytoplasm and nucleus of epidermal target cells. By altering the gas pressure, the depth of DNA penetration into the skin layers can be modulated. Lower pressures result in vaccine deposition on the skin surface and in the stratum corneum with reduced transfection efficiency, whereas higher pressures propel a greater proportion of particles into the underlying dermis which may be associated with increased local reactogenicity.

Employing gene gun technology has been shown to generate humoral and cellular immune responses in humans at plasmid doses of less than 10 μg. Up to 1,000-fold more DNA is required to elicit an equivalent response with intramuscular needle injection, as most of the injected DNA remains extracellularly where it is prone to rapid degradation by nucleases (11, 12).

One of the most promising approaches for increasing the immunogenicity of DNA vaccines is to combine them in a two-modality prime-boost regimen wherein a vaccine prime is given with DNA, followed by a viral vector encoding the same antigen or by a protein boost matching with the antigen encoded by the DNA prime (13, 14). While various boosts are effective with a DNA prime, the sequence of administration is important, with DNA being the optimal prime (14).

In principle, the safety of medicinal products has to be evaluated for each specific product individually and should not be addressed on a generic basis (15). This is of even greater importance given the multiple and distinct approaches to develop DNA vaccines. However, this chapter intends to reflect on some general and important aspects to be considered for such medicinal products. While focusing on biolistic administration of DNA, the considerations are not confined to this approach but also relate to other modes of administration.

From a regulatory perspective, recombinant DNA may fall under different definitions for a medicinal product. In Directive 2009/120/EC (16), it is specified that administration of recombinant DNA to the human body has to be considered as a gene therapy medicinal product (GTMP; (16)). However, the definition of GTMPs excludes vaccines against infectious diseases, irrespective of whether such vaccines are administered for prophylactic or therapeutic purposes. Hence, a recombinant DNA vector, for example, used for the vaccination against an infectious disease such as HIV-1, will fall under the definition of vaccines. However, if the recombinant DNA vector is used for therapeutic tumor vaccination, the product falls under the definition of a GTMP (16). In this review, we will discuss the safety aspects of DNA vector vaccination strategies based on the intended mode of action of the medicinal product, irrespective of the regulatory classification as vaccine or GTMP.

2. Safety Aspects of DNA Vaccination

Safety is an important aspect in the evaluation of a medicinal product, including DNA vaccines, with the safety profile being assessed in the context of the benefits and the efficacy of the product. The crucial and decisive evaluation of a medicinal product is based upon the assessment of its benefit-risk profile, i.e., the decision whether the benefits of the product are sufficient to outweigh its risks for the intended patient population (17). In general, DNA vaccination offers several potential advantages, including:

- No risk of infection,
- No reversion to virulence,
- Potential for combining multiple epitopes or diverse immunogens into a single preparation allowing simultaneous immunization for several diseases,
- No interference from preexisting or vector-induced antibodies,
- Expression of native antigen in situ,
- Capacity to induce both humoral and cell-mediated immune responses,
- Stability at ambient temperature which may eliminate the need for a cold chain,
- Simplicity of design,
- Flexibility of construction,
- Reproducible large-scale production,
- Ease of storage and transport,
- Long shelf life,
- Relatively low costs.

On the other hand, a number of basic safety concerns have been raised with respect to DNA vaccines, including

- Integration of the plasmid DNA into chromosomes of vaccinees' somatic cells,
- Germline integration allowing vertical transmission,
- Induction of autoimmune responses,
- Formation of anti-DNA antibodies,
- Induction of immunological tolerance,
- Direct or immune-mediated cell and tissue damage at the site of administration,
- Immunomodulation,
- Adverse effects resulting from spread to nontarget tissues,
- Impact of plasmid DNA release into the environment.

These issues basically relate to the mode of administration, the quality profile, and the inherent nature of the plasmid DNA, which is considered to be the active substance in the final medicinal product, and will be discussed in the following sections. However, as for all medicinal products, additional issues may arise from the specific manufacturing or administration of a specific DNA product.

3. Insertional Mutagenesis

Insertional mutagenesis is one of the major concerns raised in connection with vector-based gene transfer approaches, including biolistic administration of DNA. Aside from the intended action, i.e., to serve as a template for transcription/translation of a prophylactically or therapeutically active molecule, the applied plasmid DNA may interact with the host chromosomal DNA and subsequently integrate into the genome of cells of the recipient. Such incorporation can occur by random integration or as a result of homologous recombination mediated by homologous sequences present in the plasmid and the host genome. Depending on the integration site, insertion may be phenotypically silent, could be mutagenic or potentially carcinogenic if the integration event results in the activation of an oncogene or deactivation of a tumor suppressor gene. This may, in the worst case, lead to cell transformation and subsequently to tumor formation. Such a situation became evident in two clinical trials where retroviral vectors had been employed for the ex vivo genetic modification of autologous hematopoietic stem cells for the treatment of children suffering from X1-type severe combined immunodeficiency syndrome

(SCID-X1; (18, 19)). In 5 of the 20 children treated, insertion of the retroviral vector in or near the proto-oncogenes LMO2, BMI1, or CCND2, respectively, in progenitor cells led to clonal expansion of T cells with development of an acute T cell leukemia between 23 and 68 months post treatment (20). However, in contrast to plasmid DNA vectors, entry of the viral DNA into the nucleus and subsequent integration into the host genome constitutes an essential step in the mode of action of retroviral vector-mediated gene delivery. Thus, these observations are not to be regarded indicative for a similar risk potential being associated with the administration of plasmid DNA vaccines.

In in vitro transfection experiments, it has been shown that integration of plasmid DNA into the host cell genome is possible even in the absence of considerable nucleic acid sequence homology. Despite this fact, the in vivo frequency of chromosomal integration events, after intramuscular inoculation of plasmid DNA, has been shown to be very low (21, 22). Virtually all of the injected plasmid was found extra-chromosomally (22). The low integration frequency of plasmid DNA might also be due to the fact that the transfected target cells such as myocytes and dendritic cells are mostly nondividing. It was estimated that between 1 and 30 copies of plasmid were present in an integrated form per μg of genomic DNA, which corresponds to approximately 150,000 diploid cells, suggesting a very low integration frequency in vivo.

The potential for chromosomal integration of plasmid DNA in vivo may be influenced not only by the route of administration but also by the method of administration and formulation. When employing electroporation or particle-mediated delivery methods, a fraction of the administered DNA directly enters the nucleus (22), as electroporation is increasing plasmid tissue levels and plasmid association with genomic DNA. This may increase the likelihood of an integration event. Using a specific and sensitive PCR method, four independent integration events were detected in DNA from muscle of mice after administration of DNA by electroporation (21).

DNA-based vaccines often suffer from low immunogenicity, and thus additional measures like biolistic or other administration methods or co-administration/co-expression of cytokines are implemented to increase DNA uptake of cells or to improve the immune response to be elicited, which may also increase the risk of chromosomal integration of the administered DNA.

Based on the nonclinical and clinical experience available so far, there is no evidence that direct nuclear entry of plasmid DNA leads in fact to significant levels of integration. Several thousands of patients have so far been treated with plasmid vectors, but the concern of insertional mutagenesis has not been substantiated, as no such event has been observed yet (23). However, clinical long-term follow-up data, encompassing suitable measures for detection of such rare integration events and its related risks, may further weaken this concern.

Biodistribution and persistence of the DNA plasmid are important parameters when evaluating the risk for chromosomal integration. In a recent review, Faurez and colleagues (24) summarized data on plasmid vector persistence obtained in various animal models, indicating that DNA can persist for more than 2 years in mice, and for at least 70–90 days in fish, 54 days in sheep, and 28 days in rats. Any measures limiting the persistence of the plasmid DNA in vivo may contribute to alleviate the concern of insertional mutagenesis, but on the other hand may reduce a long-term treatment effect, especially when vectors are employed for other than vaccination purposes. The actual WHO guideline for example states that if ≤ 30 copies of DNA per 100,000 cells persist after 60 days, a further integration assessment may not be necessary (25). In contrast, the FDA guidance document concludes that integration studies are warranted only when plasmid persists in any tissue of any animal at levels exceeding 30,000 copies per μg of host cell DNA (~150,000 cells) (26).

From a regulatory point of view, the risk of insertional mutagenesis should already be addressed at the level of plasmid design, e.g., by avoiding sequence elements showing homology to genomic DNA or being possibly associated with an oncogenic potential. These aspects are further discussed below, in the quality section of this chapter. To evaluate the risk of insertional mutagenesis of a DNA vector to be used as the active ingredient of an investigational medicinal product, an appropriate nonclinical testing program should be set up. The extent of this program should be scientifically justified and will crucially depend on the risk profile of the specific product, considering the plasmid design, the quality of the product, the method of administration, the target population, the clinical dose and indication, as well as the results from other nonclinical studies with the product. European Medicines Agency (EMA) also released specific guidance to address post-marketing long-term follow-up of patients treated with advanced therapy medicinal products (27, 28). These documents may also serve as guidance to establish follow-up measures for patients treated with DNA vaccines in order to adequately monitor the long-term safety profile of these medicinal products including the analysis of insertional mutagenesis.

4. Biodistribution

Biodistribution and persistence of the plasmid DNA depend on the route and method of administration. In mice, plasmid DNA could be detected in many organs shortly after intramuscular injection like blood, liver, spleen, lung, lymph nodes, kidneys, and even gonads (29). After several weeks, plasmid DNA was cleared from all organs but was still detectable at the site of injection. In another study, a faster clearance from the injection site has been observed. Here it has been shown that 30 min after injection DNA concentration had decreased significantly at the site of administration. After 90 min, less than 1% of the foreign DNA persisted (12). Other investigators used radiolabeled plasmid DNA to study biodistribution after intramuscular injection. More than 90% of the plasmid had been cleared from the injection site within 3 h of administration (30). Thus, the extent of biodistribution studies needed is dependent on the formulation, dose, schedule, and route of administration of the particular product.

The risk of germ line transmission of plasmid DNA should be carefully addressed in nonclinical studies. In a first step, it has to be investigated whether the respective plasmid DNA is per se distributed to the gonads as this increases the risk for potential integration into the genome of germ line cells. After injection of plasmid DNA into rat fetuses, distribution and integration of plasmid to fetal organs including germ cells was described (31). However, the plasmid used in this study was carrying recombinant retrovirus sequences and thus may not be representative for a typical DNA vaccine. Although plasmid DNA has been detected in gonads shortly upon injection into mice muscle, integration of the plasmid DNA into germ line genome has not been observed (29). Moreover, the concentration of plasmid DNA decreased over time and the low level of plasmid DNA in the gonads which was rapidly cleared, suggested that the risk of germline transmission may be low (32).

Nevertheless, biodistribution studies should address any potential localization of the plasmids in the gonads. A specific guidance document for GTMPs is available from the European Medicines Agency (33). This document may also be regarded as a relevant guidance to address potential germ line integration of DNA vaccines. It provides a step-wise algorithm for a testing program depending on positive findings of the gene delivery vector in gonads in initial nonclinical studies.

5. Induction of Autoimmune Responses and Formation of Anti-DNA Antibodies

Another safety concern is the potential induction of autoimmunity following administration of DNA vaccines. Several possible mechanisms have been proposed (34). Autoimmunity might arise through cell-, antibody- or complement-mediated destruction of vaccine antigen-expressing host cells leading to the release of cell components theoretically capable of inducing autoimmune responses, including anti-DNA or anti-nuclear antibody formation. However, such cell decay events also occur in the course of microbial infections, following conventional vaccination or during normal tissue remodeling processes. Considering the low number of foreign antigen-expressing cells following DNA vaccination, it seems unlikely that immune-mediated cytolytic effects would pose any increased risk of inducing autoimmune phenomena.

Further, the intrinsic immune stimulatory activity of unmethylated CpG motifs in the plasmid backbone or co-expression of immune modulatory cytokines could favor the development of autoimmune responses (35). Studies of serum samples from humans vaccinated with DNA showed slight increases in anti-DNA autoantibody levels. Autoantibodies against native double-strand DNA (dsDNA) are considered a hallmark of systemic autoimmune diseases such as systemic lupus erythematodes (SLE), isotype and titre correlating with disease activity (36, 37). On the other hand, low levels of nonpathogenic anti-dsDNA antibodies of different type and specificity to those found in systemic autoimmune disease are present ubiquitously in humans (38). It is generally agreed that not all anti-dsDNA antibodies play a pathogenetic role. The autoantibody titres observed in DNA vaccine recipients were well below the level associated with the development of autoimmune disease (14).

However, though no evidence has been found for an association of DNA vaccination with induction of autoimmunity, it appears advisable, that vaccinees are monitored for the production of auto-antibodies, especially if improved DNA delivery technologies or immune-enhancing plasmid designs are employed.

6. Induction of Immunological Tolerance

Another safety aspect is potential induction of immunological tolerance. A number of factors have been identified that may favor induction of immune tolerance to a specific antigen, including very small antigen doses which persist over a period of time, no antigen processing by professional APCs, or processing by cells lacking MHC class II or co-stimulatory molecules, and age of the vaccinee (39). The amount of antigen produced after DNA immunization is thought to be small, with antigen expression persisting for up to several weeks or months (14). In addition, the predominant cell type transfected following intramuscular or intradermal DNA vaccination are myocytes and keratinocytes, respectively, which are nonprofessional APCs that do not express co-stimulatory or MHC class II molecules (40). Thus, the possibility exists that DNA vaccination may result in immunological tolerance rather than immunity, especially in vaccinated neonates because of the immaturity of their immune system. This has been demonstrated in a study in neonatal mice vaccinated with a DNA vaccine encoding the plasmodial circumsporozoite protein (41). The animals were unable to mount an antigen-specific humoral or cell-mediated immune response, and they remained unresponsive when revaccinated at an older age. On the other hand, a tolerogenic potential of DNA vaccines may be exploited as a strategy to reestablish self-tolerance in order to prevent or treat autoimmune disease.

7. Local Reactions

A further safety issue is local tissue damage or reactogenicity at the site of electroporation or particle-mediated DNA delivery. A certain level of transient and reversible tissue damage has been observed in the electroporated muscle area which was found to be dependent upon the electric field strength (42). Particle-mediated epidermal DNA vaccine delivery has resulted in mild to moderate local skin reactions including discomfort at the injection site, erythema, edema, skin discoloration, and itching (43). Less frequent reactions were petechiae, minor bruising, and scabs. However, local reactions/tolerance has to be addressed during the development of a DNA vaccine.

8. Environmental Risks

Clinical use of medicinal products that contain genetically modified organisms (GMOs) may harbor a potential risk for the environment, as active ingredients may be excreted by the patient into the environment thereby causing potentially harmful effects on the ecosystem as well as on human health. Depending on national requirements (e.g., DNA is considered to be a GMO and thus subjected to the deliberate release regulations), an environmental risk assessment for DNA has to be included in the clinical trial application. This assessment should include shedding data, plasmid DNA stability studies, sequence homology information, integration studies, and information about the presence of genes that could be favorable or harmful for other organisms (44). Information on how to perform an environmental risk assessment could be found in the EMA guideline on scientific requirements for the environmental risk assessment of GTMPs (45).

9. Quality of DNA-Based Vaccines and its Influence on Biosafety

Above, crucial safety issues were identified which have to be considered during the clinical development of a DNA vaccine. In the following section, the main safety concerns related to product quality, like DNA-plasmid design, process-related and product-related impurities, excipients used as a carrier for a biolistic mode of application, and product potency are discussed. As for all biologicals, a validated and consistent manufacturing process including adequate in-process controls and quality control assays is an integral part of the product quality. Excipients of the medicinal products have to be justified and controlled like for other medicinal products. An excipient specific for DNA vaccines administered biolistically are the particles used as carrier for the plasmid DNA. In most cases gold particles are used for this purpose. Regarding such carriers for biolistic administration, sterility, purity, and particle size uniformity have to be demonstrated. Of course, biological activity (or inertness), nontoxicity to cells and its impact on nucleic acids, as well as the pharmacokinetic properties of the carrier particles are important aspects that should be considered during nonclinical and clinical analysis of the medicinal product. In terms of manufacturing GMP-grade plasmids for human use, several guidelines addressing the quality issue of DNA-based products are available (Table 1).

DNA vaccines are commonly based on recombinant bacterial plasmids or their derivatives. Beside the intended therapeutic transgene expression cassette(s) consisting of a viral or eukaryotic enhancer/promoter, the gene of interest and a suitable termination signal, plasmids also contain a bacterial origin of replication and a selection marker.

Various selection systems have been developed for the amplification of plasmid DNA in bacteria. Expression of antibiotic resistance genes driven by a prokaryotic promoter are widely used as a selection marker. Although the probability is low, there is a remaining risk associated with the use of antibiotic resistance genes, as they might be transferred to the endogenous microbial flora of the treated individual (46). Thus, selection markers possibly interfering with the efficacy or safety of clinically applied antibiotics in the intended patient population should be avoided (47). In addition, antibiotics such as penicillin and other β-lactam antibiotics used during manufacture of the active substance, harbor the risk for immune sensitization or for anaphylaxis in case residual amounts are present as contamination in the final product. The use of kanamycin or neomycin resistance genes is recommended when an antibiotic resistance marker is required, which may be considered acceptable by regulatory authorities. These aminoglycoside antibiotics are not extensively used in the treatment of bacterial infections in the clinical setting.

Table 1
Selection of basic guidelines addressing quality, nonclinical and clinical issues in the development of DNA vaccines

Guidance for DNA vaccines	CHMP/308136/07
Guidelines for assuring the quality and nonclinical safety evaluation of DNA vaccines	WHO Technical Report Series No 941, 2007
Ph. Eur. chapter 5.14., "Gene transfer medicinal products for human use"	Pharmacopeia 7.0, 01/2010:51400
Quality, Preclinical and Clinical Aspects of Gene Transfer Medicinal Products	CPMP/BWP/3088/99
Revision of the note for guidance on the quality, preclinical and clinical aspects of gene transfer medicinal products	CHMP/GTWP/234523/09
Quality, nonclinical and clinical aspects of live recombinant viral vectored vaccines	CHMP/VWP/141697/09
Nonclinical testing for Inadvertent Germline transmission of Gene Transfer Vectors	EMEA/273974/05
Nonclinical studies required before first clinical use of gene therapy medicinal products	CHMP/GTWP/125459/2006
Follow-up of patients administered with gene therapy medicinal products	CHMP/GTWP/60436/07
Guidance for Industry: Considerations for Plasmid DNA Vaccines for Infectious Disease Indications	FDA, CBER, November 2007

However, irrespective of which antibiotic is used, validation studies should demonstrate an adequate capability of the purification process for clearance of residual antibiotics in the product.

In fact, antibiotic-free selection strategies, of which varieties are currently under development, are preferably used to resolve this issue. For example, a mutant bacteria strain which is unable to synthesize an essential amino acid can be complemented with the plasmid carrying the gene which provides for its synthesis, besides the expression cassette for the antigen (48). Other approaches are based on RNA/RNA antisense interaction involving the naturally occurring RNA derived from the origin of replication of the plasmid. This means, that the bacterial strains are modified, so that the plasmid replication inhibitor RNA I could suppress the translation of a growth essential gene by RNA-RNA antisense reaction. The concerned essential gene is then modified so that a repressor protein (e.g., tetR) would hamper its expression. Only in the presence of plasmid and, hence, RNA I, the repressor gene will be silenced and thus relieves the expression of the essential gene, allowing bacterial growth (49). Another antibiotic-free selection system is the operator-repressor titration (ORT). Here an essential chromosomal gene is controlled by the operator/promoter lac-operon of the *E.* The lacI repressor binds to the chromosomal operator and cell growth is prevented unless an inducer such as IPTG is present (50–52). The latter system has already been used for the manufacturing of DNA vaccines applied in clinical trials (53, 54).

In addition, DNA vectors totally devoid of prokaryotic elements have been developed. One example is mini-circle DNA, which is derived from a parental plasmid carrying two recombination sequences flanking the expression cassette containing the gene of interest. Through an intra-molecular recombination process the expression cassette is released from the plasmid backbone and is circularized (55, 56). The development of linear dumbbell-shaped expression cassettes, containing hairpin oligodeoxynucleotides at their ends for protection, is another approach. The respective DNA expression cassettes are generated in vitro either by polymerase chain reaction amplification or by endonuclease processing (57, 58). However, despite these technological achievements, diverse aspects for the production of antibiotic-free plasmid DNA still need to be addressed to develop robust processes for a high quality and pharmaceutical large-scale production of DNA vaccines required for commercial use.

Apart from these advanced technologies, in standard manufacturing processes a plasmid backbone is still present and is made up of bacterial DNA which contains immune-stimulatory sequence motifs. Since they (may) contribute to activate the innate and adaptive immune system they may modulate the immune response elicited by the antigen and in specific cases may be crucial for the therapeutic efficacy of the DNA vaccine. Unmethylated CpG DNA stimulates dendritic cells, natural killer cells, B cells and monocytes/macrophages (59, 60). Depending to some extent on the route and form of administration, CD4+ T helper cells (Th1) and CD8+ T lymphocyte responses are preferentially induced (61). The CpG motifs are recognized by Toll-like receptor 9 molecules, which induce a series of immune stimulatory cytokines including interferons and chemokines (62, 63). The presence of such CpG motifs in the inoculated plasmid DNA may contribute to the efficient immunization by DNA similar to the effect of adjuvants in subunit vaccines (63).

However, under rare circumstances an overactivation of the innate immune system might also elicit an immune response against self-antigens and might lead to autoimmune phenomena (64, 65). After replacing the cytosine residues of CpG islets with 5-methyl cytosine a loss of immune stimulatory effects can be observed, suggesting that lack of methylation of CpG motifs in the concerned bacterial and plasmid DNA is crucial for that effect (66). Moreover, the immunological reaction elicited by CpG sequences can vary depending upon the sequence of flanking nucleotides as well as upon the origin of the backbone. It has been demonstrated that polyG or

GC-rich oligodeoxynucleotides suppress the immune stimulatory effect of bacterial CpG (67, 68). If possible, it seems to be desirable to design the CpG motifs of the bacterial backbone of the intended plasmid to gain the desired immune stimulatory effect (69).

Genomic DNA is extremely sensitive to shearing forces. Processing of the plasmid from the production culture will usually yield fragments of bacterial DNA, which could not be readily removed during the subsequent purification steps. Thus, chromosomal DNA fragments derived from the bacterial host are a major contaminant of plasmid preparations. Some plasmid DNA preparations, for example, have been described to cause significant muscle necrosis, whereas others caused no muscle damage. The occurrence of muscle damage was shown to correlate with the quantity of bacterial genomic DNA present in the final product, but not with plasmid sequence, size or the type of gene encoded (70). The quantification of residual host cell DNA is thus of high importance, because this contamination will be present within a plasmid DNA product even when a CpG-free plasmid is manufactured. The production process should be optimized to consistently remove impurities while retaining product activity. Like for all biologicals all product- and process-related impurities should be determined quantitatively. Besides the chromosomal DNA, host cell RNA and protein are possible process-related impurities, just as cell culture reagents and additives (e.g., benzonase, BSA, antibiotics, endotoxin), and impurities resulting from the purification process (e.g., column leachables). The presence and specific amount of these impurities do contribute to and determine the safety profile of the DNA vaccine and thus have to be addressed, whenever such a product is manufactured for clinical use.

Plasmid preparations should be analyzed with respect to the endotoxin content. Plasmids manufactured in gram-negative bacteria like *E. coli* most likely are contaminated with endotoxins derived from the cell membrane of these bacteria. Since endotoxins may elicit a wide variety of pathophysiological effects, even at a low concentration, endotoxin content should be strictly limited according to the current guidelines (71).

Plasmid DNA may appear in different conformations and it has to be defined what conformation represents the active substance. The covalent closed circular (ccc) form is the most compact structure and intact molecule from which the respective transgene is potentially expressed completely and efficiently and thus in most cases will be regarded as the active substance. The other plasmid forms (e.g., multimers, relaxed monomers and linear forms) which are to some extent co-purified along with target supercoiled plasmid DNA should be regarded as product-related impurities.

However, this quality aspect also impacts on the safety profile of the product since the probability of chromosomal integration is different for the various plasmid forms; the ccc form is considered to be the one with the lowest risk of chromosomal integration (72). Thus, purification processes should be in place which ensures manufacture of supercoiled ccc plasmid DNA in almost pure form.

The analysis of the integration potential of a new plasmid in advance of any clinical trial application is endorsed by regulatory authorities, particularly when the backbone vector has not been tested before. In any case, careful design of the plasmid vector is required to minimize the risk of integration.

10. Concluding Remarks and Outlook

The first clinical trial of a DNA-based vaccine included 15 symptomatic HIV-infected patients who received a plasmid containing the viral env and rev genes via direct intramuscular injection. The vaccine was well tolerated, and patients showed elevated vaccine-induced antibodies and some CTL activity (73). Since then, a growing number of DNA vaccine candidates have entered into clinical trials for a variety of diseases, including preventive and therapeutic vaccines for infectious diseases (e.g., HIV, malaria, tuberculosis, influenza, hepatitis B and C, Ebola), as potential immune therapeutics for various types of cancer, or as immune modulators to prevent or alleviate autoimmune disease (Fig. 1). While safety

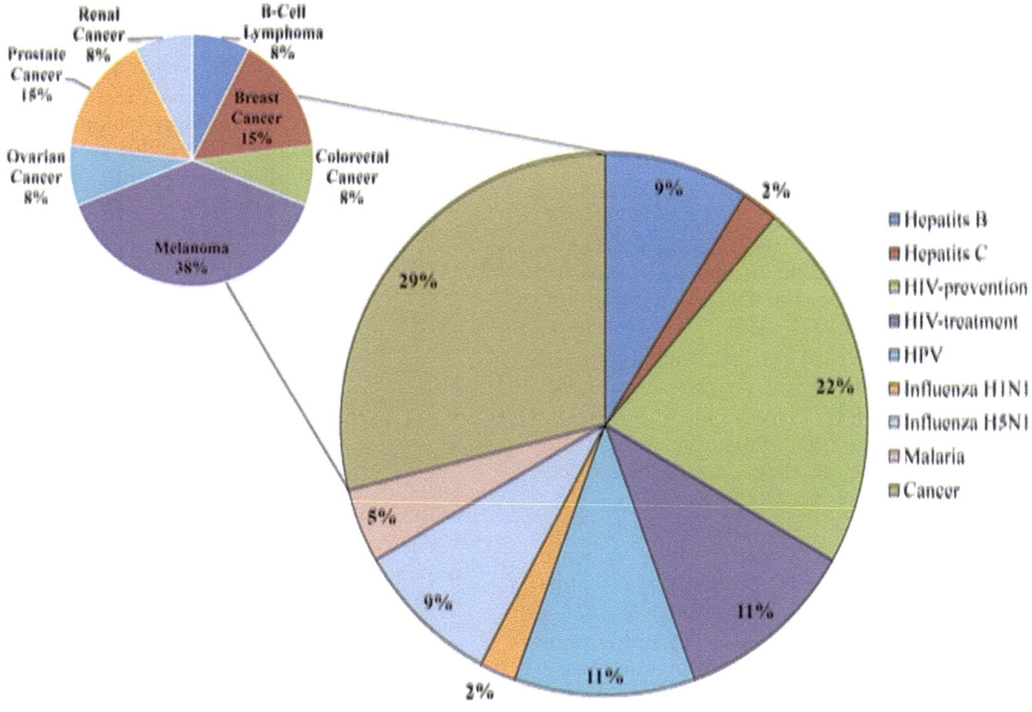

Fig. 1. Current DNA vaccine clinical trials (taken from (9) with permission of Oxford University Press).

data accumulated so far indicate that DNA vaccines are well tolerated and safe, efficacy results obtained in humans have not been satisfactory yet. A number of efforts have been made to enhance the immunogenicity and efficacy of DNA vaccines, including biolistic administration, vaccination schedules or co-expression/coadministration of immune modulating molecules as adjuvants (9). In this regard, recent studies using a heterologous prime-boost approach have demonstrated very promising results (74). Given the very good safety profile and potential to improve immunogenicity, DNA vaccine technology holds promise for future prophylactic and therapeutic vaccine development.

References

1. U.S. National Library of Medicine, National Institutes of Health, Clinical Trials database <http://clinicaltrials.gov/ct2/search>.
2. European Medicines Agency, EU Clinical Trials register <https://www.clinicaltrialsregister.eu/contacts.html>.
3. Liu MA (2011) DNA vaccines: an historical perspective and view to the future. Immunol Rev 239:62–84
4. Kendall M (2006) Engineering of needle-free physical methods to target epidermal cells for DNA vaccination. Vaccine 24:4651–4656
5. Banchereau J, Steinman RM (1998) Dendritic cells and the control of immunity. Nature 392:245–252
6. Albert ML, Sauter B, Bhardwaj N (1998) Dendritic cells acquire antigen from apoptotic cells and induce class I-restricted CTLs. Nature 392:86–89
7. Donnelly JJ, Wahren B, Liu MA (2005) DNA vaccines: progress and challenges. J Immunol 175:633–639
8. Feltquate DM et al (1997) Different T helper cell types and antibody isotypes generated by saline and gene gun DNA immunization. J Immunol 158:2278–2284
9. Ferraro B et al (2011) Clinical applications of DNA vaccines: current progress. Clin Infect Dis 53:296–302
10. Otten G et al (2004) Enhancement of DNA vaccine potency in rhesus macaques by electroporation. Vaccine 22:2489–2493
11. Satkauskas S et al (2001) Slow accumulation of plasmid in muscle cells: supporting evidence for a mechanism of DNA uptake by receptor-mediated endocytosis. Mol Ther 4:317–323
12. Barry ME et al (1999) Role of endogenous endonucleases and tissue site in transfection and CpG-mediated immune activation after naked DNA injection. Hum Gene Ther 10:2461–2480
13. Dale CJ et al (2006) Prime-boost strategies in DNA vaccines. Methods Mol Med 127:171–197
14. Gurunathan S, Klinman DM, Seder RA (2000) DNA vaccines: immunology, application, and optimization. Annu Rev Immunol 18:927–974
15. Schüle S et al (2010) Regulatory requirements for clinical trial and marketing authorisation application for gene therapy medicinal products. Bundesgesundheitsblatt Gesundheitsforschung Gesundheitsschutz 53:30–37
16. 2009/120/EC, The Commission of the European communities. Off J Eur Comm L 242 15(9)2009, p. 3–12. <http://ec.europa.eu/health/files/eudralex/vol-1/dir_2009_120/dir_2009_120_en.pdf>.
17. 2001/83/EC. The European Parliament and the Council of the European Union. Off. J Eur Comm. L311, 67–126 (2001). <http://ec.europa.eu/health/files/eudralex/vol-1/dir_2001_83_cons2009/2001_83_cons2009_en.pdf>.
18. Gaspar HB et al (2004) Gene therapy of X-linked severe combined immunodeficiency by use of a pseudotyped gammaretroviral vector. Lancet 364:2181–2187
19. Cavazzana-Calvo M et al (2000) Gene therapy of human severe combined immunodeficiency (SCID)-X1 disease. Science 288:669–672
20. Fischer A, Hacein-Bey-Abina S, Cavazzana-Calvo M (2010) 20 years of gene therapy for SCID. Nat Immunol 11:457–460
21. Wang Z et al (2004) Detection of integration of plasmid DNA into host genomic DNA following intramuscular injection and electroporation. Gene Ther 11:711–721
22. Ledwith BJ et al (2000) Plasmid DNA vaccines: investigation of integration into host cellular DNA following intramuscular injection in mice. Intervirology 43:258–272

23. Kutzler MA, Weiner DB (2008) DNA vaccines: ready for prime time? Nat Rev Genet 9:776–788
24. Faurez F et al (2010) Biosafety of DNA vaccines: new generation of DNA vectors and current knowledge on the fate of plasmids after injection. Vaccine 28:3888–3895
25. WHO. World Health Organization: Guidelines for assuring the quality and nonclinical safety evaluation of DNA vaccines; WHO Technical Report Series No 941, 2007, Annex 1. <http://www.who.int/biologicals/publications/trs/areas/vaccines/dna/Annex%201_DNA%20vaccines.pdf>.
26. Food and Drug Administration: Guidance for Industry: Considerations for plasmid DNA vaccines for infectious disease indications, 2007 <http://www.fda.gov/biologicsbloodvaccines/guidancecomplianceregulatoryinformation/guidances/vaccines/ucm074770.htm>.
27. Committee for the Medicinal Product for Human Use (CHMP): Guideline on safety and efficacy follow-up - risk management of advanced therapy medicinal products (EMEA/149995/2008) <http://www.ema.europa.eu/docs/en_GB/document_library/Regulatory_and_procedural_guideline/2009/10/WC500006326.pdf>.
28. Committee for Medicinal Products for Human Use CHMP/GTWP/60436/07. Follow-up of patients administered with gene therapy medicinal products (Draft). EMEA, London. 2008. <http://www.emea.europa.eu/pdfs/human/genetherapy/6043607endraft.pdf>.
29. Manam S et al (2000) Plasmid DNA vaccines: tissue distribution and effects of DNA sequence, adjuvants and delivery method on integration into host DNA. Intervirology 43:273–281
30. Bureau MF et al (2004) Intramuscular plasmid DNA electrotransfer: biodistribution and degradation. Biochim Biophys Acta 1676:138–148
31. Gallot D et al (2002) Systemic diffusion including germ cells after plasmidic in utero gene transfer in the rat. Fetal Diagn Ther 17:157–162
32. Schalk JA et al (2006) Preclinical and clinical safety studies on DNA vaccines. Hum Vaccin 2:45–53
33. 273974/2005. EMEA/CHMP: Guideline on non-clinical testing for inadvertent germline transmission of gene transfer vectors <http://www.emea.europa.eu/pdfs/human/swp/27397405enfin.pdf>.
34. Donnelly JJ et al (1997) DNA vaccines. Annu Rev Immunol 15:617–648
35. Mor G, Eliza M (2001) Plasmid DNA vaccines. Immunology, tolerance, and autoimmunity. Mol Biotechnol 19:245–250
36. Pavlovic M et al (2010) Pathogenic and epiphenomenal anti-DNA antibodies in SLE. Autoimmune Dis 2011:462841
37. Manson JJ, Isenberg DA (2006) The origin and pathogenic consequences of anti-dsDNA antibodies in systemic lupus erythematosus. Expert Rev Clin Immunol 2:377–385
38. Isenberg DA et al (2007) Fifty years of anti-ds DNA antibodies: are we approaching journey's end? Rheumatology (Oxford) 46:1052–1056
39. Ichino M et al (1999) Factors associated with the development of neonatal tolerance after the administration of a plasmid DNA vaccine. J Immunol 162:3814–3818
40. Fioretti D et al (2010) DNA vaccines: developing new strategies against cancer. J Biomed Biotechnol 2010:174378
41. Mor G et al (1996) Induction of neonatal tolerance by plasmid DNA vaccination of mice. J Clin Invest 98:2700–2705
42. van Drunen Littel-van den Hurk S, Hannaman D (2010) Electroporation for DNA immunization: clinical application. Expert Rev Vaccines 9, 503–517
43. Dean HJ (2005) Epidermal delivery of protein and DNA vaccines. Expert Opin Drug Deliv 2:227–236
44. Anliker B, Longhurst S, Buchholz CJ (2010) Environmental risk assessment for medicinal products containing genetically modified organisms. Bundesgesundheitsblatt Gesundheitsforschung Gesundheitsschutz 53:52–57
45. ERA. Committee for the Medicinal Product for Human Use (CHMP). Guideline on scientific requirements for the environmental risk assessment of gene therapy medicinal products. Doc. Ref. EMEA/CHMP/GTWP/125491/2006. EMEA, London, 2008. < >.
46. Droge M, Puhler A, Selbitschka W (1998) Horizontal gene transfer as a biosafety issue: a natural phenomenon of public concern. J Biotechnol 64:75–90
47. 5.14. European Pharmacopeia, 7th Edition, Supplement 6.3. Chapter 5.14. Gene Transfer Medicinal Products for Human Use. 01/2008:54100, correct 6.0.
48. Degryse E (1991) Stability of a host-vector system based on complementation of an essential gene in Escherichia coli. J Biotechnol 18:29–39
49. Mairhofer J et al (2008) A novel antibiotic free plasmid selection system: advances in safe and efficient DNA therapy. Biotechnol J 3:83–89

50. Cranenburgh RM et al (2001) Escherichia coli strains that allow antibiotic-free plasmid selection and maintenance by repressor titration. Nucleic Acids Res 29:e26
51. Garmory HS et al (2005) Antibiotic-free plasmid stabilization by operator-repressor titration for vaccine delivery by using live Salmonella enterica Serovar typhimurium. Infect Immun 73:2005–2011
52. Williams SG et al (1998) Repressor titration: a novel system for selection and stable maintenance of recombinant plasmids. Nucleic Acids Res 26:2120–2124
53. Hanke T, McMichael AJ (2000) Design and construction of an experimental HIV-1 vaccine for a year-2000 clinical trial in Kenya. Nat Med 6:951–955
54. Hanke T et al (2002) Lack of toxicity and persistence in the mouse associated with administration of candidate DNA- and modified vaccinia virus Ankara (MVA)-based HIV vaccines for Kenya. Vaccine 21:108–114
55. Darquet AM et al (1997) A new DNA vehicle for nonviral gene delivery: supercoiled minicircle. Gene Ther 4:1341–1349
56. Kay MA, He CY, Chen ZY (2010) A robust system for production of minicircle DNA vectors. Nat Biotechnol 28:1287–1289
57. Schirmbeck R et al (2001) Priming of immune responses to hepatitis B surface antigen with minimal DNA expression constructs modified with a nuclear localization signal peptide. J Mol Med (Berl) 79:343–350
58. Lopez-Fuertes L et al (2002) DNA vaccination with linear minimalistic (MIDGE) vectors confers protection against Leishmania major infection in mice. Vaccine 21:247–257
59. Stacey KJ, Sweet MJ, Hume DA (1996) Macrophages ingest and are activated by bacterial DNA. J Immunol 157:2116–2122
60. Krieg AM et al (1995) CpG motifs in bacterial DNA trigger direct B-cell activation. Nature 374:546–549
61. Jakob T et al (1998) Activation of cutaneous dendritic cells by CpG-containing oligodeoxynucleotides: a role for dendritic cells in the augmentation of Th1 responses by immunostimulatory DNA. J Immunol 161:3042–3049
62. Landrigan A, Wong MT, Utz PJ (2011) CpG and non-CpG oligodeoxynucleotides directly costimulate mouse and human CD4+ T cells through a TLR9- and MyD88-independent mechanism. J Immunol 187:3033–3043
63. Sato Y et al (1996) Immunostimulatory DNA sequences necessary for effective intradermal gene immunization. Science 273:352–354
64. Lindau D et al (2011) Nucleosome-induced neutrophil activation occurs independently of TLR9 and endosomal acidification: implications for systemic lupus erythematosus. Eur J Immunol 41:669–681
65. Yasuda K et al (2009) Requirement for DNA CpG content in TLR9-dependent dendritic cell activation induced by DNA-containing immune complexes. J Immunol 183:3109–3117
66. Ada G, Ramshaw I (2003) DNA vaccination. Expert Opin Emerg Drugs 8:27–35
67. Yamada H et al (2002) Effect of suppressive DNA on CpG-induced immune activation. J Immunol 169:5590–5594
68. Krieg AM et al (1998) Sequence motifs in adenoviral DNA block immune activation by stimulatory CpG motifs. Proc Natl Acad Sci U S A 95:12631–12636
69. Tolmachov O (2009) Designing plasmid vectors. Methods Mol Biol 542:117–129
70. Wooddell CI et al (2011) Muscle damage after delivery of naked plasmid DNA into skeletal muscles is batch dependent. Hum Gene Ther 22:225–235
71. Magalhaes PO et al (2007) Methods of endotoxin removal from biological preparations: a review. J Pharm Pharm Sci 10:388–404
72. Schleef M et al (2010) Production of non viral DNA vectors. Curr Gene Ther 10:487–507
73. MacGregor RR et al (1998) First human trial of a DNA-based vaccine for treatment of human immunodeficiency virus type 1 infection: safety and host response. J Infect Dis 178:92–100
74. Lu S, Wang S, Grimes-Serrano JM (2008) Current progress of DNA vaccine studies in humans. Expert Rev Vaccines 7:175–191

Part VI

Related Applications for Biolistic Delivery of Molecules

Chapter 28

DiOlistics: Delivery of Fluorescent Dyes into Cells

Nyssa Sherazee and Veronica A. Alvarez

Abstract

DiOlistic labeling utilizes a particle-mediated delivery system to incorporate dye into cells. Because of its random nature, this technique generates sparse fluorescent labeling which is well suited for the study of neuronal dendritic branching and dendritic spine morphology. DiOlistics is a quick, reliable and nontoxic method that can be used in combination with other techniques such as immunostaining, biolistic DNA transfection, and retrograde tracing. In this article, we describe the methods for diOlistic labeling of neurons from rodent brain slices using DiI and the imaging of neuronal and synaptic morphology using confocal microscopy.

Key words: Gene gun, Dendritic spine, Dendrite branching, Neuronal morphology, DiI, Mouse brain, Confocal microscopy, Ballistics, Perfusion

1. Introduction

DiOlistics uses a gene gun to introduce fluorescent dyes into cells which then get incorporated into the cytoplasm or the plasma membrane. This technique most commonly uses DiI (1-1′-dioctadecyl-3,3,3′,3′-tetramethylindocarbocyanine perchlorate), a lipophilic dye that will partition in and diffuse through the cell membrane providing a well-defined outline of neuronal processes which is ideal for high resolution confocal imaging.

The primary application of this technique has been the study of neuronal morphology in the central nervous system, but it can be applied to a variety of cell types, from living or fixed tissues, as well as in diverse species such as rodents, primates, and zebrafish (1–3). Another advantage of diOlistics is that it can be used in animals of all ages, making it suitable for developmental studies, as well as a complement to behavioral assays in adult or

Stephan Sudowe and Angelika B. Reske-Kunz (eds.), *Biolistic DNA Delivery: Methods and Protocols*,
Methods in Molecular Biology, vol. 940, DOI 10.1007/978-1-62703-110-3_28, © Springer Science+Business Media, LLC 2013

aged animals. DiOlistics is a versatile labeling technique that can be used in combination with immunostaining, to identify subpopulations of cells (4–7). Additionally, the technique can be used in combination with biolistic particle-mediated DNA delivery, and/or retrograde tracing for circuit mapping or cell identification (2, 8, 9).

The procedures of the technique are described in detail in the following sections. First, trans-cardiac perfusion is used for efficient fixation of the brain and all other tissues. Fixed brains are then sliced into sections. Bullets are prepared by coating tungsten beads with DiI and a gene gun is used to shoot the fluorescent bullets onto the brain slices, sparsely labeling cells within minutes. Incorporating immunostaining is an optional step described here and recently, improvements have been suggested for the combination of the methods (9, 10). After mounting the sections, images of fluorescently labeled neurons are acquired using confocal microscopy for detailed analysis of dendrite and spine morphology.

2. Materials

Prepare all solutions using ultrapure water and analytical grade reagents. Prepare and store all reagents at room temperature, unless indicated otherwise. Follow all waste disposal regulations when discarding waste materials.

2.1. Bullet Materials

1. Tungsten Beads, 1.7 μm diameter (Bio-Rad, Hercules, CA, USA). To prepare aliquots, suspend 6 g tungsten beads into 2 mL methylene chloride. Vortex to create a homogenous solution and pipette 100 μL of bead solution into individual eppendorf tubes resulting in 300 mg aliquots. Store at room temperature until needed in step 3.1.1 (see Note 1).
2. DiI (Invitrogen, Carlsbad, CA, USA).
3. Tefzel bullet tubing (Bio-Rad).
4. Polyvinylpyrrolidone (PVP): Prepare 10 mL of 10 mg/mL PVP solution dissolved in deionized water.
5. Tubing Prep Station (Bio-Rad) connected to a nitrogen tank.
6. 12 mL Syringe with tubing adaptor, made of flexible tubing with slightly larger diameter than Teflez tubing.
7. Tubing cutter (Bio-Rad).

2.2. Perfusion Materials

1. Ketamine/xylazine anesthetic cocktail: 100 mg/mL ketamine and 20 mg/mL xylazine.

2. Butterfly needle with blunted tip. Use a Dremel-like tool to smooth the pointed tip of the needle.
3. Pump or gravity system used for perfusion.
4. Scissors.
5. Chilled container.

2.3. Solutions

1. Phosphate Buffer Solution (PBS): 137 mM NaCl, 2.7 mM KCl, 10 mM Na_2HPO_4O, 2 mM KH_2PO_4O. To make 1 L, add 8 g NaCl, 200 mg KCl, 1.44 g Na_2HPO_4O, and 240 mg KH_2PO_4O to H2O. Use NaOH to adjust pH to 7.2–7.4.
2. Fixative Solution: 4% (w/v) paraformaldehyde and 4% (w/v) sucrose in PBS. Weigh paraformaldehyde under the chemical hood and mix ingredients in a large beaker. Heat and stir to solubilize, but do not boil. If needed, add NaOH pellets to solubilize and bring to pH neutral once in solution. Filter before using. Good for about 3 weeks if stored at 4°C.
3. Sucrose solutions: 15% and 30% (w/v) sucrose in PBS.
4. Permeabilization solution: 0.01% (v/v) Triton X-100 in PBS.
5. Blocking solution: 10% (v/v) goat serum, 0.01% (v/v) Triton X-100 in PBS.

2.4. Shooting and Mounting Materials

1. Vibratome.
2. 24-Well plate.
3. Paintbrush.
4. Gene Gun and accessories (Fig. 1): o-rings, barrel liner, diffusion screens, cartridge holders, cartridge extractor tool (Bio-Rad). Modify diffusion screens (see Note 2) by removing the radial poles with pliers as shown in Fig. 2.
5. Isopore membrane filter paper 3.0 μm pore size (Millipore, Billerica, MA, USA).
6. Helium gas tank.
7. ProLong Gold Antifade mounting medium (Invitrogen).
8. Slides: 25 mm x 75 mm x 1 mm and coverslips: 18 mm x 18 mm x 1 mm.
9. Nail polish.

2.5. Evaluation Materials

1. Rocking platform.
2. Primary and fluorescence-conjugated secondary antibody.
3. Confocal microscope equipped with a 561 nm laser (LSM 520 META, Zeiss, Germany).

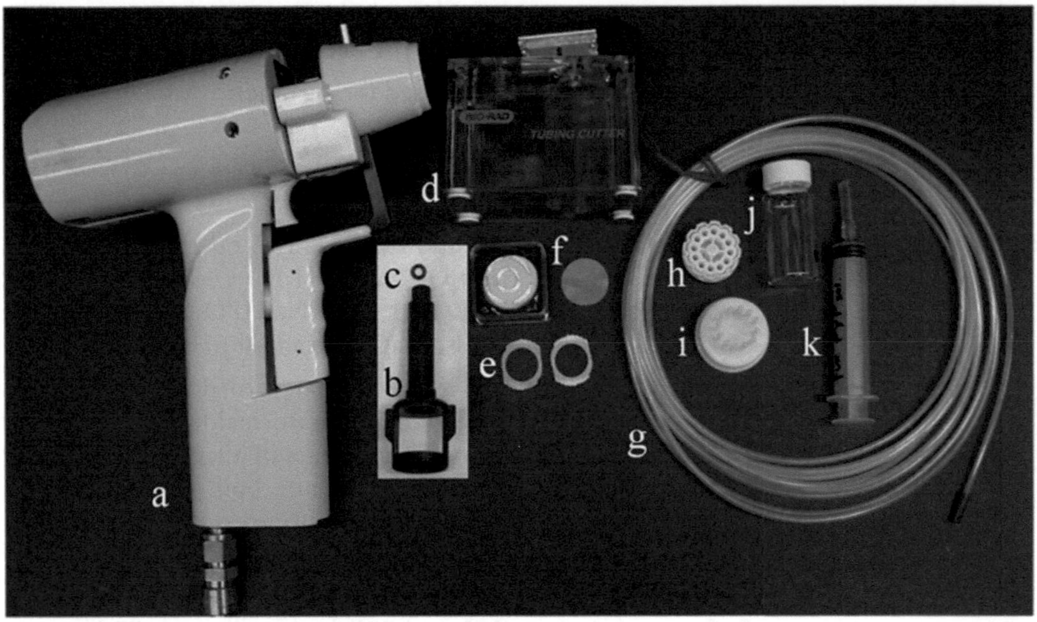

Fig. 1. Shooting Materials: (**a**) Gene gun, (**b**) Barrel liner, (**c**) O-ring, (**d**) Tubing cutter, (**e**) Modified diffusion screens, (**f**) Isopore membrane filter paper picture with box, (**g**) Teflez tubing, (**h**) Cartridge holder, (**i**) Cartridge extractor tool, (**j**) Scintillation vial, (**k**) Syringe with tubing adapter.

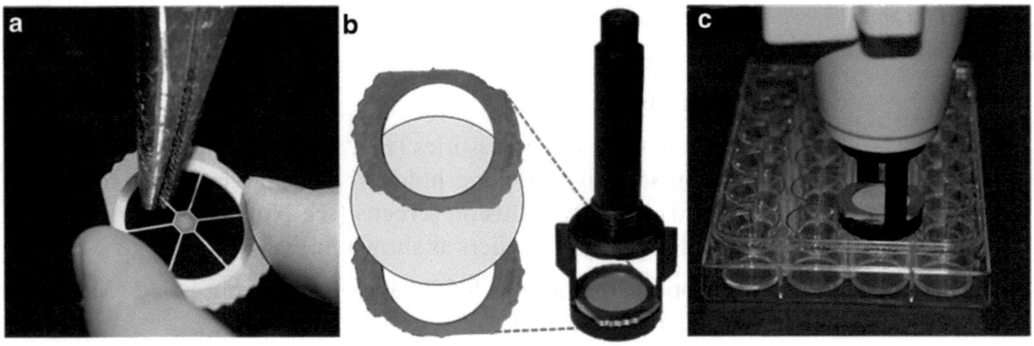

Fig. 2. (**a**) Diffusion screens are modified by removing the radial poles with pliers. (**b**) A filter membrane should be placed between two diffusion screens and then inserted at the bottom of the barrel liner, twisting to ensure a secure fit. (**c**) When shooting, the barrel liner should be centered on the well and pressed flat onto the well plate.

3. Methods

3.1. Preparing DiI/ Tungsten Bead Bullets

1. Resuspend a 300 mg tungsten aliquot in 100 µL of methylene chloride and cover quickly to prevent evaporation.
2. Sonicate for 3 min to disrupt any bead clumps.
3. Combine 13.5 mg DiI with 450 µL of methylene chloride for a final concentration of 3 mg/100 µL.

4. Coat beads with dye by pipetting the bead solution onto a glass slide and adding the DiI solution on top. Mix thoroughly using pipette tip until methylene chloride evaporates and beads turn light grey.
5. Place on a piece of wax-coated weigh paper under the glass slide and use a razor blade to scrape beads off the glass slide and dice into a fine powder.
6. Scrape beads onto weigh paper and funnel beads into a 15 mL conical tube.
7. Add 3 mL of water and sonicate in water bath for 10–30 min (see Note 3).
8. Cut Teflez tubing slightly longer than the length of the tubing prep station. Use a 12 mL syringe with tubing adaptor to pass the PVP solution through the Teflez tubing and then expel it out. Reuse solution to coat more tubing if several batches of bullets are prepared simultaneously and discard PVP solution after use.
9. Vortex the beads at a low speed to produce a homogenous solution and simultaneously use the syringe to pull the bead solution through the tubing so beads are evenly distributed throughout the tubing (see Note 4).
10. Feed the tubing through the prep station and wait about 1 min to allow the beads to settle. Use the syringe to gently remove the water so that only the beads remain (see Note 5).
11. Spin the tubing in the prep station and dry with nitrogen gas until water droplets are no longer visible, approximately 10–20 min.
12. Cut bullets with tubing cutter into 13 mm bullets and store in a container, such as a scintillation vial. Wrap vials in foil to protect from light. Bullets can be prepared in advanced and stored for several months.

3.2. Trans-Cardiac Fixation and Brain Extraction

This procedure is preferred when possible because it allows for reliable and efficient fixation. Alternatively, a block of fresh tissue can be fixed before or after slicing (3).

1. Anesthetize the mouse with an intra-peritoneal injection of ketamine/xylazine solution (0.1 mL/20 g mouse) (see Note 6).
2. Make a sagittal incision along the midline exposing the liver and the chest. Cut through the diaphragm and ribs, along the sternum toward the mouse's left side. Clamp sternum with hemostats and flip above the mouse's head to hold chest cavity open, exposing the heart.
3. Using a small scissor, make a small incision into the left ventricle and insert a butterfly needle with blunted tip (see Note 7).

4. Use a small scissor to cut the vein coming into the right atrium (see Note 8).
5. Perfuse 10–15 mL of cold PBS at a speed of 2.5 mL/min (see Note 9). A pump or gravity can be used in this step.
6. Perfuse about 20 mL of paraformaldehyde solution at a speed of 2.5 mL/min. Once the animal is stiff, stop perfusion (see Note 10).
7. Decapitate the mouse and gently remove the brain.
8. Place brain in chilled container (e.g., glass scintillation tube) with 10–20 mL 4% paraformaldehyde solution for 30 min (see Note 11) on ice.
9. Wash in PBS for 5 min on ice, three times.
10. Incubate brains for 10 min in 15% sucrose solution followed by 10 min in 30% sucrose solution to improve neuronal morphology (see Note 12).
11. Keep at 4°C in container filled with PBS until needed.

3.3. Cutting the Brain Slices

1. Glue the brain to a vibratome platform to allow brain sectioning at the desired orientation. The orientation should be determined based on the region of interest and the morphology of the cells.
2. Add cold PBS to cover the brain and keep solution and chamber on ice (see Note 13).
3. Cut sections of 200–300 μm thickness with a sharp blade.
4. Transfer slices with a brush to individual wells of a 24-well plate containing PBS (see Note 14).

3.4. DiI Staining

1. Remove PBS from wells and place slices in the center of the wells using a paintbrush.
2. Prepare gene gun: Load bullets into cartridge holder, assemble diffusion screens and filter paper into the barrel liner (Fig. 2) and insert both into gene gun (see Note 15).
3. Press the barrel liner flat onto the surface of the 24-well plate (Fig. 2) and shoot once in each well at 120–180 psi helium gas pressure (see Note 16).
4. Quickly wash slices with PBS, two times. For the third wash, cover slices with foil and place on a rocking platform for 30 min.
5. Mount slices: Gently place slice on glass slide and add 8–10 μL ProLong Gold Antifade mounting media on top avoiding air bubbles. Carefully place a coverslip over the slice and leave at room temperature overnight for mounting media to dry (see Note 17).

6. Seal edges of the coverslip with nail polish after mounting media has dried. Then, store slides at 4°C (see Note 18).

3.5. Immunostaining

Immunostaining is an optional addition to this method and, when used, it should be performed after DiI shooting and before mounting (see Note 19).

1. Permeabilization: Incubate slices in permeabilization solution (300–500 μL per well) for 15 min at room temperature (see Note 20).
2. Blocking: Incubate slices in blocking solution for 30 min at room temperature on a rocking platform.
3. Primary antibody: Dissolve primary antibody in blocking solution (e.g., 1:1,000) and incubate slices in 300–500 μL per well for 1–2 h on a rocking platform (see Note 21).
4. Wash slices with PBS for 5–15 min, three times on a rocking platform.
5. Secondary antibody: Dissolve fluorescence-conjugated secondary antibody in blocking solution (e.g., 1:1,000) and incubate slices in 300–500 μL per well for 30 min on a rocking platform.
6. Wash slices with PBS for 5 min, four times on a rocking platform.
7. Mount slices as described in the "DiI staining" section.

3.6. Confocal Microscopy

A confocal microscope equipped with a 561 nm laser (LSM 520 META, Zeiss, Germany) is used to visualize fluorescent labeling (Fig. 3).

1. Acquire images (512 × 512 pixels, 8 bit) by averaging two frames at a speed of 7 (scale 0–10).
2. Set pinhole size at 1Airy Unit (AU) for each objective. The size in microns will vary depending on the objective and wavelength.
3. Adjust the detector's gain to achieve saturation only in the brightest pixels. In this case, the detector's gain was set around 700. The use of digital gain and offsets is not recommended; increase laser power instead if needed.
4. To capture the entire cell morphology, acquire a z-stack of images with a 20× objective (air, *NA*= 0.8). Pinhole size is 2.0 μm and z-stack interval is set to 1.4 μm (0.7 times the pinhole size). Recommended laser power is 4–15%, but optimal power might vary depending on the microscope and the fluorescence intensity.
5. To image dendrites, a 63× objective (water, *NA*= 1.2) is recommended and a digital zoom of 2 can be added. Pinhole size

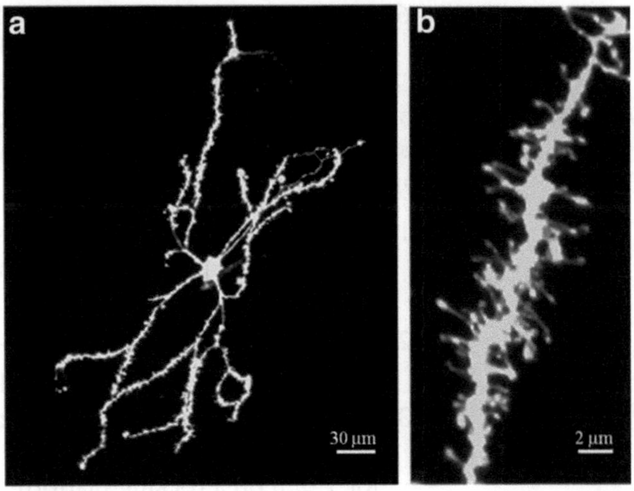

Fig. 3. (**a**) Confocal image of a labeled medium spiny neuron of the mouse nucleus accumbens obtained with a 20× objective. The high intensity of the fluorescent labeling and the low background is suitable for the study of dendritic branching and neuronal morphology. (**b**) Confocal image of a labeled tertiary dendrite obtained with a 63× objective, displaying the diverse morphology of dendritic spines.

of 1AU is 1.0 µm and z-stacks of images are collected at a 0.7 µm interval. Recommended laser power is 40% but optimal power might vary with the microscope and fluorescence intensity (see Note 22).

4. Notes

1. The optimal diameter of the tungsten bead depends on the size of the targeted cell soma. In this particular case, we use 1.7 µm diameter beads for striatal medium spiny neurons which have a cell body size of 15–20 µm. Smaller cells might require smaller diameter beads such as 1.1 µm.
2. In this method, diffusion screens are only used to hold the isopore membrane filter. The membrane is the primary diffuser of tungsten beads and also a filter of bead clumps.
3. Dicing dye-coated beads with a razor and sonication are used to minimize the formation of bead clumps that can commonly occur due to the lipophilic properties of the dye, which would create large fluorescent areas and disrupt the sparse labeling of individual neurons.
4. Try to minimize the number of air bubbles in the tubing to ensure coated beads are evenly distributed.

5. If beads are being removed with the water, wait longer for the beads to settle and remove the water more slowly.
6. Another anesthetic can be used ensuring mouse is unresponsive to tail pinch before proceeding.
7. Blunting the butterfly needle prevents unwanted punctures in the heart.
8. This will allow for the drainage of blood and perfused solutions.
9. During this stage, the liver should turn from dark red to pale pink and the fluid drainage should run clear. If this is not happening, the butterfly needle may need adjusting and/or PBS may need to be perfused for a longer period of time.
10. When perfusing multiple animals, rinse the tubing with PBS before beginning the next perfusion.
11. Do not fix for a longer period of time because it will affect labeling. Over-fixation will disrupt the cell membrane integrity causing DiI to leak out of the cell.
12. Another option is to incubate with 30% sucrose solution overnight at 4°C.
13. Keeping solution cold and the chamber on ice will keep the brain firm for more even slicing. In addition, because the tissue is mildly fixed, keeping it cool will maintain the integrity of the slices.
14. When deciding the appropriate thickness of the slice, it is best to ensure all neuronal processes are intact. When processes are cut, dye will leak out of the cell, making imaging more difficult. It is important to consider the morphology of cells you are labeling. For example, when labeling multipolar cells, thicker slices should be considered to minimize the number of dendrites cut. Furthermore, due to short fixation time, slices thinner than 150 μm may be soft and difficult to handle.
15. Make sure that the barrel liner has the rubber o-ring in the back. The o-ring makes a tight seal with the gene gun and results in more consistent helium pressure during shooting.
16. It is important to keep the same side of the slice that was shot facing up throughout the process. This will ensure that labeled neurons will be within focal distance of the objective. Objectives with high magnification and numerical aperture are ideal for fluorescent confocal imaging, but have a short focal distance, which limits the depth that can be imaged.
17. It is best to mount slices on the same day, as the integrity of the slices might not be well maintained overnight.
18. Slides can be used at least 6 months to a year if stored in the dark at 4°C. DiI is light sensitive and long-term exposure to light will cause the fluorescence to fade.

19. Recently, two modifications have been suggested to this procedure which includes the use of a cholesterol-specific detergent, digitonin (8), and the use of a DiI derivative, Cell Tracker CM-DiI (10).
20. Do not use higher concentrations of detergent as this will dissolve DiI and significantly reduce the fluorescent staining. Try to minimize the time in detergent as extensive incubations will also reduce the DiI labeling. If extended incubations are necessary, keep in PBS rather than in blocking solution.
21. If overnight incubations are necessary, remove detergent from blocking solution.
22. When studying dendritic spine density and morphology, we recommend systematic sampling of dendrite order (primary, secondary, tertiary, etc.) because spine density might vary across different order dendrites. For example, image collection might be restricted to tertiary dendrites. The total number of images collected will depend on the experimental requirements and the degree of variability within a neuron, and across neurons and animals.

Acknowledgments

We would like to thank Fumi Ono's laboratory for providing access to his confocal microscope. Our research is funded by the intramural program of NIAAA and NINDS.

References

1. Connaughton VP, Graham D, Nelson R (2004) Identification and morphological classification of horizontal, bipolar, and amacrine cells within the zebrafish retina. J Comp Neurol 477:371–385
2. O'Brien J, Lummis SC (2004) Biolistic and diolistic transfection: using the gene gun to deliver DNA and lipophilic dyes into mammalian cells. Methods 33:121–125
3. Seabold GK et al (2010) DiOLISTIC labeling of neurons from rodent and non-human primate brain slices. J Vis Exp 41. doi:10.3791/2081
4. Dobi A et al (2011) Cocaine-induced plasticity in the nucleus accumbens is cell specific and develops without prolonged withdrawal. J Neurosci 31:1895–1904
5. Grutzendler J, Tsai J, Gan WB (2003) Rapid labeling of neuronal populations by ballistic delivery of fluorescent dyes. Methods 30:79–85
6. Kim Y et al (2009) Methylphenidate-induced dendritic spine formation and DeltaFosB expression in nucleus accumbens. Proc Natl Acad Sci U S A 106:2915–2920
7. Lee KW et al (2006) Cocaine-induced dendritic spine formation in D1 and D2 dopamine receptor-containing medium spiny neurons in nucleus accumbens. Proc Natl Acad Sci U S A 103:3399–3404
8. Matsubayashi Y, Iwai L, Kawasaki H (2008) Fluorescent double-labeling with carbocyanine neuronal tracing and immunohistochemistry using a cholesterol-specific detergent digitonin. J Neurosci Methods 174:71–81
9. Neely MD, Stanwood GD, Deutch AY (2009) Combination of diOlistic labeling with retrograde tract tracing and immunohistochemistry. J Neurosci Methods 184:332–336
10. Staffend NA, Meisel RL (2011) DiOlistic labeling of neurons in tissue slices: a qualitative and quantitative analysis of methodological variations. Front Neuroanat 5:14

Chapter 29

Protein Antigen Delivery by Gene Gun-Mediated Epidermal Antigen Incorporation (EAI)

Sandra Scheiblhofer, Uwe Ritter, Josef Thalhamer, and Richard Weiss

Abstract

The gene gun technology can not only be employed for efficient transfer of gene vaccines into upper layers of the skin, but also for application of protein antigens. As a tissue rich in professional antigen presenting cells, the skin represents an attractive target for immunizations. In this chapter we present a method for delivery of the model antigen ovalbumin into the skin of mice termed epidermal antigen incorporation and describe in detail how antigen-specific proliferation in draining lymph nodes can be followed by flow cytometry.

Key words: Gene gun, Gold particles, Epidermal antigen incorporation, Epicutaneous immunization, Cell transfer, C57BL/6, OT-II, CFDA SE

1. Introduction

Whereas the gene gun technology has been extensively employed for inoculation of upper skin layers with antigen encoding plasmids, experimental studies using this method for epidermal transfer of protein molecules remain exceptional (1, 2). For successful delivery of antigen into the skin, circumvention of the outermost layer, the stratum corneum, is essential (3). Tape stripping represents a widely used method to break this barrier; however, the amount of antigen taken up following this procedure remains modest. On the other hand, technologies such as tattooing (4) or the use of microneedles (5) rather target the dermal layers than the epidermis.

Theoretically, delivery of protein antigen to the epidermal compartment results in uptake of the molecules by specialized antigen presenting cells (Langerhans cells) residing in this skin layer,

Stephan Sudowe and Angelika B. Reske-Kunz (eds.), *Biolistic DNA Delivery: Methods and Protocols*, Methods in Molecular Biology, vol. 940, DOI 10.1007/978-1-62703-110-3_29, © Springer Science+Business Media, LLC 2013

followed by migration of these cells from the periphery to secondary lymphoid organs, where they can present the acquired antigen to T cells. The role of Langerhans cells vs. dermal dendritic cells in initiation of different types of immune responses is currently a topic of intense investigations (6–8). Specifically targeting antigen to different skin layers therefore represents an important tool to study skin immunology.

In this chapter, we present a protocol for epidermal antigen incorporation (EAI) by gene gun and a method to indirectly demonstrate the presence of antigen presenting cells harboring the respective antigen in draining lymph nodes. The day before EAI, wild type C57BL/6 mice received fluorochrome-labeled lymphocyte preparations from Ovalbumin (OVA) transgenic OT-II donors by intravenous injection. EAI is performed on shaved and depilated skin by application of OVA in aqueous solution, followed by a gene gun shot with empty gold bullets (Fig. 1). At a helium discharge pressure of 300–400 psi, gene gun bullets are predomi-

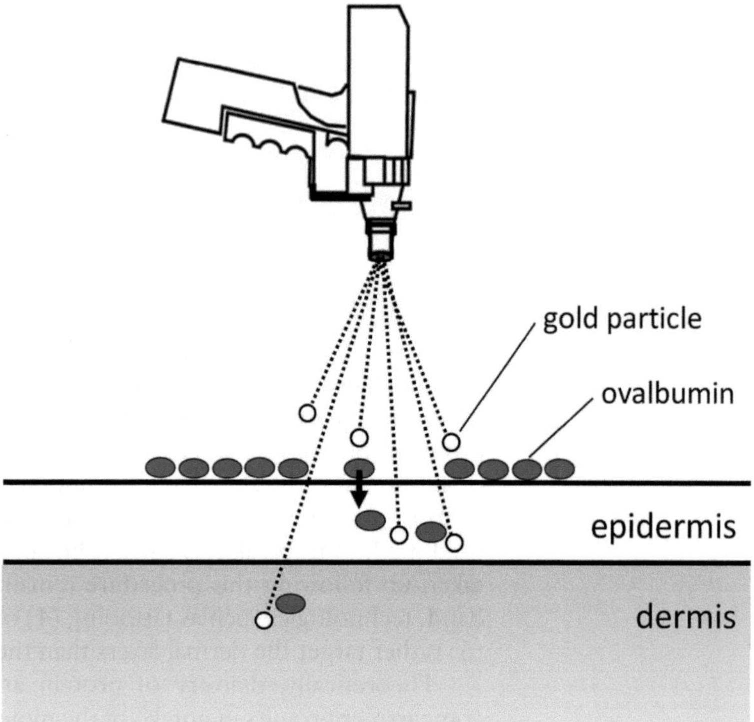

Fig. 1. Epidermal antigen incorporation (EAI). Aqueous ovalbumin solution is painted onto the skin and left to dry. Subsequent gold particle bombardment using a gene gun transfers antigen predominantly into the epidermis and to a lower degree into the dermal compartment of the skin.

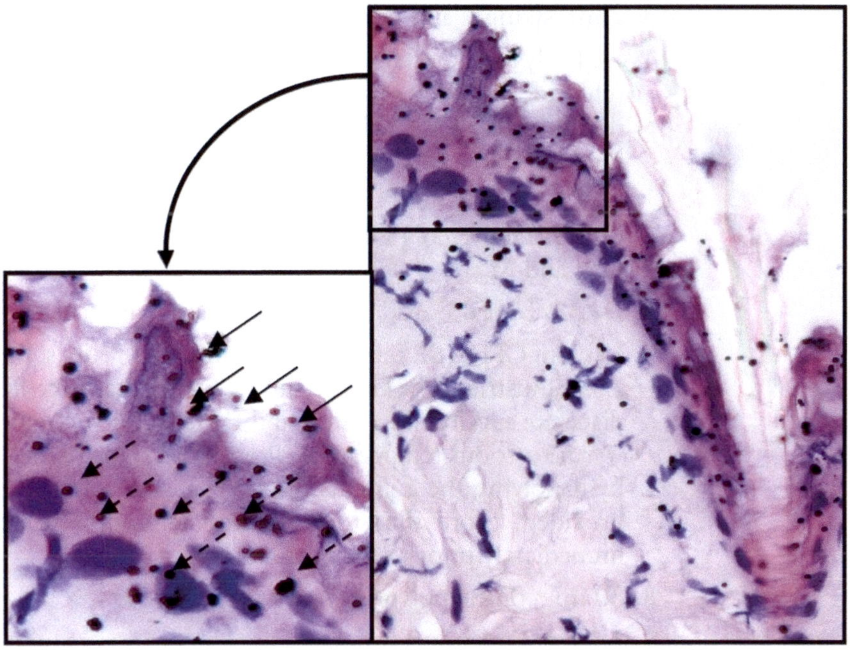

Fig. 2. Representative section of bombarded skin area after EAI. The majority of bullets are found in the stratum corneum (*solid arrows*) and the epidermal compartment (*dotted arrows*). Only a limited number of bullets are detectable in superficial areas below the epidermis (Reprinted from (2), Copyright 2007, with permission from Elsevier).

nantly deposited in the stratum corneum and the epidermis (Fig. 2). OVA-transgenic T cells from donor mice express a Vα2/Vβ5 T cell antigen receptor (TCR) specifically recognizing a C-terminal OVA epitope (AA323-339) (9) and proliferate upon contact with antigen presenting cells. As the donor lymphocytes have been stained with the fluorochrome CFDA SE prior to transfer, transport of the incorporated antigen to secondary lymphoid organs and local antigen presentation can be indirectly measured by monitoring proliferation of transgenic T cells via flow cytometry. Targeting different areas of the body, i.e., the abdomen, the ears, or the hind feet, leads to antigen-specific proliferation restricted to different draining lymph nodes.

2. Materials

2.1. Mice

1. Female C57BL/6 mice.
2. Female OVA transgenic OT-II mice (see Note 1).

2.2. CFSE-Staining and Cell Transfer

1. Dissecting set.
2. Cell strainer: 40 μm (BD Falcon, Schwechat, Austria).

3. 2 mL syringes and pestel.
4. ACK lysing buffer (Ammonium-Chloride-Potassium (**K**) (chloride)): 8.29 g/L NH_4Cl, 1.00 g/L $KHCO_3$, 0.0372 g/L disodium EDTA • $2H_2O$ dissolved in deionized H_2O. Adjust the pH value to 7.2–7.4 by addition of HCl, if necessary. Sterilize the solution by filtration or autoclave.
5. DPBS: Dulbecco's Phosphate Buffered Saline (PBS) without Ca^{2+} and Mg^{2+}.
6. DPBS/3% FBS: Add 3% Fetal Bovine Serum to DPBS.
7. Vybrant CFDA SE cell tracer kit (Molecular Probes, Invitrogen, Lofer, Austria): The kit contains CFDA SE (carboxy-fluorescein diacetate succinimidyl ester) and DMSO. Allow the product to warm to room temperature. Prepare a 10 mM stock solution by dissolving the contents of one vial in 90 μL of the DMSO provided. Store the stock solution at −20°C or prepare it immediately prior to use (see Note 2).
8. 50 mL conical polypropylene tubes.
9. 1.5 mL Eppendorf tubes.
10. 37°C warm water bath.
11. 1 mL syringes with 27 gauge needles.
12. Mouse restrainer with tail access.

2.3. EAI

1. Electric shaver.
2. Depilatory cream for sensitive skin.
3. Helios™ Gene gun (Bio-Rad, Hercules, CA, USA).
4. Gene gun bullets containing empty gold particles of 1.6 μm diameter (see Note 3).
5. Recombinant OVA: grade V (Sigma-Aldrich, Vienna, Austria).
6. PBS.
7. Anesthesia equipment: Isofluorane inhalation anesthesia is the preferred method; if no anesthesia machine for rodents is available, use Ketaminehydrochloride/Xylazin: Add 1 mL of a 10% Ketaminehydrochloride solution (Narketan, Vétoquinol AG, Blep, Switzerland), and 0.5 mL of a 2% Xylazine solution (Rompun, Bayer Health Care, Vienna, Austria) to 3.5 mL endotoxin-free ultrapure water.
8. 37°C heating pad or hot water bag.

2.4. Analysis by Flow Cytometry

1. Dissecting set.
2. Hanks' Buffered Salt Solution with Ca^{2+} and Mg^{2+}.

3. Cell strainer: 40 μm (BD Falcon, Schwechat, Austria).
4. 50 mL conical polypropylene tubes.
5. Eppendorf tubes.
6. 2 mL Syringes.
7. FACS buffer: 3% FBS, 0.05% NaN_3, 2 mM Ethylenediaminetetraacetic acid (EDTA) in PBS.
8. PE-labeled anti-mouse-Vβ5.1/5.2 antibody (BD PharMingen) (see Note 4).
9. PerCP-labeled anti-mouse-CD4 antibody.
10. FACSCanto II flow cytometer (BD Biosciences, Schwechat, Austria).

3. Methods

3.1. Preparation of Mice

The day before EAI, mice have to be shaved and depilated at the respective skin area on their abdomen, ears, or hind feet.

1. For EAI at the abdomen hold the animal at the neck and backside with one hand, and carefully shave its abdomen with an electric shaver in an area of approx. 4×2 cm.
2. Under isofluorane anesthesia (see Note 5) apply depilatory cream with a soft brush to the shaved area. The duration of exposure has to be adjusted to the mouse skin (see Note 6).
3. Gently but thoroughly remove the depilatory cream with a sponge containing hand-hot water and dry the skin with a paper towel.
4. For EAI at the ear pinna or the hind footpad, remove extensive hair with a small pair of scissors; then use the depilatory cream as described for abdominal EAI, but with a shorter exposure time.
5. Let the animals rest for some hours before transfer of lymphocytes from transgenic donor mice as described under Subheading 3.2.

3.2. CFSE Labeling and Cell Transfer

1. Prewarm DPBS and DPBS/3% FBS to 37°C in a water bath.
2. Sacrifice an OT-II donor mouse by cervical dislocation and aseptically remove the spleen and lymph nodes (see Note 7).
3. Gently, mince the spleen with the plunger of a 2 mL syringe at the bottom of a petri dish, and the lymph nodes with a pestel in an Eppendorf tube to prepare cell solutions.

4. Let the cell solutions settle in Eppendorf tubes for 5 min and take the monodisperse supernatants to get rid of connective tissue and other particulate matter (see Note 8).
5. For lysis of erythrocytes, add pooled preparations from spleen and lymph nodes to 7 mL ACK lysis buffer (RT) in a 14 mL polypropylene tube, and mix thoroughly by inverting the tube about 5 times.
6. Incubate at RT for a maximum of 10 min, add 6 mL DPBS (RT) and centrifuge at 260 g for 10 min.
7. Aspirate the supernatant; resuspend the pellet in 10 mL DPBS (RT) and centrifuge again for 10 min at 260 g.
8. Let the frozen CFDA SE stock solution come to room temperature or prepare it freshly as described under Subheading 2.2, step 6.
9. Dilute the CFDA SE stock solution to 5 μM in prewarmed DPBS.
10. Aspirate the supernatant of the last washing step and gently redissolve the pellet in DPBS containing the fluorescent dye (see Note 9). Incubate at 37°C in a water bath for 15 min and gently shake the tube every 5 min.
11. Add 40 mL fresh prewarmed PBS/3% FBS to stop the staining and pellet the cells.
12. Resuspend the pellet in fresh prewarmed PBS/3% FBS and incubate for another 30 min to ensure complete modification of the probe (see Note 10).
13. Centrifuge the cells for 10 min at 260 g, and gently resuspend the pellet in PBS.
14. Count the cells and adjust the concentration to 4×10^6/200 μL with PBS (see Notes 11 and 12).
15. Warm up the tails of recipient mice in the warm water bath for approximately 30 s, put the animal in a rodent restrainer and intravenously inject 200 μL of the cell solution per mouse (see Note 13).

3.3. EAI

The day following shaving/depilation of the skin and transfer of transgenic T cells, mice receive recombinant OVA in solution via EAI.

1. Prepare a 50 mg/mL OVA solution in PBS and use 5 μL thereof for EAI.
2. Anesthetize the mice by isoflurane anesthesia or intraperitoneal injection of Ketamine/Xylazine.

3. Apply 5 μL of the OVA solution to the shaved/depilated area at the back side of the ear pinna, the abdomen, or the upper side of the hind foot (see Note 14).
4. Gently allocate the solution over the respective skin area using a pipette tip. Let the solution completely dry under an infrared light lamp (see Note 15).
5. After evaporation of the liquid, bombard the skin with 1.6 μm empty gold particles using a Helios™ Gene gun at a helium discharge pressure of 400 psi. For EAI at the ear pinna, it is recommended to use a stencil (see Note 16).
6. If you are using injection of Ketamine/Xylazine for anesthesia, put the mice on a heating pad or warm water bag during awakening.

3.4. Flow Cytometry

Antigen-specific proliferation in draining lymph nodes can be analyzed at various time points after EAI (see Note 17).

1. Prechill FACS buffer on ice.
2. Kill the mice by cervical dislocation and dissect the draining lymph nodes (see Note 18).
3. Transfer the lymph nodes into an Eppendorf tube containing 1 mL HBSS.
4. Mesh the lymph nodes through a cell strainer with the plunger of a 2 mL syringe into a 50 mL polypropylene tube.
5. Rinse the cell strainer with 10 mL HBSS.
6. Centrifuge the cells at 260 g for 10 min.
7. Dilute the antibodies in chilled FACS buffer (see Note 19):
   - Anti-mouse-Vβ5.1/Vβ5.2—PE or anti-mouse-CD45.1-PE (see Note 4).
   - Anti-mouse-CD4-PerCp.
8. Resuspend pellets in 100 μL staining solution.
9. Incubate for 10 min on ice in the dark.
10. Centrifuge cells at 260 g for 10 min.
11. Resuspend cells in chilled FACS buffer.
12. Repeat steps 10 and 11.
13. Analyze samples in a flow cytometer: Gate on CD4 positive T cells and plot CFSE (FITC channel) vs. Vβ5.1/Vβ5.2 (or CD45.1). Donor cells can be distinguished from recipient cells by expression of the Vβ5 chain of the transgenic T cell receptor (see Note 20). Proliferating T cells can be identified by a reduction of CFSE dye intensity. Individual populations of lower CFSE-staining resemble the number of cell divisions (Fig. 3).

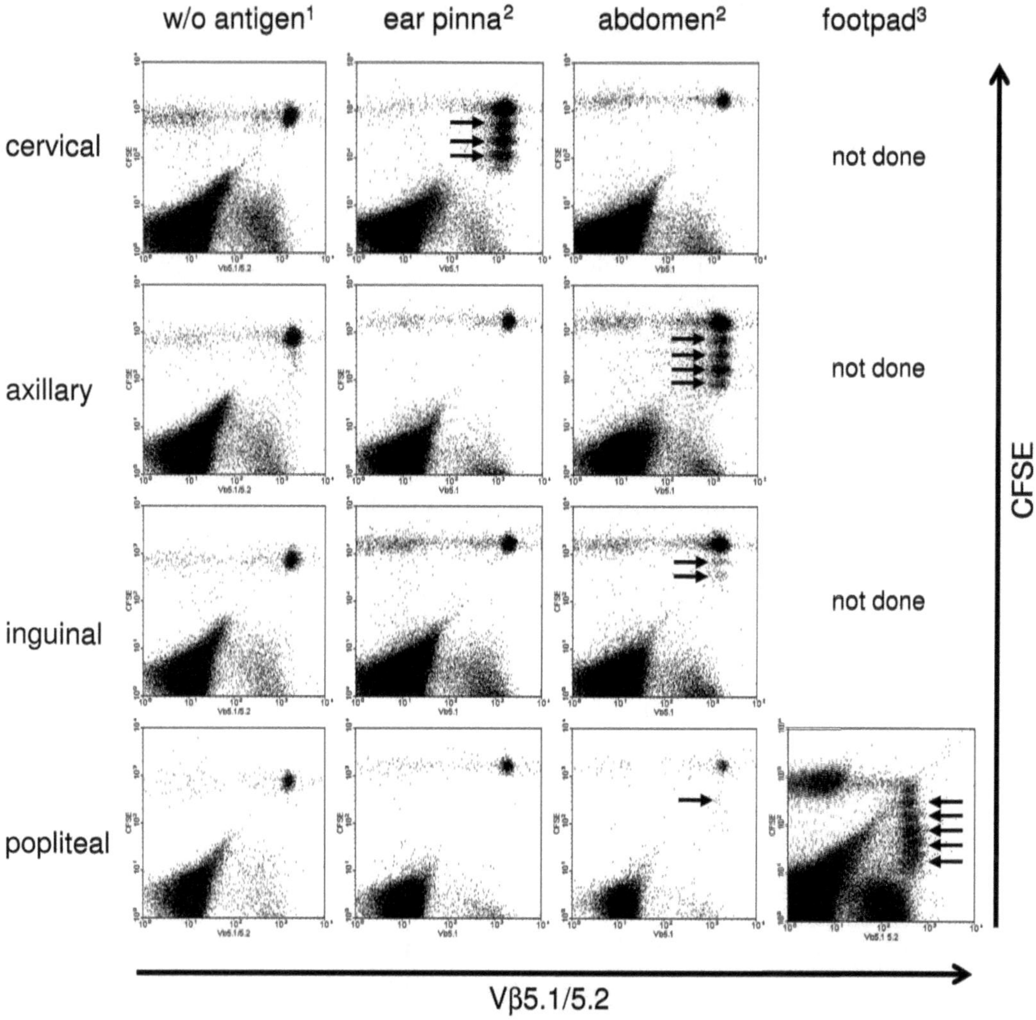

Fig. 3. In vivo proliferation of adoptively transferred OT-II T cells after EAI. Splenocytes from OT-II transgenic mice were labeled with CFSE and transferred into C57BL/6 wildtype recipients. After 24 h, ovalbumin was applied by EAI to the ear pinna, abdomen, or upper side of hind feet, and draining lymph nodes (cervical, axillary, inguinal, popliteal) were harvested at indicated time points. Monodisperse lymph node cell solutions were prepared, stained with anti-mouse-CD4/PerCp and anti-mouse-Vb5.1/5.2/PE, and analyzed by flow cytometry. Plots have been gated on CD4 positive cells. Populations resembling cell divisions of proliferating OT-II cells are marked by arrows. Representative data of individual mice are shown. (*1*) mouse has received one gene gun shot on the depilated abdomen without prior antigen application. Lymph nodes were harvested 144 h post EAI. (*2*) 5 μL of ovalbumin [50 mg/mL] has been painted on the skin prior to gene gun application. Lymph nodes were harvested 48 h post EAI. (*3*) 5 μL of ovalbumin [50 mg/mL] has been painted on the upper side of the hind foot prior to gene gun application. Lymph nodes were harvested 72 h post EAI.

4. Notes

1. Especially for long-term experiments, it is recommended to cross OT-II mice with C57BL/6 CD45.1 mice and use the F1 offspring for cell transfer. Thereby, discrimination of CD45.1

donor from CD45.2 recipient cell populations is facilitated. For further details see also Notes 4 and 20.

2. Alternatively, other cell tracking dyes that emit light at different wavelengths can be used, if the flow cytometer is equipped accordingly. For example, PKH26 (Sigma-Aldrich) can be excited with a 488 nm blue laser and emits light in the PE channel, while Cell Proliferation Dye eFluor® 670 (eBioscience) can be excited with a 633 nm red laser and emits light in the APC channel. CFSE, PHK26, and Cell Proliferation Dye eFluor® 670 can all be used for long-term experiments. For short-term experiments (up to 7 days), various CellVue® dyes (Sigma-Aldrich) with a broad color range are available.
3. As described in detail in other chapters of this book, prepare gene gun bullets by coating the inner side of a tefzel tubing with gold microcarriers of 1.6 μm diameter. In contrast to standard protocols, skip the step in which plasmid DNA is added.
4. If CD45.1/OT-II mice are used as donors, use PE-labeled anti-CD45.1 antibody.
5. Alternatively inject 80 mg Ketamine and 8 mg Xylazine per kg body weight intraperitoneally. For a 25 g mouse this corresponds to 100 μL of the solution described under Subheading 2.4, step 7.
6. Mouse skin is more sensitive to depilation than human skin. Depending on the used product, an incubation time of 30–60 s is sufficient.
7. Depending on the age of the mouse and the skills of the experimentator 5×10^7 to 1×10^8 cells can be obtained from one donor, sufficient for transfer of 10–20 recipient mice.
8. You can also mesh spleen and lymph nodes through a 40 μm cell strainer. Then rinse the strainer with DPBS, pellet the cells by centrifugation at 260 g for 10 min and proceed with erythrocyte lysis.
9. It is recommended to use a 50 mL polypropylene tube for this incubation step because this enables to gently mix the lymphocytes during the staining period and to directly add a large volume of fresh prewarmed DPBS/3% FBS to stop the staining and extensively wash the cells.
10. CFDA SE passively diffuses into cells, where it remains colorless and nonfluorescent until cleavage of its acetate groups by intracellular esterases, which leads to formation of highly fluorescent, amine-reactive carboxyfluorescein succinimidyl ester (CFSE). The succinimidyl ester group reacts with intracellular amines, forming fluorescent conjugates that are well retained, whereas excess unconjugated reagent and by-products passively diffuse to the extracellular medium and can be

removed by washing. In labeled cells, the fluorescent conjugates are retained during development, meiosis, and in vivo tracing. The labeling is not transferred to adjacent cells in a population, but inherited by daughter cells after cell division. The number of cell divisions can be directly determined as the fluorescence intensity halves with every cell division.

11. Smaller cell numbers (10^6 and less) can be applied. Using smaller cell numbers results in a more complete activation of the transferred cells, but reduces the overall number of proliferated cells. This may lead to detection problems, especially when analyzing small lymph nodes.
12. You can optionally check proper CFSE labeling by analyzing a small aliquot in a flow cytometer before cell transfer. The freshly labeled cells will be extremely bright in the FITC and PE channel. At this time point, no proper compensation is possible. At later time points in the experiment, cells will lose some of their brightness and maintain a level of staining that can be compensated.
13. Gently shake the vial containing the cells in between intravenous injections so that they will not settle at the bottom of the vial. Be careful not to inject any air bubbles as this could cause air embolism.
14. Include appropriate controls in your experiment such as mice receiving a gene gun shot without prior application of protein (negative control), or intradermal/footpad injection of OVA (positive control).
15. The mouse skin is hydrophobic. The protein solution can be evenly distributed on the application area by moving the droplets over the skin with a pipette tip several times.
16. Use a small piece of cardboard and cut an opening of approximately 10 mm × 1 mm. Pull the ear through the slot using forceps and spread the pinna on the cardboard.
17. You can expect approximately two cell divisions per 24 h. Analyzing cells at late time points can make the discrimination of donor and recipient cells difficult. In this case, CD45.1+ donor mice are recommended.
18. For EAI into the ear pinna, the draining lymph nodes are the cervical lymph nodes, for abdominal EAI the axillary or inguinal lymph nodes (depending on the exact localisation of the area used for application of the protein solution and the gene gun shot), and for EAI into the hind foot the popliteal lymph nodes. Helpful hints for exact localization of lymph nodes in mice can be found in a publication by Harrel et al. (10).
19. Refer to manufacturers' instructions for recommended antibody concentrations or perform pre-titration experiments for

optimal antibody dilutions. If no pre-titration experiments have been done, use antibodies [0.5 μg/μL] at a 1:100 dilution.

20. Approximately 10% of T cells of C57BL/6 mice also express the Vβ5.1/Vβ5.2 allele. Therefore, donor cells cannot be distinguished from recipient cells after longer periods of proliferation. In this case use CD45.1+ donor cells.

References

1. Maddelein ML et al (2002) Amyloid aggregates of the HET-s prion protein are infectious. Proc Natl Acad Sci USA 99:7402–7407
2. Weiss R et al (2007) Epidermal inoculation of Leishmania-antigen by gold bombardment results in a chronic form of leishmaniasis. Vaccine 25:25–33
3. Kalluri H, Banga AK (2011) Transdermal delivery of proteins. AAPS PharmSciTech 12:431–441
4. Pokorna D et al (2009) Vaccination with human papillomavirus type 16-derived peptides using a tattoo device. Vaccine 27:3519–3529
5. Weldon WC et al (2011) Microneedle vaccination with stabilized recombinant influenza hemagglutinin induces improved protective immunity. Clin Vaccine Immunol 18:647–654
6. Brewig N et al (2009) Priming of CD8+ and CD4+ T cells in experimental leishmaniasis is initiated by different dendritic cell subtypes. J Immunol 182:774–783
7. Ritter U, Osterloh A (2007) A new view on cutaneous dendritic cell subsets in experimental leishmaniasis. Med Microbiol Immunol 196:51–59
8. Stoecklinger A et al (2011) Langerin+dermal dendritic cells are critical for CD8+ T cell activation and IgH gamma-1 class switching in response to gene gun vaccines. J Immunol 186:1377–1383
9. Robertson JM et al (2000) DO11.10 and OT-II T cells recognize a C-terminal ovalbumin 323–339 epitope. J Immunol 164:4706–4712
10. Harrell MI et al (2008) Lymph node mapping in the mouse. J Immunol Methods 332:170–174

INDEX

A

B

C

D

E

F

G

Stephan Sudowe and Angelika B. Reske-Kunz (eds.), *Biolistic DNA Delivery: Methods and Protocols*, Methods in Molecular Biology, vol. 940, DOI 10.1007/978-1-62703-110-3, © Springer Science+Business Media, LLC 2013

MIX
Papier aus verantwortungsvollen Quellen
Paper from responsible sources
FSC® C105338

FSC
www.fsc.org

If you have any concerns about our products,
you can contact us on
ProductSafety@springernature.com

In case Publisher is established outside the EU,
the EU authorized representative is:
Springer Nature Customer Service Center GmbH
Europaplatz 3, 69115 Heidelberg, Germany

Printed by Libri Plureos GmbH
in Hamburg, Germany